America's Forgotten Poet-Philosopher

SUNY series in American Philosophy and Cultural Thought
———————
Randall E. Auxier and John R. Shook, editors

America's Forgotten Poet-Philosopher

The Thought of John Elof Boodin in His Time and Ours

Michael A. Flannery

Published by State University of New York Press, Albany

© 2023 State University of New York

All rights reserved

Printed in the United States of America

No part of this book may be used or reproduced in any manner whatsoever without written permission. No part of this book may be stored in a retrieval system or transmitted in any form or by any means including electronic, electrostatic, magnetic tape, mechanical, photocopying, recording, or otherwise without the prior permission in writing of the publisher.

For information, contact State University of New York Press, Albany, NY
www.sunypress.edu

Library of Congress Cataloging-in-Publication Data

Name: Flannery, Michael A., 1953– author.
Title: America's forgotten poet-philosopher : the thought of John Elof Boodin in his time and ours / Michael A. Flannery.
Description: Albany : State University of New York Press, [2023] | Series: SUNY series in American philosophy and cultural thought | Includes bibliographical references and index.
Identifiers: LCCN 2023009051 | ISBN 9781438495712 (hardcover : alk. paper) | ISBN 9781438495736 (ebook) | ISBN 9781438495729 (pbk. : alk. paper)
Subjects: LCSH: Boodin, John Elof. | Philosophy, American. | Philosophy, Modern. | Science—Philosophy. | Metaphysics.
Classification: LCC B945.B574 F63 2023 | DDC 191—dc23/eng/20230726
LC record available at https://lccn.loc.gov/2023009051

10 9 8 7 6 5 4 3 2 1

*To Catherine Tallen
for her perennialist spirit*

Portrait of John Elof Boodin, ca. 1925. Courtesy of Carleton College.

Contents

Acknowledgments	ix
Prologue: Who Was John Elof Boodin?	xi
Introduction: Science and Its "Mad Clockwork of Epicycles": The Key to Understanding Boodin	1
Chapter 1. Boodin's Time	17
Chapter 2. Pragmatic Realism: Boodin's Metaphysics	45
Chapter 3. Evolution, 1925	67
Chapter 4. A Theological Trilogy	107
Chapter 5. *The Social Mind*: Boodin's Sociology of Spirit	137
Chapter 6. Boodin in a Hostile World	171
Epilogue	203
Appendix	213
Glossary of Important Terms	217
Notes	225
Bibliography	263
Index	283

If Aphrodite did rise in her full-formed beauty from the mists of the sea, it was not altogether owing to the potencies of salt water. There was also the cosmic genius of Zeus.

—John Elof Boodin, *Cosmic Evolution*, 35

Acknowledgments

First and foremost, I must thank Randall Auxier, professor of communications and philosophy at Southern Illinois University, Carbondale, for first urging me on to follow my impulse and examine John Elof Boodin more fully. It soon became a labor of love that was spurred ever onward by his continual encouragement and sustained interest. I am also deeply grateful to John S. Haller Jr., emeritus professor of history and medical humanities at Southern Illinois University, Carbondale, for reading earlier manuscript versions of this book and making many helpful comments and suggestions. I have benefitted from John's scholarship and perceptive grasp of American intellectual history for many years, and this project has been no exception. Finally, I have to thank my understanding and longsuffering wife Dona who endured another of her husband's obsessions and attended to life's many details while I followed Boodin down his poetic philosophical trails. She allowed me the luxury of inhabiting the stars while she stayed fixed on this terrestrial plane. Despite all the collegial and personal support I've received throughout this project for which I am sincerely grateful, all errors of omission or commission remain my own.

Prologue

Who Was John Elof Boodin?

Before examining the thought of John Elof Boodin it will be helpful to have a general sketch of his life, covering his origins, his immigrant experience, his education, his professional career, and the significant influences bearing upon them. Although this book is not intended to be biographical in any real sense, it must be recognized by intellectual and cultural historians that every concept ever formulated in the past came from a real life actually lived. The historian is doing neither the craft nor the reader any service by lapsing into the antiquarianism of a name-and-date recitation nor by reifying ideas as entities without owners; either extreme is simply bad history. Barbara Tuchman who knew better than most how to write good history in an engaging style acknowledged all biography (even the shortest) as "a prism of history . . . [that] attracts and holds the reader's interest in the larger subject." Without it, she adds, "the central theme wanders, becomes diffuse, and loses shape. One does not try for the whole but for what is truthfully *representative*."[1] It is in that spirit this life of our subject is offered.

Early Years

His origin was as obscure as his legacy, born in the parish of Pjetteryd amidst the Småland highlands of southern Sweden on September 14, 1869, the middle of a large household of ten children. His mother Kristina (his father's second wife) and Elias Nilsson provided the warm, nurturing environment lacking in the cold and often inhospitable climate that had forged the character of these strong, self-reliant people since the Nordic Bronze

Age. He died in Los Angeles, California, on November 14, 1950, a *very* different time and place. As mentioned, this is *not* the story of his life; that has already been ably told by Charles H. Nelson.² It is, however, an attempt to reveal the remarkable intellect that graced this planet during that period. Boodin himself described his life as emerging "from the darkness like a meteor," a celestial rocket that "makes a transient trail and disappears into the darkness again."³ In the following pages I will endeavor to explain the brilliance of that meteoric mind that deserves more than the passing notice of a falling star.

The first sixteen years of young John's life were circumscribed by the daily rhythms of a harsh but predictable farm life (see the appendix). Raised in reverent piety dictated by confessional Lutheranism, the boy learned the harsh realities of agricultural life softened by an idealism drawn from a land steeped in ancient history and myth. As Boodin described it, "I was nurtured on the Bible and the legends of the place. There were groves and alters of the old Stone Age and there was the much older background of the trolls that lived once in the hills and in the bordering lake."⁴ It was an early life well suited to stimulating the exacting habits of mind of a philosopher and encouraging the emotional transports of the poet.⁵ Boodin would become both.

The first to notice Boodin's brilliance was Pastor C. A. Bomgren and his successor Claes Sjöfors of the local church. It was the Sjöfors family who took him in and prepared him for the gymnasium (the equivalent of a college preparatory school) in Uppsala. This ended John's agrarian life and launched him toward the trajectory of a remarkable academic career. After two years at the Fjellstadt school, Boodin probably heard of the Swedish-American communities emerging with strong ties to Swedish Lutheran pastors, and Pjetteryd had an exceptionally large emigration to America in the last half of the nineteenth century. In the summer of 1887 Boodin booked steerage to America and landed in Colchester, Illinois, a coal-mining town about ninety miles northwest of Springfield. The peculiar migration to this out-of-the-way Midwestern location is probably suggested by the fact that his brother Johannes located there with a thriving community of about 100 to 150 fellow Swedes. The realism John learned on the farm was now reinforced by spending two years as a blacksmith, all the while honing his skills in English and adding as finances and circumstances allowed to his education. In 1890 he completed his teacher's training at the college in Macomb, Illinois, and two years later entered the University of Colorado. Shortly thereafter he had an opportunity to work with the Episcopal Church and the Swedish immigrants in Minneapolis, and so from 1893 to 1894 he found himself at the University of Minnesota. Boodin describes this as an important time for his development. It was here that he was introduced to the social sciences by the university's influential William Watts Folwell and to the psychology of William James by James Rowland Angell, who would later serve as president of Yale. Both of these men would have a profound impact on Boodin's career.

Go East Young Man!

Although Horace Greely had advised America's youth to go West, Boodin had his sights set in the opposite direction. For a young man with the scholarly turn of mind now so in evidence with Boodin, the future lay in the settled academies of New England—Dartmouth, Yale, Harvard, and so forth. While Boodin yearned to attend Harvard, for the immediate future Providence, Rhode Island, would do nicely, and a new and extremely fertile period opened for the aspiring student when he headed east to Brown University. It was there that James Seth introduced him to philosophy. Here his remarkable academic pedigree began; it was 1894.

At Brown, Boodin started a small Philosophical Club and it is there that he met "the man whom I wanted most of all for a friend,"[6] Harvard professor William James. There was good reason, for James's reputation preceded him. He had helped make psychology (then typically taught as a branch of philosophy) an independent discipline with his massive, two-volume *Principles of Psychology* (1890). In that landmark work James established psychology as an experimental science emphasizing the importance of a well-founded phenomenology of experience. It influenced Edmund Husserl, the founder of modern phenomenology. John Dewey, another of pragmatism's leading lights, singled out James's discussion of conception, discrimination and comparison, and reasoning in *Principles*—not his later *Pragmatism* (1907)—as most influencing him in his abandonment of Kantian idealism for functionalism.[7] By the time Boodin met James the Harvard luminary was well on the way toward establishing himself among America's leading exponents of pragmatism, an idea he carried forward from Charles Sanders Peirce and developed more fully into a theory of experiential and empirical truth. Although Peirce coined the term in 1878, James acknowledged his debt and took the idea in these new directions. More will be said about pragmatism later. Besides his reputation for pragmatism, James would publish in just a few years his most widely read and bestselling title, *Varieties of Religious Experience* (1902), a book that would launch modern religious studies. But perhaps James became most revered for taking philosophy out of the abstract ivory tower and into our daily lives; this is what Boodin most wanted to befriend, and rightly so. As James's biographer has so well stated:

> In place of the mythological world of fixed ideas, James has given us a world of hammering energies, strong but evanescent feelings, activity of thought, and a profound and restless focus on life now. For all his grand accomplishments in canonical fields of learning, James's best is often in his unorthodox, half-blind, unpredictable lunges at the great questions of how to live, and in this his work sits on the same shelf with Marcus Aurelius, Montaigne, Samuel Johnson, and Emerson. James's best is urgent, direct, personal, and useful.[8]

But ironically the young Boodin, instead of being struck dumb by the awed presence of his idol, used the occasion of their meeting to challenge James. James had just published "Is Life Worth Living?" in the *International Journal of Ethics* (October 1895) and Boodin's recently formed philosophy club used the occasion to invite America's leading pragmatist to respond to members' reactions to the paper. James did respond, but no critique was more thorough and telling than Boodin's. He argued that James provided no solid criteria for values and instead "made all our superstitions, day dreams and air castles equally sacred with the best established truth."[9] James took Boodin's criticisms to heart, gave him the sobriquet "Orator," and launched an enduring friendship with the young and eager scholar. Boodin believed that James's interest was sparked not only by his strong and compelling intellectual engagement with James but also by the fact that he was "different," an immigrant background with an unusual accent, "surrounded with an air of mystery." It was through James that Boodin learned the heart of pragmatism: "How ideas must terminate in concrete experience in order to be declared true."[10]

Throughout their association up until James's death on August 26, 1910, Boodin was treated as much as a potential convert to the cause of pragmatism as a friend. Perhaps this is because James saw so much promise in Boodin's work. When the latter's *Time and Reality* was published in 1904, James called it "a masterly piece of work, both in thought and style, and represents a synthesis vast enough, and original enough to give you an admitted master's place."[11] As time passed, James increasingly pressed Boodin to join the pragmatist ranks, an invitation from which Boodin remained less resistant than aloof. Boodin, it seems, preferred the role of the uncommitted critic and commentator rather than partisan participant. Speaking of James's passion for pragmatism, his biographer has written:

> James's ambitions for pragmatism were breathtaking. When he compared the pragmatist movement to the Protestant Reformation, he was not being ironic. Just as he was launching the lectures in October 1906 [the Lowell Lectures on pragmatism], he wrote to Giovanni Amendola, one of Papini's Italian pragmatists, "I think that pragmatism can be made—is not Papini tending to make it?—a sort of *surrogate* of religion, or if not that, it can combine with religious faith so as to be [a] surrogate for dogma." Saving his best shot for last, as he had in Edinburgh [at the Gifford Lectures, 1901–1902], he closed his pragmatism lectures by saying, "On pragmatistic principles, if the hypothesis of God works satisfactorily in the wider sense of the word, it is true." This is not the Protestant God, not the Christian God, not monotheism.[12]

Boodin would have much more to say about God and religion, but it is worth pointing out in this context something very important about our subject of interest, namely that Boodin had his influences but he was always his own man. We will see later on as we examine Boodin's ideas that his understanding of God and pragmatism was not James's. James was surely an influence on Boodin's thought, but it never impressed itself upon him like a carbon copy. In fact, if we are to find a stronger influence on Boodin we must look elsewhere.

On to Harvard and Royce

In 1897 Boodin was able, with the assistance of Brown University psychology professor Edmund B. Delabarre and a Hopkins scholarship obtained with the help of Harvard's Bussey professor of theology Charles Carroll Everett, to enter Harvard, at that time the epicenter of American philosophy. It is a measure of Boodin's intellect and commitment to his studies that virtually every faculty member wherever he attended was immediately impressed with his abilities. The only person who apparently needed convincing was the senior faculty member of Harvard's philosophy department, George Herbert Palmer, who feared that the Swedish immigrant might not have sufficient proficiency in English and perhaps more subliminally might not "fit in" to the culture at Harvard Yard. Assuaged of Palmer's concerns, Boodin entered a new and important chapter of his education. Here he would be proximate to American philosophy's best and brightest, not only William James but also major figures such as Hugo Münsterberg, George Santayana, and most significantly Santayana's mentor, Josiah Royce.

When Boodin arrived at Harvard, Royce had been in Cambridge for five years. He had already established himself as America's foremost absolute idealist with *The Religious Aspect of Philosophy* (1885). John McDermott has aptly described Royce as one who "combines the approach of a preacher with that of an extraordinary intellectual virtuoso."[13] And Boodin would see Royce at the height of his career and influence. Royce's argument for an absolute mind—the Absolute—with all finite minds forming a basis for truth over error was what James's pragmatism was designed to respond to. For him, thought could pick its objects purely behaviorally and experientially. But it would be wrong to say that Royce has nothing in common with pragmatism. In fact, pragmatism deserves a more nuanced approach than it is often given. It is more accurate to see pragmatism consisting of two strands: one characterized as "radical empiricism" as promoted by James and Dewey, and a second "idealist" current followed by Peirce and Royce.[14] (Another way of dividing pragmatism will be discussed in chapter 6.) Thus, when Boodin came to Harvard he found himself in the midst of these two distinct manifestations of pragmatism. It is no wonder then that Boodin's ideas would become blended as it were. Boodin's biographer sees "the realistic and scientific aspects of James's pragmatism [that] permeated Boodin's idealistic metaphysics" as "paradoxical," but seen in the light of these two variations on the pragmatic theme perhaps less so.

In any case, when Boodin chose to work on his dissertation—"The Concept of Time"—under Royce, he was coming under the influence of America's best philosopher of religion.[15] More importantly, while James was Boodin's soulmate, Royce provided the greatest influence over Boodin's own developing metaphysic through the oppositional character of Royce's idealism. However, far from creating a chasm between mentor and protégé, Boodin's stimulating influence and insightfulness prompted Royce to confess to Boodin as he launched his career in the isolation of Grinnell College in Iowa (1900–1904) that it was the settled opinion of the faculty that he "was the best metaphysical mind that had come to Harvard so far."[16] Royce would remain Boodin's friend and professional ally until the former's death in 1916. When his student was searching for a publisher for what would become his book *Truth and Reality*, based upon his dissertation, Royce told George Platt Brett Sr. of Macmillan Publishing in a letter dated January 2, 1911:

> Professor Boodin is a rising man, with a good record and standing as a philosopher,—as strong a man as any of his age in the West. I have advised him to get himself into book form as soon as possible. His essays, so far, are predominantly although not extremely technical. But he is clear and wholesome, and ought to win a good hearing. I should advise your taking up the question of his proposed book. And I commend him to your earnest attention.[17]

This is high praise coming from one of America's most noteworthy living philosophers. It seems as though Brett took Royce's recommendation to heart; Macmillan would become Boodin's book publisher for the remainder of his career.

Given the intimate relationship between Royce and Boodin, it is worth examining the two a little more closely before moving on to a discussion of his subsequent years at the University of Kansas and the University of California at Los Angeles. Although Boodin always remained an independent thinker, it would be unrealistic to think that someone as closely connected to his early training as Royce would have no effect. For Royce the Absolute is manifested in humanity itself and everyone can (at least potentially) be assured of fulfilling this positive significance in their lives by being a part of a community expressed in the will of the Absolute. It has been noticed that in this there is a trend toward process philosophy, different from the methodologies of idealism but nonetheless reminiscent in the foundational work of Alfred North Whitehead and Charles Hartshorne.[18] Here Boodin's emergentism suffused throughout his "pragmatic realism and cosmic idealism" forms a synthesis between Royce and James catalyzed through "creativity" that somehow leaves one with the impression that there is always more of Royce than James in Boodin's pragmatism.[19] Although James thought that Royce's Absolute was a form of theological determinism that was unnecessary under pragmatic testing, we shall see that Boodin's *A Realistic Universe* (1916), *Three Interpretations of the Universe* and *God and Creation* (1934), and *Religion of Tomorrow* (1943) develop a unique and productive response to this question as if in answer to a dialogue between the two men that had been raging in his head since his Harvard days. All of this would be cast within a strong sense of community no doubt drawn especially from Royce in *The Social Mind* (1939), Boodin's important contribution to sociology. Of course, Boodin always gave science a close reading, and no scientific issues bearing upon these issues were rifer than the nature of evolution, both cosmological and biological. Here Boodin gave his response in *Cosmic Evolution* (1925) that would address ideas only touched on by Royce and James.

Academic Life: An Assessment, 1904–1967

Boodin would hammer out these works at three institutions. The first was at the University of Kansas beginning in 1904. His time at Kansas was troubled by campus politics and run-ins with the administration. Continued difficulties marked his departure, effectively begun with a leave of absence in which he spent 1912 to 1913 in "exile" in Cambridge, Massachusetts. In the fall of 1913, Boodin accepted an invitation to join the faculty at Carleton College from his good friend Donald Cowling, the institution's president. This period has been aptly described as one of "loyalty and deferred ambition." As Boodin's Kansas experience was marked by strife, his years at Carleton, located just south of Minneapolis, were tranquil, and the atmosphere must have seemed congenial since Boodin had worked with Swedish immigrants through St. Mark's Episcopal Church in that city some twenty years before. Despite these advantages, Boodin's ambitious career and burgeoning publication record had him yearning for a larger, higher profile academic setting. He spent 1927–1928 as a visiting professor at the University of Southern California where he was reunited with an old friend, Charles Henry Rieber. Rieber became dean of the College of Arts and Letters at what was then the Southern Branch of the University of California (now UCLA). The school was growing fast, and it was almost certainly through Rieber that Boodin was offered a position to join the philosophy faculty there in 1928. He would remain there for the rest of his career.

What influence did place have in Boodin's thought? He was certainly aware of it and said quite explicitly, "My philosophy has thus taken form on the prairies of the great Middle West—the heart of old America—with its strong common sense and conservative social philosophy. The expansiveness of nature, with its unimpeded vistas and yet its sense of homely friendliness, must, it would seem, in the course of years get into a man's spiritual outlook. So far as my social relations are concerned, they have mostly been with the Western students whose hearts and minds have been as expansive as the prairies."[20] One thing is sure: Boodin's ideas were shaped by distinctly American patterns of individualism tempered by community responsibility unfamiliar to Europeans constrained by ancient institutions hardened by centuries of tradition and custom. Swedish though he was, Boodin's mind was shaped and transformed by his experiences in America, from his blacksmithing days in Illinois to his life on America's western-most boundary of Los Angeles. Yet it would be good to remember that Boodin's ability to synthesize pragmatic realism with communal realism was probably, in part, due to his experience as a poor immigrant. As Nelson points out, if Boodin had a special knack for such a creation it was perhaps because he was a product of "the traditional peasant community and its mystical *Weltanschauung*."[21] Perhaps that was "the exotic" that James saw in Boodin.

In his own day Boodin was not ignored. As we have seen, his brilliance was quickly recognized by every one of his teachers; relationships thus formed were strong and lasting. He was elected president of the Western Philosophical Association fairly early in his career, and with nearly sixty articles and eight books to his credit, it would be inaccurate to say that Boodin's voice couldn't get a fair hearing. Indeed, Boodin did influence some important people in American intellectual life. One was America's poet laureate Robert Frost, who, it is said, read everything Boodin wrote and considered him the only worthwhile philosopher.[22] Another was John Steinbeck, who was introduced to Boodin's philosophy through Richard Albee, one of Boodin's students at UCLA. Steinbeck was drawn to Boodin's ideas developed in his *Cosmic Evolution* and *The Social Mind* (1939). Boodin helped Steinbeck form his "phalanx theory" based on the analogy of the Greek military strategy of gaining strength through close formation. For Steinbeck, writing during a time of great socioeconomic upheaval, the "group-man" could as a phalanx of collective strength work toward great social change, especially as workers struggling against the harsh "survival of the fittest" capitalistic ethos or as refugees caught in the throes of military oppression. Boodin's ideas would find their way into Steinbeck's *Tortilla Flat* (1935), *In Dubious Battle* (1936), *The Moon is Down* (1942), and most notably *The Grapes of Wrath* (1939).[23] Boodin didn't influence all of Steinbeck's work, but he had a significant impact on a major—perhaps the most important—portion of it.

Despite this positive attention, Boodin himself always felt neglected. He once complained to Arthur O. Lovejoy, "The only philosopher of standing who has taken pains to study my books is [Alfred] Hoernlé, and he has pleaded to deaf ears!"[24] Lovejoy's own response to Boodin was to prove his point by largely ignoring him. Boodin's exasperation was not entirely unfounded. R. F. Alfred Hoernlé, a devoted follower of the idealist philosopher Bernard Bosanquet and passionate champion of racial equality in South Africa, generally found Boodin's work congenial to his inclinations and did indeed point out that Boodin's work was being unfairly neglected. "The old saying that 'a prophet is without honor in his own country' seems to apply . . . to Professor John Elof Boodin," he charged in his review of *Cosmic Evolution* in 1927. "At least, it appears," he added, "that American philosophers have hardly accorded to Boodin's philosophical writings the credit which they deserve. His successive volumes . . . mark the development of a worldview of distinct individuality, in the expression of which Boodin has displayed a singular combination of philosophical thoroughness with strong poetic imagination and solid scientific knowledge—surely, an equipment which should have earned him wider recognition than has actually fallen to his lot."[25] By the time of his death Boodin still found his place among the panoply of American philosophers, but sparingly and not without criticism. Andrew Reck gave more complete consideration to Boodin in his *Recent American Philosophy*,[26] but just a few years later in the monumental and compendious work *The Encyclopedia of Philosophy*, he had this to say: "Boodin's philosophy never won a wide audience in its own time, and after World War II it receded from attention. Relativity theory in physics had raised difficulties with regard to Boodin's categories of space and time, and the retreat from holistic concepts in the life sciences had weakened the plausibility of his system. When contrasted with more recent forms of scientific philosophy, Boodin's thought seems to be vague and unwarranted, but also imaginative and speculative."[27] However accurate this assessment was when it was written over fifty years ago, I believe this conclusion needs sweeping and thorough revision. Important changes have occurred in the history, philosophy, and practice of both the physical and biological sciences since Reck offered this dismissive appraisal. It is the job of the remainder of this book to explain precisely what those changes have been and why they suggest a revival of Boodin's thought. In fact, Reck's consigning Boodin's obscurity to the "march of science" has a certain triumphalist tone that doesn't ring true. Surely other factors explain Boodin's erasure from American intellectual life.

The Anatomy of Obscurity: A Reassessment

There are three reasons for Boodin's decline and obscurity, none of which can be attributed to advances in physics and/or biology. First is the unfortunate experience of Boodin's early connection with George Herbert Palmer at Harvard. Palmer was one of those faculty members better known for their sycophantic relationship to university administration than their devotion to teaching or scholarship. There are few who have spent any time in the academy who haven't met yes-man manipulators like Palmer. Arthur Child, a former student of Boodin's at UCLA, provided this revealing glimpse into Boodin's early life at Harvard and its impact upon his teaching career:

> The careers of the Harvard doctors in philosophy were managed by George Herbert Palmer, and crossing Palmer could be fatal to a career. A great snob, long since recognized as a total non-entity, Palmer was one of those non-entities with a grasp of, and hold on, administration. Re-reading Santayana's autobiography [George Santayana had received his PhD under Royce and subsequently joined the Harvard faculty], *Persons and Places*, I find him remarking that Mrs. Palmer ". . . (unlike her husband), inspired me with immediate confidence and respect"—which suggests a good deal about Santayana's colleague. I take this to be the bitterness of Boodin's life: that a position at an important Eastern university, and above all Harvard itself, being in Palmer's conviction inappropriate for a Swedish immigrant, Palmer recommended him off to the Midwest [Grinnell College], where Palmer thought he belonged but where he never acquired the intellectual influence he might have exerted in the East.[28]

In addition to this inauspicious start was the timing of Boodin's career. His active period began with the publication of his revised dissertation, *Time and Reality*, as a monographic supplement to *The Psychological Review* (October 1904); it ended with the publication of *Religion of Tomorrow* (1943). Just a glance at this period exposes the severe distractions—two world wars and the Great Depression—facing America during Boodin's most active years. These certainly did not contribute to a widespread interest in the musings of a philosopher first in America's heartland of the Midwest then in the very young UCLA just beginning to establish itself. In fact, German philosopher Alfred Hoernlé, one of the few consistent supporters of Boodin's work, who (among other things) is remembered for providing, along with his wife, an English translation of mystic Rudolph Steiner's most important work, *The Philosophy of Freedom* (1916), openly admitted that the exigencies of both wars negatively impacted first Boodin's *A Realistic Universe* and then *The Social Mind*:

> *Inter arma silent leges*: would that reviewers, too, could be silent in times of war! It is a misfortune for Professor Boodin's book that it happened to be published in the month of the outbreak of the present war. This coincidence has largely robbed it of the attention it deserves, at least in all English-speaking countries involved in the war. And the war, too, is responsible for the fact that this review appears two years after the publication of the book. An apology is due, not only to the Editor and the readers of MIND, but, above all, to Professor Boodin himself.[29]

But it wasn't just the bad luck of being caught up in world events beyond one's control. A third factor relevant to Boodin's obscurity is the inopportune confluence of place and profession. The center of American philosophy was neither in the Midwest nor in the West (not to say it was entirely absent—Hans Reichenbach would build his career and UCLA's philosophy department with a brand of philosophizing very different from Boodin's, and in the Midwest the so-called Chicago pragmatists led by John Dewey along with George Herbert Mead [a name we'll encounter later] and others had already influenced American intellectual life through the 1930s). But by the post–World War II years pragmatism was on the wane and considered passé by the generation of scholars following Boodin's death. The centrifugal force came not from the West or the Midwest but from what has been called the "institutional nexus" of Harvard and Oxford in the postwar years up through the 1970s. "Hostile to metaphysics and certainly to the notion that philosophy was a guide to wisdom," writes Bruce Kuklick, "analysis rather looked at philosophy as an activity that clarified ordinary talk and the structure of science."[30] How effectively it did either one is debatable. Boodin's biographer puts it bit more pointedly: "The piecemeal and sectarian precision of modern philosophy was inadequate to the task [of addressing Boodin's great passion for metaphysics] precisely because it neglected all contexts other than linguistic or those associated with a narrow theory of knowledge."[31]

As Randall Auxier has observed, philosophy took a linguistic turn and became "reduced to the consideration of word-puzzles or to dialectical word-play."[32] However much the new generation of philosophers wanted to "clarify" ordinary talk, it surely became an exercise that made philosophy utterly irrelevant to the lives of most ordinary people. Thinking itself especially clever and far beyond the dated concerns of the pragmatists like Peirce, James, and Dewey, and certainly too "sophisticated" to be further bemused by Royce's Absolute or the personal idealism of George Holmes Howison or William Ernest Hocking, the profession lost its focus and turned into an increasingly incestuous self-absorption. "In the chaos," writes Auxier, "numerous important voices were lost to subsequent thinkers and scholars, including Royce's and two of his better students, Hocking and John Elof Boodin. Our understanding of pragmatism has been grievously diminished by this loss of memory, by the accidents of history, and by the fickleness of philosophical fashion."[33] It is the recovery of that memory that animates the present effort and the reintroduction of this unique and in some senses prescient voice back into general accessibility in the public arena.[34] Indeed far from consigning Boodin to Reck's ash heap of history, it has been suggested that it would be possible to refurbish much of Boodin's thought by rewriting it "in the present-day idiom" and in so doing present him to a new and appreciative academic and popular audience. As an American intellect, he can be considered an original and perceptive mind of the first order even exceeding that of Hocking.[35] I agree, and had I not been introduced to Boodin by and through Auxier's presentations of this engaging poet-philosopher, this book might never have been written, although I "discovered" him quite by accident searching for material related to Alfred Russel Wallace. More on him later.

But I believe there is an additional reason for a book on Boodin's thought and his place in American intellectual history. Boodin has particular relevance for today. Fifty years ago, John K. Roth gave a list of pressing issues relevant to the philosophy of Josiah Royce in this "age of crisis": the ecology of our planet, foreign entanglements, race relations, turmoil over educational responsibilities, and growing uncertainties over our moral values and religious life. As an undergraduate at what was then Northern Kentucky State College, I remember them well, and it is a remarkable—if rather sad—commentary that all of these are as pertinent today as when Roth presented them (some even more so!). Here we see perhaps Boodin's closest affinity to Royce, namely the sense of community and its unity of consciousness. "Especially in times of crisis like ours," Roth pleads, "there is a need to try to see things whole, to hope and strive for a way to affirm the basic intelligibility, meaningfulness, and goodness of our lives, both individually and communally. Without such effort, man's future is not very bright." We need most desperately "to become constructive metaphysical thinkers."[36] This was written in reference to Royce, but it equally applies to his student, Boodin. In this sense Boodin is keenly relevant to today, certainly more than the linguistic prattle of philosophers or the reductionist infatuations of so many of their colleagues who think (if I may borrow and alter a phrase from Whitehead) that all philosophy is but footnotes to Democritus. I have here in mind our all too easy degeneration into "sick" communities; treating scientists as a new priesthood, from Richard Dawkins's atheistic fundamentalism (really just the flip side of religious fundamentalists like Ken Ham) and the destructive "universal acid" of Daniel Dennett's neo-Darwinism to E. O. Wilson's scientism and Stephen Hawking's declaration that philosophy is "dead."[37] Symptomatic thinking for sociopathic times.

What we can learn from Boodin is a renewed sense of the importance of sound reasoning built into a comprehensive philosophy—a thoroughgoing metaphysic if you will—that is once again relevant to our daily lives. The totality of Boodin's metaphysic is epistemological, cosmological, and sociological *creation*. It is inherently non-reductionist, at once teleological and inherently religious so long as it is understood that the teleology is genuinely purposeful (not some "teleonomic" invention) and religious without degenerating into some ugly caricature of *sola scriptura* Bible thumping or hidebound reversion to doctrine. This is not Boodin. What *is* Boodin is something he called *pragmatic realism*. He presented it in *Truth and Reality*, one of his earliest works, as putting science into metaphysics in recognition of reality *as real*. It puts into practice the principles of Peirce and James in no uncertain terms by adopting an outward looking "truth attitude."[38] There is, of course, much more to Boodin than this, but this is a good starting point. This, then, establishes the need for Boodin's reappraisal and points toward some directions it might take.

Moving Forward (in History)

If my attempt to answer "Who was John Elof Boodin?" has raised more questions than it answered, good. The remainder of this book will attempt to answer those questions by carefully examining Boodin's thought and by bringing it into meaningful dialogue with his peers, past and present. In the chapters that follow I hope to introduce an old philosopher to a new audience and perhaps revitalize some aspects of American thought in need of some restoration. That, however, comes with a caveat: I do so not as a philosopher but as a historian. I have to believe that this would have met with Boodin's approval since he valued the historical approach highly and it suffuses his own writing, including one thoroughly historical work, *Three Interpretations of the Universe*. If part of the problem has been a loss of memory, then who best to restore it than someone whose stock-in-trade is the past? Restoring some of that memory by giving Boodin a renewed voice and a fair hearing should be a good start. I cannot rewrite Boodin and I have no intention of trying, but I do hope to give in the coming chapters a complete appraisal of his ideas, placing them in their historical context as well as comparing and contrasting them with other ideas that competed with Boodin's at the time. Here, caution is in order. Herbert Butterfield was far too perceptive a historian to ignore the "Whiggish" tendency "to praise revolutions [scientific or otherwise] provided they have been successful, to emphasize certain principles of progress in the past and to produce a story which is the ratification if not the glorification of the present."[39] We must follow Butterfield's lead in our appraisal of Boodin; in fact, we must disentangle Boodin from the Whiggish tendencies to which philosophers are as prone as historians.

We begin that task with an introductory chapter on science. As his definition of pragmatic realism would suggest, Boodin always gave science—whether in physics or biology—a close reading, and it is impossible to understand much of his philosophy without first understanding the scientific times in which he lived and the issues with which he grappled and ultimately synthesized into his own thought. It is important in this regard to reject the notion that science is somehow an inexorable march of progress. It is not now nor has it ever been. Here too we may proceed with some helpful advice from Butterfield and realize that little will be accomplished "if we merely pay attention to the new doctrines and take note of the emergence of the views we now regard as right. It is necessary on each occasion to have a picture of the older systems—the type of science that was to be overthrown."[40] This was Reck's mistake and we shall not repeat it here. In fact, we shall find that Boodin was often not on the "right" side of science but that today, whether in biology or cosmology, "right" has new meaning, a "*new*" founded upon concepts initiated long ago by Plato, about which more will be said in the introduction.

It is also important to understand the intellectual trends affecting science itself. The tremendous lure of positivism, its close twentieth-century cousin logical positivism, and its attendant scientism was definitely on the rise in the first half of that century, and not just within the ranks of the academy (though it seems to be its epicenter). While Boodin's virtual disappearance from American intellectual life has, as we have already seen, many sources, it cannot be fully understood without an appreciation of the blinding authority bestowed upon science during Boodin's career. The resulting reductionism, positivism, scientism, and materialism were all anathema to Boodin, and he maintained the courage of his convictions to refuse to yield to them. Like the pragmatists generally, Boodin used science as a verification principle, as a surgical scalpel to cut away extraneous material in order to see what philosophical positions really meant; positivists used verification as a guillotine to cut off any and all philosophy. For this reason a detailed discussion of this process must be explained and critiqued before an examination of Boodin's thought can begin in earnest.

Once completed, that effort starts with understanding Boodin's handling of time, which formed the basis of his epistemology. While other material becomes pertinent in this regard, the two main sources are *Time and Reality* (1904) and *Truth and Reality* (1911), though years later Boodin's 1937 faculty lecture, published as *Man in His World* (1939) would provide perhaps his most succinct and mature temporal statement. After establishing Boodin's epistemology, chapter 2 can develop Boodin's metaphysic, something James was never able to accomplish. Here we will want to look at *A Realistic Universe* (1916). In chapter 3, evolution exposes itself in three ways at a particular time, all in 1925: Darwin's theory; the discovery of the Taung child in South Africa, ostensibly vindicating Darwin's simian-to-hominid prediction; and finally, Boodin's very different *Cosmic Evolution*. Chapter 4 takes up Boodin's theology with discussions of his *Three Interpretations of the Universe, God: A Cosmic Philosophy of Religion* (both published in 1934), and *Religion of Tomorrow* (1943). As already mentioned, sociology was an important subject for Boodin, especially as it relates to the nature of persons and community. Chapter 5 covers *The Social Mind* (1939), which comprises his major treatment of the subject. In chapter 6 Boodin's defenders and critics old and new are examined and scrutinized to place him in context. Besides this broad outline, much filling in along the way will be included as needed with Boodin's extensive body of journal literature, including the extremely valuable *Studies in Philosophy: The Posthumous Papers of John Elof Boodin* (1957) compiled and edited by his UCLA colleague and friend Donald Ayers Piatt. The book concludes with an epilogue that suggests three broad areas of Boodin's relevance for today—science, history/philosophy, and modern culture.

But before moving on to the introduction it might be emphasized here that rather than a working biography or philosophy this is essentially a history, namely a history of Boodin's thought. As a work of history, the reader is warned against what C. S. Lewis called the error of chronological snobbery, an unwarranted privileging of "the intellectual climate common to our own age" based on the assumption that all previous views have been outdated and discredited on the sheer basis of their replacement. These "newer is better" prejudices, Lewis reminds us, "are likeliest to lurk in those wide-spread assumptions which are so ingrained in the age that no one [or *almost* no one] dares to attack or feels it necessary to defend them."[41] In the coming pages a number of well-entrenched icons will be challenged, and the reader is simply asked not to succumb to the temptation Lewis warns against. Some ideas deserve replacement or at least modification; others have been cast aside too hastily in the face of mere fashion. We must decide which is the case.

With this in mind, we might begin by referencing what history meant in Boodin's epistemology. He wrote: "History must be regarded, then, as our ideal construction on the basis of present symbols, which represent a factual order, now real only for us. Its justification is a practical one. In appropriating the institutional or accumulated life, we come to consciousness of ourselves, we come to understand *our* world better and anticipate better its behavior, though the music and discord of the past have been merged into the movement of the present."[42]

The goal of this book is to achieve the same. Hopefully by its conclusion the reader will see in the rich dimensions of Boodin's thought—rescued from an unfortunate collective amnesia—its relevance for our "present strata" in order to enhance our own consciousness, in short, "to understand *our* world better."

I must add by way of postscript that the frequent quotations throughout the coming pages have a purpose, namely to let the voices of these thinkers express themselves as they expressed them to Boodin and as Boodin expressed them to us. We must hear for ourselves the ideas directly, whether from Whitehead or James or Bergson or Boodin or Boodin's commentators, friends or foes. Quotation here is not a shortcut, it is a projection, an amplification of voices through time spoken directly and *immediately* to us. Neither is quotation an abrogation of interpretation. We can do both.

Introduction

Science and Its "Mad Clockwork of Epicycles"

The Key to Understanding Boodin

Because Boodin always gave science such a close reading in his work, some preliminary comments on the state of science in the first half of the twentieth century are in order. If there is one single feature of science worth noting in Boodin's lifetime, it is the rise of positivism, both as a theoretical and as an operative principle that would, for many, become the reigning *Weltenschauung* underlying a host of isms that became empirical masks for a priori ideologies about the "proper" relationship between science and society. The problematic nature of the greatest of those isms—Darwinism—has been revealed by the careful work of historian John C. Greene. Almost the entire positivist program was also built upon the fiction of science as a cumulative building-block of progress, a corrective provided by Arthur Koestler and Thomas Kuhn. Once these thorny and controversial issues are examined, Boodin's difficult but courageous position as a vocal non-reductionist in a reductionist age can be seen.

Understanding Science in Boodin's Day and Ours

As we have seen, Boodin's active career lasted thirty-nine years (from 1904 to 1943). This means that he lived through three signal events—properly considered revolutions—in science: first, the rise of the Darwinian theory of evolution with the publication of *On the Origin of Species* on November 24, 1859, and its subsequent synthesis to incorporate modern genetic science

1

in the 1930s and 1940s; second, Einstein's special theory of relativity first proposed in a paper published June 1905 in the *Annalen der Physik* and proven in a series of confirmations from 1914 to 1916; and third, the rise of quantum theory beginning with Max Planck's "quantum of action" idea in 1900, followed by Einstein's "quantum theory of the solid state" in 1906, added to with Niels Bohr's "atom-model" that evolved over a period of time between 1913 and 1924, eventually leading to Werner Heisenberg's award of the Nobel Prize in physics for the discovery of quantum mechanics in 1932 and to Irwin Schrödinger's Nobel Prize the next year, in the committee's words, "for the discovery of new productive forms of atomic theory." Boodin would carefully follow these developments virtually as they occurred.

Such a summary invites the naive assumption that science proceeds linearly, like building blocks of seminal discoveries stacked one upon the other rising to a great intellectual edifice of human achievement. We might all dismiss this as the exuberance commonly displayed by over-achieving adolescents at the local high school science fair but for the fact that it is frequently displayed by "experts" acting as ambassadors in their respective fields to the public—E. O. Wilson, Richard Dawkins, and Stephen Hawking are examples. And if the bestseller lists are any indication, they have a large and enthusiastic audience. All of them have had distinguished careers: Wilson in myrmecology (the study of ants), Dawkins in evolutionary biology, and Hawking in theoretical physics and cosmology. All of them are examples of good scientists doing bad philosophy. One particularly egregious example is the enormous bully pulpit given to "Bill Nye the Science Guy" in his TV series funded by the National Science Foundation and aired on PBS. His principle expertise appears to lie in his ability to tie a bowtie. While most of it is pretty innocuous, he is sure to tell kids everywhere in his intro song that "science rules!" Reason and logic found new lows when Bill Nye debated Ken Ham at Ham's Creation Museum (more properly called a fundamentalist theme park) on February 4, 2014, becoming every thinking person's game of the weak.

These more recent examples notwithstanding, this attitude has a long history; much of our understanding of science today is rooted in conceptual ideas born in the seventeen century, but its cultural embodiment came two hundred years later. In fact, the term *scientist* was coined by William Whewell when the British Association for the Advancement of Science convened at Cambridge on June 24, 1833. At that meeting William Whewell pushed aside Kant's declaration that philosophy was the "Queen of the Sciences" and crowned astronomy with that honor, agreeing with Samuel Taylor Coleridge that "natural philosopher" was "too wide and lofty" a term. Something less speculative and concrete was needed, "by analogy with *artist*," he suggested, "we may form *scientist*."[1] It was an interesting shift in perspective that not only sought to relegate the older designation of natural philosophy to the history books but also revealed an attitude toward the discipline devoted to the pursuit of wisdom to be too vague and impractical for his tastes. As significant as this was, the infatuation with science and especially with the promise of scientific progress for our purposes might be said to have begun with Auguste Comte.

The Rise of Positivism

Despite its distinctly French genesis, positivism found a small but influential group of adherents in Britain, first with Richard Congreve, who under Comte's spell openly espoused the "Religion of Humanity," started a positivist church in its name, and was openly declared head of the British arm of the movement by Comte himself in 1857. Others followed, such as Edward Spencer Beesly, John Henry Bridges, George Earlam Thorely, Frederic Harrison, John Stuart Mill, and George Henry Lewes. These British apostles of positivism were following the ideas of Auguste Comte, born in Montpellier, France, in 1798. Comte studied at the Ecole Polytechnique in Paris and befriended Saint-Simon. His initial motivation was to demonstrate that philosophy was becoming absorbed by the sciences. He eventually built this into a scientific religion that even proposed a new positivist calendar of 558 "Great Men" to replace the saints of the liturgical calendar. It was to be a "world-wide faith growing out of philosophy, with positive [hence *positivist*] knowledge at the root."[2] Comte's positivism had two aspects: the first, science validated by a hard verificationism; the second, a system of ethics best considered as secular humanism. Seeing the past as a series of organic developments, Comte always displayed an element of historicism. He accordingly saw humanity as emerging in three stages of development: from superstition and otherworldliness (the *theological* period); then what he called a "fiction of abstractions" (the *metaphysical* period); and finally the organization of knowledge into rational categories, systems, and disciplines (the *scientific* period). Put another way, human society—Comte-invented sociology—moved from legends and myths to the ideas of Plato and Aristotle to the Laws of Nature (duly capitalized to indicate their ontological significance). For positivists, then, all philosophical propositions (especially metaphysical ones) were simply meaningless, not worthy of serious consideration. All "legitimate" inquiry could be reduced to scientific inquiry with science itself pointing the way toward virtually limitless improvements in the human condition. As such, it is really a comprehensive system best labeled as *scientism*.

This is important because the theory of evolution as presented by Charles Darwin, not only in his *Origin* but also his *Descent of Man* (1871), is really at heart a doctrine of positivism. The theory itself simply states that all life developed from common descent with the diversification of species by means of natural selection. Evolution itself wasn't new. Comte de Buffon gave an account in his *Histoire naturelle* (1749–1789) as did Jean Baptise Lamarck in *Philosophie Zoologique* (1809) and *Histoire naturelle des animaux san vertèbres* (1815). Even Darwin's grandfather Erasmus contributed a deistic version of evolution with his *Zoonomia* (1794–1796). Robert Chambers tilled the intellectual soil of Britain with transmutationism in his *Vestiges of the Natural History of Creation* (1844). Chambers's speculative and quirky work invited considerable criticism, but it also got everyone thinking and talking about transmutation. It sparked another naturalist, Alfred Russel Wallace, on his own extensive journey (first in South America, later in the Malay Archipelago) that would result in his collateral discovery of natural selection that he shared and shocked Darwin with in a long letter from the remote island of Ternate. When Darwin received it on June 18, 1858, he exclaimed, "I never saw a more striking coincidence, if Wallace had my M.S. sketch written out in 1842 he could not have made a better short abstract! Even his terms now stand as Heads of my Chapters." But Darwin's theory with natural selection as the evolutionary drivetrain was unique. As one historian explains, with natural selection, "differential death rates caused by purely natural factors created new species. God was superfluous to the process."[3] More importantly, as Curtis Johnson convincingly argues, Darwin sought to enthrone chance as an operative principle in biology, a principle that gave evolution "a single meaning to Darwin from beginning to end."[4] For Darwin, chance meant something much more than simply an "unknown cause" but rather an unknowable cause "not [due to a] lack of human understanding but rather a lack of directing rational agency."[5] Although this would develop over time, the atheistic implications were understood early on (as early as the spring of 1838 in his personal notebooks and a bit later in his "Old & Useless Notes"). Here the idea is expressed that life sprang from inorganic matter as a purely "contingent" result of responses to the immediate agencies of chemical catalysts responding stochastically to heat, light, and other atmospheric conditions. Darwin would reiterate this idea in a letter to Joseph Hooker in 1871:

> It is often said that all the conditions for the first production of a living organism are now present, which could ever have been present. But if (& oh what a big if) we could conceive in some warm little pond with all sorts of ammonia & phosphoric salts—light, heat, electricity & c present, that a protein compound was chemically formed, ready to undergo still more complex changes, at the present day such matter would be instantly devoured, or absorbed, which would not have been the case before living creatures were formed.[6]

Darwin, it seems, had settled on this "big if" thirty-three years earlier in his private notes. As Johnson concludes, "Here, at last, we find Darwin's dangerous idea: the pure chance origin of life on this planet. Once chance governs the beginning, chance can govern all the way down. God has been ushered out the door."[7] As for morals, those too are wholly naturalistic, derived from sympathies evolved from our social instincts.[8] If we are to "be moral," we need look no further than our selves. Taken on the whole this is science wrapped in the dogmas of positivism.

Associating Darwin with positivism should not be seen as something new and avant-garde. Darwin's positivism has considerable historiographic precedence. Neal C. Gillespie discusses Darwin's positivism at length, which he has defined as "that attitude toward nature that became common among men of science . . . which saw the purpose of science as the discovery of laws which reflected the operation of purely natural or 'secondary' causes. It typically used mechanistic or materialistic models of causality," and he rejected any supernatural or teleological factors that it regarded as beyond the scientific pale.[9] Intellectual historian John C. Greene agrees with Gillespie as have others.[10] In his theory of social evolution, Greene calls Herbert Spencer a "Darwinian before Darwin" with his emphasis on population pressure and survival of the fittest. "Both [Spencer and Darwin] were powerfully influenced by the positivistic faith of nineteenth-century science as the sovereign key to knowledge of reality."[11]

Given Darwin's strong inclination toward positivism one would think both he and his outspoken apostle, Thomas Henry Huxley, would have been drawn to the British Comteans. After all, both had almost limitless faiths in science and were equally wary of traditional religion. But Huxley was a vocal opponent of positivism and serves as a good example of why neither he nor Darwin ever joined with Comte's British allies. Huxley was opposed only to the Comtean window dressing added to its more substantive program of hard verificationism, rejection of metaphysical speculation, and insistence that knowledge can be reduced to the senses. Huxley had no patience with a "Religion of Humanity," a positivist priesthood, or what he derisively called "Catholicism *minus* Christianity."[12] As historian Neal Gillespie puts it, for Huxley, "It was not enough to drive out the old ideas. Their advocates had to be driven out of the scientific community as well. . . . In order for the world to be made safe for positive science, its practitioners had to occupy the seats of power as well as win the war of ideas."[13] It would do no good to build a new scientific order upon old ecclesiastical trappings and notions of a humanistic religion. In this sense, Huxley's opposition to the positivists of his day made perfect sense. This notwithstanding, English positivist Frederic Harrison, who sparred with Huxley on several occasions, emphasized their many points of agreement and even offered to bestow Darwin's bulldog honorary membership in the positivist movement. Huxley, for his part, was willing to accept Harrison's olive branch "from the plenipotentiary of latter-day positivism" only if offered under better terms, presumably without the absurd Comtean baggage.

Getting "Caught on the Scientific Horn of the Positivist Dilemma"

Much more will be said concerning Darwin's evolutionary theory in chapter 3 (including its advance with the neo-Darwinian synthesis in the 1930s and 1940s), but more immediately, within the context of the present discussion and as an idea, positivism faces at least four serious dilemmas. The first is that the positivistic faith in science claims to be based upon some idea of hard commitment to reality but cannot escape its own naiveté on the subject. Science simply is not a discipline of steady or cumulative advance. Thomas Kuhn, for example, has stressed science as a noncumulative process. In effect, these revolutions in our understanding of nature remake the scientific discipline anew. Kuhn argues that scientific disciplines rest upon paradigms (i.e., "universally recognized scientific achievements that for a time provide model problems and solutions to a community of practitioners") and that accumulating anomalies push these paradigms to the breaking point. The ensuing crisis in any one of them at a given time develops into a full-fledged revolution in which the old order collapses and a new one emerges, transforming the nature of the discipline itself under a fresh paradigmatic structure.[14] Interestingly, Kuhn sees Darwin's theory as a "very nearly perfect" example regarding progress through resolution of a revolution. Greene is less convinced. He sees Kuhn's model as "sealed off" from significant outside issues of politics, power, and national cultures. He is doubtful that Darwinism even fits the Kuhnian idea of a revolution since Darwin's theory arose not amidst pressing anomalies within biology but instead from other outside influences, especially in geology, to wit, Lyell's uniformitarianism. Kuhn also fails to consider counter-paradigms from Buffon or Lamarck.[15] Kuhn does have something valuable to say about the noncumulative nature of science, but it is also a limited monistic, non-contextual model of scientific change.

But we didn't need Kuhn to tell us that science was not a cumulative enterprise. The fascinating polymath, Arthur Koester, told us as much several years before:

> In fact, we have seen that this [scientific] progress was neither "continuous" nor "organic." The philosophy of nature evolved by occasional leaps and bounds alternating with delusional pursuits, *culs-de-sac*, regressions, periods of blindness, and amnesia. The great discoveries which determined its course were sometimes the unexpected by-products of a chase after quite different hares. At other times, the process of discovery consisted merely in the cleaning away of the rubbish that blocked the path, or in the rearranging of existing items of knowledge in a different pattern. The mad clockwork of epicycles was kept going for two thousand years; and Europe knew less geometry in the fifteenth century than in Archimedes' time.[16]

The positivist faith in science was a curious one indeed, not one that could instill a great enthusiasm for the future of humanity. Was the technocratic world that produced the Bomb, global pollution, eugenic genocide, and urban clutter one to instill confidence in positivists' hope for science? Was it really better than the theological world it supposedly improved upon? Koestler thought not. "A puppet of the Gods is a tragic figure," he admitted, but "a puppet suspended on his chromosomes is merely grotesque."[17]

A second positivist dilemma is its insistence that philosophy and metaphysics are meaningless while they boldly—and apparently blindly—express a myriad of metaphysical positions. This was particularly noted with regard to the so-called logical positivists, Comte's twentieth-century permutation especially in the Vienna Circle founded by Moritz Schlick in 1924 that included Gustav Bergmann, Rudolf Carnap, and Herbert Feigl. Alfred North Whitehead in his *Adventures of Ideas* chided scientists and even some philosophers for avoiding metaphysics by saving "the importance of science by an implicit recurrence to their metaphysical persuasion that the past does in fact condition the future."[18] Similarly, Philip Paul Weiner complained that

> neo-positivists [logical positivists] who define the class of possible operations by reference to future acts of verification can do so only by assigning some metaphysical status to time. The necessity of a temporal distinction in distinguishing operationally actual from possible operations presupposes an absolute assumption about the temporal character of existence and knowledge. This implicit reference to a constant flow of events falls in the class of materially certain and necessary truths. How can this assertion be made compatible with the neo-positivist doctrine that all necessary propositions refer only to the tautologies of discourse and not to events?[19]

As Brian G. Henning has wryly observed, "One can choose one's metaphysic, but one cannot choose not to have a metaphysic."[20]

A third dilemma in positivism relates not just to its rejection of metaphysics but its demand for a *conclusive verification* that is impossible to meet. This is not an undesirable goal, but realistically it is simply unrealizable in all instances. Hilary Putnam's favorite example is the statement that "there are no extraterrestrials in the universe" is not amenable to conclusive verification; there might be some verification if ever we contact them or they choose to contact us, but if that statement is true, it will never be conclusively verified. Yet the statement itself should not be regarded as meaningless. In fact, Putnam uses the positivists notion of verification to distinguish them from the pragmatists. "In short," writes Putnam, "for the positivists, the whole idea was that the verification principle should *exclude* metaphysics (even if they were mistaken in thinking that their own ideas were simply scientific and not metaphysical), while for the pragmatists the idea was that it should *apply to* metaphysics, so that metaphysics might become a responsible and significant enterprise. There is all the difference in the world between these attitudes."[21]

Finally is what Greene calls "the scientistic horn of the positivist's dilemma." If science is so narrowly defined so as to exclude all value judgments, all nonempirical statements, then how is *anything* to be regarded as valuable or important? In fact, why should there be any passion for systematic inquiry at all? If, on the other hand, science can serve as a reliable guide to human destiny, which one do we choose: Julian Huxley's humanism, Comtean humanism, Watson and Skinner's behaviorism, Galton's eugenics, Marx's communism, Freud's psychoanalysis, Jung's analytic psychology, and so on? As Greene concludes, "In the ensuing struggle the central idea of science as an enterprise in which all qualified observers can agree as to what the evidence proves vanishes from sight. Thus, whichever horn of the dilemma the positivist takes, science is the loser."[22] In effect, science confronts its own scientism.

In the end, Darwin's attachment to chance as a real force in nature and the constant pull of positivism caught him up near the end of his life in a rather sad nihilism. He told William Graham less than a year before his death, "You have expressed my inward conviction, though far more vividly and clearly than I could have done, that the Universe is not the result of chance. But then with me the horrid doubt always arises whether the convictions of man's mind, which has been developed from the mind of the lower animals, are of any value or at all trustworthy. Would any one trust in the convictions of a monkey's mind, if there are any convictions in such a mind?"[23]

Science versus Scientism and the Pragmatists' Answer

Of course, the problem being delineated here is not science but scientism. The pursuit of systematic inquiry into natural phenomena has yielded wondrous benefits in many ways, and it is as old as human endeavor itself. All the pragmatists appreciated science, Boodin not the least of them. But as a human endeavor it is not omnipotent. When it engages in boundary transgressions into other areas of human inquiry—philosophy, theology, the arts—problems arise. We have been discussing this in the previous section primarily in terms of biology, but it can be seen in physics as well (another discipline that Boodin followed closely). Unfortunately, this seems least appreciated by some of the most popular physicists who are most ready to discard the one discipline most capable of making sense of the physical world's data. The renowned Richard Feynman was always scornful of philosophy, the late theoretical physicist/cosmologist Stephen Hawking has proclaimed that "philosophy is dead," and astrophysicist Neil deGrasse Tyson considers it "useless." Of course, there are important exceptions like Werner Heisenberg, who was no stranger to philosophy having read Plato's *Timaeus* as a youth hiking through the Bavarian Alps. In answer to the question "Who was right the atomist materialists or the platonic realists?" he said, "I think that modern physics has definitely decided in favour of Plato. In fact these smallest units of matter are not physical objects in the ordinary sense; they are forms, ideas which can be expressed unambiguously only in mathematical language."[24] Also, theoretical physicist Freeman Dyson has escaped the sickness of scientism.[25] Not so with these high-profile nay-sayers who have lodged themselves within our popular culture as authoritative spokesmen for science. Feynman's books are still readily available and read, Hawking's status is iconic, and Tyson's presence seems ubiquitous on major media outlets, from PBS's *NOVA* to his own television series *Cosmos: A Spacetime Odyssey* and most recently his *StarTalk* guest interview show currently aired on the National Geographic channel. Nobel laureate in physics Frank Wilczek, despite his eloquent portrayals of the universe in *A Beautiful Question* (2015) and *Fundamentals* (2021), clearly believes that if science is not omnipotent it is surely potentially omniscient to any question that reasonably "makes sense." This is just another variation on scientism since what "makes sense" is, for Wilczek, cast within his magisterium of science and it alone.

Sociologist Stanley Aronowitz knows better. Science is never an enterprise divorced from the society in which it resides. He writes: "As scientific discourse permeates state and civil society, scientific culture spills over beyond the laboratory. Business dares make no decisions that are not grounded on mathematical calculation that provides projections; legislators enact laws based on 'data' generated by scientifically trained experts."[26] Science today is subsumed by financially and politically powerful agencies sporting three-letter acronyms like the NIH, NSF, CDC, and FDA, all (with the exception of the NSF) well under way during Boodin's lifetime. In this manner society itself becomes subservient to technocratic elites, and however innocent and well-intentioned their motives, it soon finds itself dominated by an ethos of scientism. As Aronowitz explains, "At issue is the claim of enlightenment science to certainty and its refusal to acknowledge its own discourse as a form of ideology."[27] As the physicists described above search for a "unified theory" that can dispense with the untidy issues raised outside their field, Aronowitz wisely points out, "The facts do not speak for themselves and, through this door, marches religion and other metaphysical doctrines as well as philosophy."[28] It is this door so many high-profile physicists and biologists wish to close.

In some ways there is nothing new here. Those who would deign to speak on behalf of today's science are quick to chide their medieval forebearers like John Buridan, Nicholas of Autrecourt, Thomas Bradwardine, Duns Scotus, and Roger Bacon of being constrained by the superstitions of the Church; they believe too readily in the myth of science and religion warfare and suffer from presentism, conveniently forgetting that their science suffers from its own constraints. "The popular image of the medieval Church as a monolithic institution opposing any sort of scientific speculation is clearly inaccurate. . . . But the price of having a rich sponsor," writes James Hannam, "is having to bend to their interests and avoid subjects they find controversial. Modern scientific researchers competing for funds from big companies have exactly the same problem. The Church allowed natural philosophers a much wider dispensation than many corporate interests allow their researchers today."[29] Then as now, we need to be reminded that science is a method and an approach that is not a system of ultimate beliefs; it cannot be decontextualized or reified into an idol. Science, as historian Steven Shapin reminds us, is "never pure."[30] When it sports an "ism" it becomes even less pure, not because isms are necessarily bad—they stand or fall on their own merits—but because there is no surer sign of a boundary transgression between science and philosophy as when the former becomes an ism. Boodin surely has his isms but they are part of his overall philosophical outlook and not part of an allegedly "purely" scientific credo.

This leads to one of the untoward consequences of scientism—its role in what Hilary Putnam has productively critiqued as the fact/value dichotomy. It is, in fact, according to Putnam, the dangerous legacy of logical positivism: the notion that values are *subjective*, beyond the scope of rational argument. Putnam insists that there can be responsible inquiry into value questions. This is, of course, of immense practical importance in affirming certain normative values in society. Here a number of philosophers of science must claim a certain responsibility for the problem. Putnam's masterful summation brings us back to the pragmatists and is worth quoting at length:

> I have argued that my pragmatist teachers were right: "Knowledge of facts presupposes knowledge of values." But the history of the philosophy of science in the last half century has largely been attempts—some of which would be amusing, if the suspicion

of the very idea of justifying a value judgment that underlies them were not so serious in its implications—to evade the issue. Apparently any fantasy—the fantasy of doing science using only deductive logic (Popper), the fantasy of vindicating induction deductively (Reichenbach), the fantasy of reducing science to a simple sampling algorithm (Carnap), the fantasy of selecting theories given a mysteriously available set of "true observation conditionals," or, alternatively, "settling for psychology" (both Quine)—is regarded as preferable to rethinking the whole dogma (the last dogma of empiricism?) that facts are objective and values are subjective and "never the twain shall meet." That rethinking is what pragmatists have been calling for for over a century. When will we stop evading the issue and give the pragmatist challenge the serious attention it deserves?[31]

Some may be confused by William James's call for radical empiricism, thinking it has some relationship to an undue—perhaps even a scientistic—attachment to science. But as James Woelfel has explained, James "remains a fresh and timely voice on a range of contemporary issues" that include "an open-ended attentiveness to the richly nuanced concreteness and complexity of experience that refuses to reduce either the human phenomenon or the universe itself to the boundaries set by the sciences."[32] For James, *radical* empiricism was not radical in its reductionism at all but instead in its thoroughgoing commitment to *experience*, indeed an ontology of pluralistic "pure experience." Unlike those who would sacrifice values on the altar of science, James rescues values from such an inglorious fate. He fears an all-embracing science that threatens "the diminution of man's importance." The practical import of this is that "James promises us an outlook that will enable us to hold on to both our love of fact and our confidence in our 'human values,'" write Ruth and Hilary Putnam, "and do so without transcendentalizing those values that they become ineffectual."[33]

Another pragmatist response to scientism can be found in John Dewey. Although Dewey's appreciation for scientific method sometimes associates him with positivism or even scientism, some regard this as an unfair and superficial reading. More will be said on this in chapter 6, but Deweyites insist that his appreciation of science never meant it to be the *only* valid epistemology. For Dewey, the association of what is known with what is real is a serious "intellectual fallacy." Knowledge cannot, for Dewey, be reduced to science alone; experience is pluralistic in a way that science gives but one of many pictures. Scientific method as Dewey sees it is at once experimental and fallible. Dewey's call for a "scientific treatment of morality" is not, on a Deweyite reading, positivistic or scientistic. Gert Biesta makes a case "that rather than aiming for a scientific treatment of morality, Dewey was actually articulating a moral treatment of science."[34] In this Dewey made a lot of assumptions about how science has transformed our understanding of the world and was too taken in by the Darwinian explanatory lure, but his main concern expressed as "How is science to be accepted and yet the realm of values to be conserved?" is not for Deweyites amenable to a positivistic relegation of values to subjectivity.[35] Interestingly, this was not entirely Boodin's interpretation of Dewey, which we will examine later.

Back to Boodin

It is within this complex milieu of ideas that Boodin grappled—the rise of positivism and scientism and its impact on philosophy, all of which are being keenly felt today. While confounding and marginalizing him professionally, it also gave him significant opportunities to craft an alternative philosophical view and metaphysical vision. Nothing makes you stand out better than when you are battling a clear and present danger. The pressing issues outlined in the previous section should make clear the relevance of Boodin today. We can thus see in Boodin James's pragmatism and Royce's absolutism, but he made his own way and developed his distinctive answers to our relationship with nature and the universe. Having examined all his published work, I cannot see Boodin as anything other than an underappreciated philosopher of the first rank whose poetic writing style and perceptive grasp of twentieth-century science and the history of philosophy deserves a fresh reading and a fair review. Moreover, Boodin did something James was never able to do, complete a comprehensive metaphysic. Boodin was able to creatively transform the ideas of James and Royce into something genuinely new and, I believe, important.

We are now able to begin our intellectual journey with Boodin in earnest. In the pages that follow, Boodin's ideas will be presented as completely and systematically as possible. In so doing I have not relied on Boodin's private papers and correspondence, which reside in the special collections of UCLA. If this were a conventional biography that material would be indispensable, but, as mentioned earlier, this is not a biography. Boodin's ideas as presented in the public arena are of interest here. We will begin with an examination of Boodin's epistemology in his treatment of time.

Before proceeding, however, it might be appropriate to end this introductory chapter with a sense of the poetic spirit of his writing. It comes out periodically, but initially in the opening chapter to *A Realistic Universe*, "The Divine Five-Fold Truth." Designed as a lead to his metaphysics, despite its beauty it is not expansive enough to capture the whole of his philosophy. For that "Reality as Actuality" in his *Studies in Philosophy* seems a better candidate. It appears as he left it and as Donald Ayres Piatt published it seven years after his death. It serves as an interesting expression of his life and his work as he perceived it, emphasizing those themes—community, time, space, history, nature, and the cosmos—that formed such an intimate part of his thought. Here is Boodin's love of science, of life, of faith, and of philosophy waxing most eloquent:

> I live in a community of time and space. Temporarily I am old as life. The history of life is my history. In me are the "recollections," the traces of the whole history of life; and in its general features my history recapitulates this history, though the perspective has been, in many ways, foreshortened. Walt Whitman gives this expression to his idea in poetic form in *Song of Myself*:
>
> > I find I incorporate gneiss [rock], coal, long-threaded
> >   moss, frutis, grains, esculent roots
> > And am stucc'd with quadrupeds and birds all over,
> > And have distanced what is behind me for good reasons,
> > And call anything back again when I desire it.
>
> Of course we must not expect accurate description from a poet, but his intuition is sound. We are part of the temporal community of life. The history of life is my history—a history of adaptation to the cosmos. I am as old as the hills and older. The constituents of my body were forged prospectively through the history of the earth and the sun, back to the nebula from which the sun was born. In the crucible of nature the elements of my body were formed in due proportion. In the backward view, nothing is foreign to me. And the form of the future is indicated in the history that has passed and is, as the man is

indicated in the conditions of nature in his embryonic history. The restlessness of dust is part of my inheritance as are the passions of the animals—their love and hate—and man's long groping to find a satisfactory life in union with his fellow man and nature.

I live in a community of space. The extensity of my life in space comprises the whole cosmos. I am part not only of the field of the earth with its atmosphere and of the sun of which our earth is a part, but my life includes the most distant galaxies. Cosmic radiation from the most distant parts affects my life, though unknown to me; so that my life would be different if this radiation were different. I am part of a community of space as well as of time. This whole sensed world is part of me and much that I do not sense. Sirius is part of me and I am Sirius in this perspective. The world is mine and all its glory—and tragedy. But I also have a wealth that belongs to me alone, of feeling and emotion, of will and thought; and before this richness, the magnificence of the stars pales.

And I am part of the theme, the space-time structure, of the whole which is expressed in the vast drama of time and space, which governs the entrances and exits of galaxies as well as our entrances and exits. Time is immense and space is immense but eternity is a theme that pervades time and space and gives rhythm and order to change in a contingent world. The rhythm of electrons, of atoms, of life is an expression of this universal order. And this is an order of time and not merely of space. To stop time would be an end to all music and drama. It would mean the freezing of chaos. If we could view the cosmos from the point of view of the whole, the symphonic structure of the field of fields, we would see the cosmos as the actuality which descends centrifugally, through various levels of spirit, soul, and body, thus engendering a centripetal striving towards the actuality of the field of fields. But the relation is not a simple logical relation from premises to conclusion. The relation is a creative interaction of fields within a pluralistic world—a world of contingencies where the parts may fail and often do fail of adjustment, with consequent tragedy. Without tragedy there can be no progress. Without suffering we cannot learn our failure. And the spirit of the whole must love and suffer (in a way we cannot understand) or there can be no atonement, no harmonization.

Within this drama, my personality is the intersection with the route of nature, viewed from my perspective, not only with the contemporary community, physical and spiritual, but also with the whole direction, the ought, the eternal Logos as it becomes incarnate in finite relationships. In this cosmic symphony of movement, my vocation is to realize my actuality, as I am able and ready, from my perspective in harmony with the whole.[36]

Chapter 1

Boodin's Time

The best starting point for Boodin's thought is his theory of knowledge, an epistemology presented in two works that launched his career. The first is *Time and Reality* (1904), a revision of his doctoral dissertation; the second is a much larger and more detailed work, *Truth and Reality* (1911). It is no small tribute to the enduring quality of Boodin's epistemology that nearly a century later this latter book would be selected for volume 2 of Thoemmes Press's *Early Defenders of American Pragmatism*. There he takes his place, according to series editor John R. Shook, among "all pragmatists of stature in their own right." But insofar as Boodin started his epistemology with an in-depth discussion of time, this chapter begins with a broader analysis of time within the context of current theory of his day. It should be recognized that there are three different kinds of temporality—organic time, mechanical/clock time, and physics time—and how those are delineated becomes critical; this will become apparent when Boodin's contemporaries—James, Royce, and Bergson—are compared in their handling of time. After examining these aspects of the question, both of Boodin's works are discussed in detail and conclude with critical reviews among his peers.

Escaping the "Clutches of Time"

Boodin began his theory of knowledge (epistemology) with an examination of time. Like his mentors James and Royce, Boodin was nothing if not a temporal philosopher, and as such, history, the most temporal of disciplines, was an important element of his overall thought. It should be added that one of the important goals Boodin undoubtedly got from James and Royce was the imperative of restoring wholeness to persons alienated by industrial society. One of the ways they tried to do this is to reignite a sense of the spiritual, a reconnection of body and soul. For James the means to this end was through psychology; without being religious, he nonetheless took religion very seriously. Royce, on the other hand, was essentially a religious man who expressed himself through the Absolute. Put differently, "The restoration of wholeness in such individuals in Royce's time required only the adapting of the essential structures of Pauline Christianity to the historical peculiarities of the modern age—temporalism, and specifically historicism provided the required structures for adaptation, because Royce's age believed in history when it could no longer believe in myth."[1]

But what happens when history itself becomes myth? Time as expressed—or perhaps better *experienced*—in history is hardly monolithic or static. The question posed here asks not for a philosophical explanation of time but instead for a thoroughly historical account of time. Nonetheless, the two are related and this excursion is ultimately important to understanding Boodin's approach to time. More broadly, to have a complete appreciation for temporal thought it seems essential to understand time in its manifestly historical aspect. James, Royce, and Boodin can be forgiven for not addressing this as propaedeutic to their work; the history of time as witnessed in its various technological manifestations is a comparatively recent specialty. It didn't begin until Lewis Mumford provocatively stated in *Technics and Civilization* (1934), "The clock is not merely a means of keeping track of the hours, but of synchronizing the actions of men. . . . The clock, not the steam engine, is the key-machine of the modern industrial age."[2] By the year of its publication James and Royce had long passed, while Boodin was busy developing his idea of functional realism, his concept of creative cosmology in *Three Interpretations of the Universe*, and his theology in *God: A Cosmic Philosophy of Religion*. Even as late as 1959, twenty-five years after *Technics and Civilization*, Mumford admitted his book "stands alone, an ironic monument if not an active influence."[3]

Its immediate importance for us here is that, in Rosalind Williams's words, Mumford's book "enlarges the canon of culture" by "combining sensitive analysis of forms with analyses of the social and physical realities of labor, landscape, and capital" in a way that sets our modern technical apparatus within a larger historical context that correlates those physical changes with those taking place in our minds.[4] Mumford was the first to tie technologies to *mentalités*, which in turn becomes *une leçon des choses* ("a lesson of things"). Perhaps more clearly Marc Bloch frequently referred to *représentations collectives* ("collective representations") as synonymous with *mentalités*.[5] This Annales school perspective will be probed later. But Mumford was less interested in social organization than he was in relating the evolution of technologies to the evolution of personalities, or as Williams puts it, to identities.[6]

Here the question of time takes on significance directly relevant to any examination thereof, philosophical, historical, or otherwise. No one has made a more careful study of time in its fullest historical and Mumfordian sense than Stefan Tanaka, professor and historian at the University of California, San Diego, who is a leading scholar on issues involving the social construction of time in modern societies.[7] Tanaka explains that "time need not be a metric to emplot and organize but a way to understand the world."[8] In this sense, there is not one unilateral or longitudinal time. Time generally has been used either to map out and order the means of production or to highlight those matters deemed of primary social significance. For example, in the premodern era or non-Western cultures, calendars were less used to demarcate time than to trace celestial movements, religious holidays, or liturgical cycles. In medieval Europe, social time (often local to town or region or ecclesiastical connections) took precedence over mechanical time; chronological time didn't really begin in earnest until the Enlightenment. Temporality certainly mattered to the premodern world, but not the mechanical time we take to be universal and ubiquitous today.

The classical physics embodied in the mechanical universe ushered in by Isaac Newton's *Principia* (1687) laid the foundation for the mechanical clock time we know today. Ilya Prigogine and Isabelle Stengers write, "The ambition of Newtonian science was to present a vision of time that would be universal, deterministic, and objective inasmuch as it contains no reference to the observer, complete inasmuch as it attains a level of description that escapes the clutches of time."[9] Attempts to relate time to celestial movements as Plato had, to motion as Aristotle suggested, or to the mind as proposed by Augustine were definitively replaced. As Tanaka explains: "This is the escape from the 'clutches of time.' Time changes from activity and sensibility of humans to a mechanical, clock-like movement, a putatively neutral time that is bidirectional and turns motion into a default condition. Newtonian time establishes a single system that will be used to unify the many reckoning systems into one system that, though formulated in Europe, is presented as universal. That is, the world becomes renderable mechanically and mathematically."[10]

Time was separated from any specific social meaning and became a means placing modern peoples, in Elizabeth Ermath's words, "at odds with our personal and practical situations" or so subsumes them that we can no longer tell the difference. "We see the effect of the escape from the clutches of time," writes Tanaka, "the supremacy of space over time is the dominance of the state over lived experience, of the technological apparatus over the human."[11] Clock time from Newtonian physics attempts to plot time and space in what Alfred North Whitehead calls "simple location." "This is the famous mechanistic theory of nature," he writes, "which has reigned supreme ever since the seventeenth century. It is the orthodox creed of physical science. Furthermore, the creed justified itself by the pragmatic test. It worked. Physicists took no more interest in philosophy."[12] For Whitehead, through a process of "constructive abstraction" like the clock, we can arrive at approximations of this kind, but in actuality "there is no element whatever which possesses this character of simple location." He concludes, "Accordingly, the real error is an example of what I have termed: the Fallacy of Misplaced Concreteness."[13] In fact, with the advent of relativity and quantum mechanics, this type of concreteness no longer even accords with our present understanding of time that can be seen from relativity as a dimension of the physical universe that can be stretched and warped, that from a quantum perspective is little more than a parameter plugged into an equation, or a thermodynamic arrow pointing irreversibly from past to future toward increasing entropy.[14] In any case, James correctly rejected the Newtonian fiat that it was Absolute, True, and a mathematical flow irrespective of anything external. But James replied, "We don't livingly believe in or realize any such equally-flowing time."[15]

So of the three types of time, only the premodern was truly connected to the daily lives and experiences of people. The clock time we have come to know disassociates ourselves from this by turning time into misplaced concreteness, an abstraction of uniformity measured by plots of simple—or simplified—location. The time known by science today has its own relevance but surely not one of lived experience. This is why Tanaka can call our notion of mechanical time "outmoded" and the history we construct from it, mythical.

Myth or no myth, mechanical "clock" time is at the very least an inescapable imposition on human society and must be dealt with on its own level. We needn't go back to the ancient world to see how time as lived experience transformed into the abstraction just described. In fact, we need go no further than America. For example, the Gregorian calendar established by Pope Gregory XIII in 1582 that started the new year in January instead of the Julian calendar with the year beginning in March was not adopted by the British colonies until 1752. Dates falling between January and March were expressed in a confusing system of dual numbers demonstrating the transitional nature of time. Until then one-handed lantern clocks were popular in the colonies, a testament to the fact that precise mechanical demarcation of time remained a secondary concern.[16]

But things were changing. By 1805 Eli Terry was meeting the demand for precise inexpensive time pieces by producing one thousand two-handed clocks per year. Nevertheless, America remained until 1900 primarily an agrarian nation. Here time was more cosmic and organic, more intimately related to lived experience. The days were marked not by minutes and hours but by the rising and setting sun, and life itself was marked by the cyclical rhythms of spring planting, summer growing, fall harvesting, and winter storing. The standardization of time itself—the distinctive move toward mechanical time—was largely a post–Civil War phenomenon. With no standard time zones, towns miles apart were, by the position of the sun, minutes apart. The lack of any system played havoc with railroads until, after years of discussion and argument, on Sunday, November 18, 1883, at high noon, Standard Railway Time went into effect, at that point giving daily life two times, local time and standard time. This dual arrangement brought protests against standard time. Rural communities complained that this newfangled clock time ignored all-important seasonal variations in daylight that disrupted routines on the farm. But the rising demands of the factory along with the need to regulate increasingly long-distance supply chains would brook no opposition against clock time. The *Indianapolis Daily Sentinel* noted the future of time by declaring, "The sun is no longer to boss the job. People—55,000,000 people—must now eat, sleep and work, as well as travel by railroad time." The clock's final victory was established when the Standard Time Act became federal law on March 19, 1918, the same time the US recognized its global nature by formally recognizing Greenwich mean time (years *after* Japan, Germany, and France).[17] Misplaced concreteness could now reign supreme with the locomotive and factory whistle, the hourly wage, and the myriad clock-based schedules demanded by industrial society, constant reminders of its ubiquitous power over everyone's lives. Marx's alienation of workers from the means of production and loss of autonomy were now presided over by the clock, reified and commodified in capitalist society and even given ethical value. Workers were punished for being late, labor was set to meticulous time/motion analysis by "efficiency" obsessed industrial managers like Frederick Winslow Taylor, and failure to observe mechanical time was considered tantamount to robbery. "Unfaithfulness in the keeping of an appointment is an act of clear dishonesty," observed nineteenth-century pedagogue Horace Mann. Indeed, "You may as well borrow a person's money as his time," he moralized.[18]

All this occurred with the epic change from organic time well suited to an agrarian society to mechanical time required by the industrial age. It happened in the United States within one generation, roughly from 1850 to 1918. Interestingly James, Royce, and Boodin all lived through it. Given its consequences over the lives of modern peoples, what we want to know in examining each philosophers' handling of time is not a purely philosophical exposition of time that reinstitutes it organically but one at least cognizant of mechanical time's effects and one that seeks the reconciliation of individuals to a time-ordered world in the wholeness of *lived experience* rather than mere socioeconomic production. For context, we begin with Boodin's mentors James and Royce, with some additional comments on Henri Bergson, who influenced both.

James, Royce, and Bergson on Time

William James dealt with time extensively before Royce or Bergson. His article "The Perception of Time" was published in the *Journal of Speculative Philosophy* in the fall of 1886, thirteen years before Royce's first series and fifteen years before his second series of Gifford Lectures and their subsequent publication as *The World and the Individual* (1899–1901).[19] Bergson tackled the issue first in *Essai sur les donées immédiates de la conscience* published in 1889 (translated into English under the title *Time and Free Will* [1910]) and in a 1903 article "L'Introduction à la Métaphysique" in *Revue de Métaphysique et de Morale* and much later in his *La Pensée et le Mouvant* (1934). James was proud of his article. So proud, in fact, that he brought it to the attention of Scottish philosopher George Croom Robertson, friend and editor of *Mind*, saying, "Much of it is mere compilation; but the core of the thing is a view I have nowhere seen, that our *intuited* time is only a few seconds long, and is a genuine sensation, due to a nerve-process which I try to adumbrate hypothetically."[20] James thought his treatment of time so integral to his overall development of psychology that he reprinted it verbatim as chapter 15 of his monumental two-volume *The Principles of Psychology* (1890).[21]

James, as always, wants to know how we experience time, physiologically and psychologically. Anticipating Husserl, he proposes a phenomenology (without calling it that) of time and space. While time consists of past, present, and future, we experience the present as "no knife-edge, but a saddle-back, with a certain breadth of its own on which we sit perched, and from which we look in two directions into time."[22] The unit of time of this perception is a *duration* experienced moment to moment. Here James is sometimes misread as presenting the "real present" as "so like a knife edge,"[23] but as quoted above James specifically denies this. And this distinction is important when he shows how past and future is handled. James describes it as follows:

> Thus memory gets strewn with *dated* things—dated in the sense of being before or after each other. The date of a thing is a mere relation of *before* or *after* the present, or some other thing. Some things we date simply by mentally tossing them into the past or future *direction*. So in space we think of England as simply to the eastward, of Charleston as lying south. But, again, we may date an event exactly by fitting it between two terms of a past or future series explicitly conceived, just as we may accurately think of England or Charleston being just so many miles away.[24]

It can be seen that James's analogy of the saddleback works precisely as described. A knife's edge of the *specious present* would hardly allow for the sorting of past and future that James proposes, and it is this *specious present* (amidst the saddle) that remains "the paragon and prototype of all conceived times." To coordinate James's structure of time, it might be said that for time to be truly experienced it must be through sensation whereby (borrowing from Shadworth Hodgson) it becomes the measuring tape (the *duration*), and perception, "the dividing-engine which stamps its length." Our saddle-bound present does the sorting. James certainly had little sympathy for Bergson's treatment of time as a "real" vital process, regarding it as a "patently artificial construction."[25] In that sense, for James, Bergson was no better than Newton.

Royce's time bears some similarities and some important differences from James. It may be delineated in the following passage from an address to an Association of Clergymen in 1906:

> Time then is, I should say, a peculiarly obvious instance of the necessity for defining the universe in idealistic terms—that is, in terms of life, of will, of conscious meaning. Burdened as we all are by the mere concept of the time of the clock makers and of the calendars, by the equally conceptual time of theoretical physics and of daily business, we are prone to forget that it is the human will itself which defines for us all such concepts, which abstracts them from life, and which then often bows to them as if they were indeed mere fate. If you look beneath the abstractions, you find that time is in essence the form of the finite will, and that when I acknowledge one universal world time, I do so only by extending the conception of the will to the whole world.[26]

Auxier has highlighted this as a clear and succinct synopsis of Royce's temporalism. The instrumentalism has an affinity with James, but James always resisted Royce's eternalizing his temporalism in the Absolute.[27] More importantly, Royce presents here how we relate to time in what has been identified as idealistic pragmatism. His explication of Royce can hardly be improved upon: Royce "has indicated the mode of necessity we are to use, the dialectical sense which the practical, temporal conception of the will requires for its generalization, embedded in the intentional stance, and a rejection of the over-arching necessity associated with 'fate,' i.e., treating the resulting concept as though it were the necessary ground of the process which has produced it."[28]

Most significantly, apparently Royce rejects James's philosopher's fallacy or Whitehead's fallacy of misplaced concreteness because just before the above quoted passage Royce states:

> The time of the timepiece and of mechanical science, the time of geology and of physics, is indeed, as Professor Münsterberg maintains, but an abstraction. This abstraction is useful in the natural sciences. But it has no ultimate meaning except in relation to beings that have a will, that live a practical life, and that mean to do something. Given such beings, it can be shown

that they need the conception time of mechanics or of geology in order to define their relation to nature.[29]

Here Royce mixes in some sound pragmatism with a conflation of various types of time, types of which he seems to not completely understand. As discussed in the previous section, can it be said that clock time ("the time of the timepiece") and geological time or the time of modern physics are the same? Writing prior to the great revolution in physics about to be unleashed by Einstein and a bit later by Bohr, Heisenberg, and Schrödinger, Royce can be forgiven for seeing physics as a simple and straightforward expression of Newton's mechanical universe. But even Newtonian physics transformed into clock time is not *necessary* in Royce's sense. Much less is it needed to define our relationship with nature; in fact, it could even be said to alienate us from nature. Surely the town merchant with his ever-present pocket watch was far more removed from nature than the farmer up the road who measured his workaday world by the rising and setting sun and the continuous cycling of the seasons. And yet surely both are persons "that live a practical life, and that mean to do something."

As we have seen this fallacy of misplaced concreteness obtains every time mechanical time is universalized as a reified dogmatic concept. In other words, the fallacy of misplaced concreteness occurs whenever time clock time is treated as a ubiquitous unilateral reality. Tanaka points out that history as generally conceived—in spatial and temporal uniformities of millennia, centuries, decades, years, months, days, and minutes, all measured by calendars and clocks of *our* construction—is fostered by these simple locations. "The combination of this form of classing with history is, in my mind," he writes, "one manifestation of what Whitehead identifies as the 'fallacy of misplaced concreteness.'"[30] In a sense, then, we often ascribe temporality to peoples of many times, places, and cultures within this superimposed clockwork structure (in ways that indeed might have made little sense to them), and as such this "disorientation inverts Foucault's 'violence of time,' where the expectation of translation and transmutation shifts to our, not their, responsibility."[31] Whitehead wanted to banish the old ghosts of misplaced concreteness from all philosophical discussion—"mere awareness, mere private sensation, mere emotion, mere purpose, mere appearance, mere causation"—that could have no "mere" relationship with such abstractions.[32] Here *mere time* might be added, where an artificial mechanical construction of temporality is held to be uncritically *merely* the same thing as time itself. These problems arise whenever Whitehead's fallacy is dismissed or ignored.

But we still must deal with time in this mechanical aspect, just as we must still deal with awareness, private sensation, emotion, purpose, appearance, and causation. That said, mechanical "clock" time is never *mere* time. And yet Royce treats it as such.

Whatever Royce's handling of time may entail next to James's, one more important temporal philosopher warrants comparison—Henri Bergson. Bergson is considered the first process philosopher ahead of Whitehead and Hartshorne who would give time even fuller expression. James read Bergson's *Introduction to Metaphysics* in February of 1903 and applauded what he saw as his "conclusive demolition of the dualism of object and subject in perception."[33] Moreover, James saw in Bergson "a philosophy of pure experience." For him, Bergson was fresh and exhilarating—a "magician." "Nothing in Bergson is shop-worn or second hand," he said.[34]

Czech American philosopher Milič Čapek has argued that Royce and Bergson treated time and eternity in ways that "are remarkably similar."[35] This is all the more surprising given what Čapek calls "Royce's monistic idealism" and Bergson's tendency to treat temporalism if not fully dualistically "at least to the polarity of the mental and physical." Like Royce, Bergson agrees that consciousness transpires not instantaneously but within an interval or duration (recalling James's saddleback analogy). Both Royce and Bergson have a "perception of succession" that goes beyond time to eternity. Unlike other idealists who see the Absolute as timeless, Royce sees the Absolute (God) as *temporally extended* and that "time is the form of the will." Bergson agrees, though states it differently, calling his *élan vital* "pure willing" or *pur vouloir*.[36] But, Čapek significantly points out, Royce's God offers no way of becoming like Bergson or Hartshorne, as Bergson's eternity is dynamic rather than static. Unlike Bergson, Royce gets caught in the net of his own argument that cosmic temporality occurs in sequences preserved in the "all-embracing Eternal Now." He falls prey to the fallacy of misplaced concreteness by intentionally or not "suggesting erroneously the atomistic character of temporal events. We thus sink into the fallacy of simple location which artificially isolates each event from its temporal context—its ancestors, constituting its past and its not yet existing virtualities, constituting its future."[37] For Bergson time *is* duration experienced not as moments but continuously. Bergson also speaks of time as a "force" and a force that makes free will possible; it is "the horizon of the inner life."[38] Royce and Bergson both use the example of experiencing a melody, but Royce insists we do not "abolish the succession of tones" nor does "the divine mind" whereas Bergson sees the melody to have meaning needs to be brought into a compositional whole. The practical importance of this is that reality cannot be reduced to a succession of analyzable points in a static fixed universe (the tendency of science) but instead as dynamic, progressive, and creative (contra scientific reductionism). In this sense Bergson paved the way for the richness of all the process philosophers—for example, Alfred North Whitehead, R. G. Collingwood, Ilya Prigogine, Charles Hartshorne, Nicholas Rescher—all of whom reject the reductionist materialism of science given over to scientism.

And here is where this long side excursion into the thought of James, Royce, and Bergson on time has left us in regard to Boodin. Understanding the contributions of James and Royce in particular we can see the ideas on which Boodin cut his philosophical teeth. Here Boodin's dependence upon his mentors becomes obvious, but more importantly, his unique handling of time developed into an overall epistemology leading toward a coherent metaphysic that stands out in bold relief. If James introduced Boodin to the exhilarating and thoroughly refreshing pool of pragmatic thought, it was Royce who would take him into the deeper waters of metaphysics in ways James never could. It should also not be forgotten that Royce was a pragmatist and in his own way committed to process thinking too, and his interactions with James mutually fertilized their conviction that experience turned into ideas/actions that *mattered* was a central principle of their philosophies.[39] And yet, as we shall see, Boodin would retain the wisdom of his mentors while simultaneously transforming it into a philosophy that was truly his own. In terms of his own temporalism, Bergson's reference to the new technology of the cinema caught Boodin's eye and formed an opportunity for examination and critique beyond both James and Royce.

From the Temporal to Truth: Boodin's Epistemology

Time was the starting point in Boodin's philosophy. He used time as the foundation of an epistemology that would lead to truth. Boodin says his conception of time first came to him in 1897 as he closely studied Hegel generally and in particular Royce's "The Conception of God" originally given as an address in the fall of 1896 at the University of California, Berkeley, at the invitation of George Holmes Howison. Here Boodin would critique his mentor's idealism and come to his own unique conception of time as a "creeping in" concept rather than serial because, as he says, "it creeps into all our calculations and makes all our systems of truth unstable."[40]

He begins not with Royce but with Kant, who he insists "has lost sight entirely of the character of succession and emphasizes merely the simultaneity or timelessness of the time-form."[41] But Kant's view is too detached and cannot show any correspondence between form and succession. Time is too fundamental to divorce it from what Boodin calls "the behavior of the real."[42] Real time is expressed in its process not in any serial character, but in *fleetingness* rather than duration of process. Boodin views time as a series intrinsic to us, something *we* impose on time—it is not given to us. But the world is not static and the serial nature of time can be abstracted from a pluralistic world in the flux of processes. But if series is relative to the time-process, what may be said of eternity? Actually, this serial nature of time is only a "type of the eternal" since "the eternal is after all an abstraction from a real time-process to which its significance is relative. It is only in its abstractness that is timeless. The eternal or the world of significance is a derivative of a habit-taking time-process and shows its relativity to this." It holds a quantitative structure "necessitated by the social demand for description and common action."[43] This is a marked departure from Royce's more idealist time/eternity/Absolute nexus; here might be found more Jamesian influences. But, significantly, Boodin adds that time cannot be reduced to mere quantity, for time is inexpressible by our conveniences of measurement. Indeed, "The time character involves precisely the relativity or falsifying of any description which tries to exhaust the real subject-object. Time *creeps into* our world of description and negates it."[44] In the final analysis, time cannot be expressed as an ideal concept only—qualitatively or quantitatively—but must be considered a real property of a pluralistic world. In this sense, it is merely begging the question to ask if time is infinite since this presupposes time.

As part of the real world, then, time's character expresses itself in the fleetingness of experience measured in durations that give the time-nature of experience coherence and meaning. This entails discernments of reality such as in Boodin's example "the rose is red or not red" that relates time to judgment. This gives time an "absolute or dynamic non-being" aspect, and it is time that makes these discernments not only necessary but even possible. This is why the past is irreversible; the rose that is red (that particular rose at that particular time) was red and was *not* yellow, pink, or white, or some other color. Time gives us "the real negating and transforming of the world of experience for which the past-symbols stand." This is possible precisely because time is irreversible; the irreversibility of time is critical, for if it were reversible it would cease to be a process at all.[45] Thus time has a negative and positive dual character. Boodin is sometimes unclear here in expressing this negative aspect of time, but if we recur to his "the rose is red or not red" example we can see that time also acts as the Boolean operator "not"; things are fixed in a qualified way by time by excluding other possibilities of what they might have been. As such, the truth of the "red rose" is not a stable Truth about *all* roses, only *that* rose, and time has made it so. But this entails the positive aspect in which the judgment of the rose being red is even possible. Imagine a timeless world in which the Boolean operator "not" is removed and you have the chaos of endless possible roses of endless possible colors in which no judgment can be made, and a world without judgment is a world without creative direction. Thus time renders a world of discernable "facts" but facts that are destabilized by the sheer flow of time; this is what Boodin means by time making our truths unstable.[46] This negative/positive aspect of time is summarized as follows:

> As against Kant we have pointed out that series and order themselves involve ideal construction and hence presuppose time as a datum. What remains as the ultimate time character is pure negativity, not non-being as an empty ideal abstraction, indistinguishable from equally abstract being, as with Hegel; nor relative non-being in the sense of mere otherness, other being; but non-being as an ultimate aspect of reality, a dynamic principle, negating the habit structure of the world and transforming it into ever new structures.[47]

It is a characteristic of Boodin's writing that a careful and often enlightening examination of the history of philosophical concepts up to the present is integrated into his work and time is no exception. He observes that the major difficulty of previous philosophers with regard to time is their tendency to hypostatize it, whether it be Democritus's atoms, Herbart's qualities, Plato's impersonal ideas, or Hegel's Absolute. The Platonists, argues Boodin, have always been wrong in denying process and scorning the time-world in favor of eternal ideas, while Aristotelians are wrong to see the actual and potential coexistent. Such errors led the medievalists to insist that the acorn "wanted" to become an oak. Although we will have much more to say about this later, Boodin is correct in his assessment that science made a real advance in discarding these explanatory teleological categories. For him, process must be tied to practice and experience the whole history of which may be summed up, *viz.*, "The philosophical and scientific conceptions of process, both in mediæval and modern times, have been largely reverberations of the categories of the Platonizing Aristotle without his empirical insight."[48] But the idealists are no better—Hegel simply juggles Aristotle's categories, Hegel treats history as a kind of "kinescope" with the logical categories of Being, Non-being, and Becoming passing by in successive order, and even Royce makes time a correspondence with the "incomplete finite will," which is simply an imposition, "merely an ideal demand."[49] Boodin has equally pointed criticisms against the empirical reductionists, singling out the Darwinian British mathematician William Kingdon Clifford and logical positivist Ernst Mach as examples. To them, time can be ignored as we increasingly understand the laws governing the physical universe. "But," replies Boodin, "mathematical equations are not explanations. The seeming time facts cannot be ruled out of court by a mere ideal demand [scientific or otherwise]." Boodin concludes that time reflects a real dynamism expressed in real process. Thus it is significant to mention, as Randall Auxier has, that Boodin was identifying key features of process and reality in 1904 a quarter century *before* Whitehead and soon after Bergson's association of time with real dynamism, leaving him, as already noted, well positioned "as few others for the philosophical encounter with the theory of relativity and the quantum revolution."[50]

More importantly, an examination of an even earlier paper by Boodin, "The Reality of the Ideal with Special Reference to the Religious Ideal," published in 1900 along with his other early writings, makes a compelling case for Boodin as a very early modern process theist.[51] Referring to God in that essay as "the impartial and sympathetic Spectator and Coöperator" gives evidence of process-based theology that would follow Boodin for the rest of his professional life. This is not to argue for priority over Bergson, only that modern process theism in this pre-Whiteheadian period does not have a single source in Bergson alone.[52]

How Boodin's process thought could have been consistently ignored historically by philosophers can only be explained by the fact that there is a human yearning for unity and permanence that traditionally resists the mutually collaborative nature of process theology. Process theology does not find truth in the dogmatism of almighty permanence but in the flexibility of constructive and creative flux and change. Boodin prepares for his investigation of truth by ending *Time and Reality* on a theme to which he shall return again and again by declaring that "we [should] regard experience as making itself anew, as an essentially creative universe, which to some extent at least accumulates past experience into present structure, and transforms present structure into new experience."[53] More than that, he constructs this into what he calls "substantial, dynamic Non-being." This negative aspect is necessary for three reasons: (1) without it, as we have seen, no judgment is possible; (2) these judgments become intelligible only in relation to real non-being; and (3) our judgments vary and are unstable because they are based upon different strata (like the strata of geological time) necessitating different judgments. Definitionally, then, "Non-being . . . is the genus within which time is a species. Non-being . . . includes our judgments, both of relative or ideal negation and of absolute negation. The character of time coincides with this latter species of non-being. Time is absolute or dynamic non-being."[54]

In this sense, it can readily be seen that time is critical to Boodin's epistemology and why it launched all his later metaphysical investigations. Time is an ultimate attribute of reality that forms the requisite nature of all process and transformation. It is epistemically a "creeping" and "fleeting" part of our experience neither strictly serial, mathematical, nor substantive. It is non-being in the sense that it is neither a thing nor an energy. Time is, in effect, one of the ultimate attributes, the canvass upon which our judgments are sifted, weighed, and formed. Without time there can be no certitude regarding our judgments, indeed there can be no judgment at all. Without time there is no flux, no change, no knowledge. As such, a careful consideration of time was of paramount importance for building Boodin's epistemological framework. Boodin is essentially combating erroneous views of time—Bergson's substantialized time, Royce's reification of time, and various forms of misplaced concreteness—that lay at the heart of many epistemological confusions. With time thoroughly explicated, Boodin completes this line of inquiry seven years later in *Truth and Reality*.

It is, however, worth looking at the fallout stemming from this temporal portion of his theory of knowledge before proceeding. Apparently that collegial response must have been severe and widespread enough to prompt him to deliver a paper on April 21, 1905, to the Western Philosophical Association to "supplement" and clarify his position on time.[55] There he again criticized the concept of time as serial and implicitly aimed at Royce when he said that time cannot be handled by "simply dumping things together" as space, quantity, causality, or (in Royce's case) "to will." The nature of time is not mystical at all but a phenomenon that inherently tends to "negate our meanings, to make our judgments false and so to make new judgments necessary." Time is "non-being" not because it is unreal but because it possesses this power to negate—to reiterate, the rose that was red was *not* white, yellow, or something else; the day that was warm and sunny was *not* cold and damp. Again, without this negative quality to time we should have no basis to even know the rose was, in fact, red, or the day, in fact, was warm and sunny. Then, in a stroke that anticipates Whitehead, he states quite clearly his position as an early process philosopher: "But inasmuch as negation is never negation of itself, but is always transmutation, and therefore novelty, there time *process* [emphasis in the original], though not the time concept, is a very positive and very rich affair. *Process is the bearer of all reality and contains within itself the prophecy of new reality* [emphasis mine]." Boodin then defends his characterization as "absolute or dynamic non-being." The term *absolute*, he admitted, "has caused no small deal of trouble" for him. But by absolute, he explained, "I simply mean that time is a real property of our experience-world, subjective and objective, and not a derivative of being in any form, as the Hegelians would have it." He ends by stating his two quarrels with the idealists. One is that although all reality must be regarded as by and through experience, all experience is not reflective and cannot be reduced to concepts or to purposeful intent. Sometimes experience is quite literally *experienced*, as we experience joy or fear, hot or cold, pain or not pain, and so on. His second quarrel is with the idealists' incomplete or in-exhaustive ontological categories. Hegel, for example, fails to differentiate his dialectic of being from non-being. We need not a single, static system of meanings but dynamic systems that are "ever new."

Whether Boodin answered all criticisms definitively is questionable. Boyd Henry Bode and Percy Hughes both published reviews of *Time and Reality* critical of Boodin's definition of time as absolute non-being, his emphasis upon its negative or eliminative aspect, and its non-serial character.[56] It is unfortunate that Bode failed to take account or even read Boodin's supplemental comments on time. Hughes published his review before Boodin's address to the Association, so it is unclear if they sufficiently addressed his concerns. What can be said is that Boodin gave a reasonable and thorough defense of his views on time; if Bode or Hughes were dissatisfied with it, such dissatisfaction remained due to their own a priori assumptions rather than any inherent weaknesses in Boodin's exposition. It might be further suggested that Boodin's emphasis on the negative or eliminative understanding of temporality might even have been rather prescient. All this transpired long before cybernetics was invented, but Walter Rosenblith, who had worked with its founder Norbert Wiener, made the observation that teleological objects and organisms rooted in time and place require feedback (positive and *negative*) and, in fact, without error-inducing possibilities real teleological behavior is impossible.⁵⁷ We needn't be suspicious of cybernetics' fascination with machines to think that all these reductionist formulations are false; their emphasis on feedback is quite dynamic and process-related, and there is likely an important element of truth here that Boodin seems to have captured in his temporal philosophy forty years earlier.

As previously mentioned, seven years later *Truth and Reality* was published. This would not only complete Boodin's epistemology but also be preliminary to his metaphysical work, *A Realistic Universe*. It is clear that Boodin already had his metaphysic clearly in view when he wrote *Truth and Reality* since he states in his preface that it would soon follow. It would appear that Boodin had even prepared both works as one large manuscript; it was his longtime friend and colleague Donald Cowling who suggested splitting them into two separate works.⁵⁸ Wisely taking this advice, we too may deal with this epistemological treatise in its own right as a capstone to his discussion of time.

Boodin begins his epistemology with a proper sense of proportion and modesty, indeed in frailty and incompleteness. "We half-men, while we struggle and see through a glass, darkly," he writes, "should at least make our tolerance as large as our ignorance."⁵⁹ Beginning with an investigation of the human mind and intelligence, Boodin views our mind as instinctive and the product of "instinctive or organic adjustments." This sounds surprisingly reductionist, but he continues by ascribing the higher attributes of human consciousness to "a lucky structural variation or accumulation of variations, which changed the whole course of evolution by giving meaning to the process and thus establishing new survival values." By "lucky" Boodin does not mean wholly random or stochastic; he means something more. Evolutionary mechanisms "must be understood from the point of view of creative will, which through a variety of efforts, gradual culminations or sudden mutations, strives to make itself definite and individual and which give continuity to the whole process." We begin, he says, as pure organisms but are transformed in "in ideal beauty" or "ideal creation" which, in turn, have their "creative effect" upon our later development. Moreover, this evolutionary structure is built out or developed as a "spontaneous impulse," not as the mere mechanical heaping up of bricks by blind accident. The teleological implications here are unmistakable.⁶⁰ Boodin continues with an explanation of intelligence as "the capacity to learn from experience," which is closely related to four categories: (1) the perceptual (time, space, and habit, the last being a sorting of trial and error toward success); (2) reproductive imagination, a priori tendencies in our mental makeup; (3) empirical generalization, the sorting or evaluating of quantitative and qualitative series by interpretive cause and effect; and (4) idealization comprising the unity of parts within the whole, harmonious integration, distinctness or simplicity of relationships, and synthesis into universality and social sharing. These alone are not enough, however, as these must be integrated into a "law of totality" that is really "a faith that, somehow, the universe as a whole hangs together" and we can experience it without interruption. Indeed Boodin believes in a comprehensive unity that might be expressed as God, about which we cannot hypostasize into existence by wishful thinking or speculation but must rather apply as a hypothesis in meeting our experiential need.[61]

Boodin then proceeds by discussing what he calls the process and the morphology of truth. In terms of process, Boodin remains thoroughly pragmatic by seeing thought as a "moving, living will" giving to instinct its "articulate and self-conscious purpose" that at the same time "selects and guides." Here creativeness emerges from reason. While thought can at times proceed intuitively, eliminating familiar steps, it originated by and through thought itself. In any case, it must be tested by "the demands of experience."[62] The truth process is the three-part mental attributes of ideation, feeling, and will. It is, in effect, "self-realization—the whole self striving to realize a definite end—the will to know."[63]

In the morphology of truth Boodin is clear to distinguish actual thought as judgments from its logical representations in formal propositions and syllogisms. Judgments, then, transpire through differentiation and limitation whereby "significant denial" or negation allows for the specificity of discernment, like his epistemological example of the rose is red but *not* yellow, white, or pink. In addition, we need not parse thought too finely between induction and deduction; from a thought perspective the process is largely the same. What we want to know is the relationship of thought to truth, a connection that has a tenuous hold on belief, for although truth can at times mandate belief, it is certainly *not* true that belief can create truth. Belief, Boodin sees, is wrapped up in the past—our social constructions, our psychological influences, or customs and habits. Truth, by distinction, is forward-looking—anticipating consequences, correspondences, and conduct. Boodin's use of the term morphology in relation to truth should not be taken as a mere linguistic turn (the reduction of truth to a morpheme) but more as an examination of structural relationships relating to and ultimately comprising truth.

While this may be the *form* of truth, it is not its content; its content must be rooted in a "definite epistemological universe" of real truths if only partially discerned, not in a world where chance rules willy-nilly. "An ideal truth, which insists upon the impossibility of truth," Boodin pleads, "is the most irrational theory of all. Truth must believe in itself, in its possibility. And the belief in complete truth implies a belief in the teleological unity of the universe, in some manner, and so postulates internal relations."[64] Although some relations are clearly external (i.e., "not grounded in the natures of their terms"—symbols as in math, serial relations [clock time], space relations, quantitative comparisons), others *are* internal, found in the "intimacy of purposive overlapping and interpenetration." We must, in the end, let our finite experience lead us to a universe of implicit "internal relations or relations of teleological significance."[65] These forms of truth are related to certain postulates or laws of truth, which presuppose all our knowing: the law of consistency, the law of totality, the subject-object form or the necessity that knowledge be representative, and the law of finitude.[66] Boodin's epistemological convictions toward the nature of truth will not permit atomism or some psychophysical parallelism between the immaterial and the material that he likens to hydra-headed beasts any more than he can countenance agnosticism (a "dismal abyss"). Boodin's world of knowledge and truth is richer than these, but they must be tested in examination and experience. Here the will has a role to play by furnishing the motive and goal of thought or else thought would occur in a vacuum.[67] Boodin is ever resistant to the lazy lure of reductionism—the easy shortcut of explaining away rather than explaining—and affirms but limits idealism within a context of science and reason, showing that he still walks hand-in-hand with Royce's Absolute and James's pragmatism. This interesting balance is best explained by quoting him at length:

> I have faith in a higher consciousness than the human as the fulfillment of our fragmentary insight and "the final cause" of the evolutionary process. But I do not see any leading toward this mind in the infra-human world—the world of the stone and the amoeba. I must rather seek it in the supra-human reaches as the goal of our ideal striving. While mystical and esthetic intuition may seem to furnish some of us a very intimate acquaintance with such a world, I cannot see that such a faith exempts reason from dealing with it as an hypothesis and from testing it as any hypothesis is tested, through its success in simplifying and guiding experience. I do not deny the possibility of the idealistic absolute. There is certainly nothing contradictory in

the conception of such a complete, systematic experience. On the contrary, it must always figure as an epistemological ideal, even if not an ontological assumption.[68]

Boodin proceeds with an emphatically supportive exposition of pragmatism, which is his basic criterion for truth. As he has so often, he reiterates the epistemological foundation of pragmatism, "Knowledge everywhere must be based upon evidence as furnished through human experience."[69] Stated another way, "Man is the measure of all things." In this sense there is, for Boodin, nothing new in pragmatism, only the need to test all hypotheses in the scientific spirit of verification, whether those investigations happen to be natural or philosophical. But, it should be added, *not* in the hard—even destructive—verificationism of the positivists who sought to rule out philosophical propositions as meaningless but rather to evaluate what philosophical propositions might really mean. In this sense verification in the hands of the pragmatist can actually affirm certain philosophical or metaphysical positions. Boodin adds, however, an additional aspect of knowledge that he believes has been neglected by his fellow pragmatists, the epistemological importance of the *creative imagination*. This can lead to truth, a truth that (à la Whitehead) recognizes the empirical *and* the rational as well as "the finitude of truth as an adjustment to an infinite process."[70] But at the same time pragmatism has sometimes been misunderstood or even caricatured as simplistically equating the truth with the useful. Here Boodin gives many counter examples and even admits that in our "pluralistic and plastic" social world, even fiction may be better than truth. For Boodin, it *is* possible to be brutally honest to a fault; truth must be tamed by the ethical and moral contexts in which it finds itself; it must be restrained by the sense of greater good or even compassion unknown by an absolute and dogmatic veracity insisted upon only by demons. Just as truth must not be confused with usefulness, neither should pragmatism generally be seen as just another form of humanism. It will prove significant later on, but for now Boodin makes it clear that although reality is known only by the difference it makes to us as living persons, truth is not made so *because* of our humanity. Much less is truth bound by the sweeping reductionisms embraced by most humanists, oftentimes when they themselves claim to be *least* reductionist (as with Roy Wood Sellars, discussed in chapters 3 and 6). It bears particular notice that Boodin places the meaning and validity of pragmatism within "a stream of processes" of cosmic proportions that is not a "mere chance" affair but is determined by an active formal constitution of the whole universe.[71]

This being the case, it is reasonable to ask, How do we as finite and fallible creatures relate to this universe? Or, as stated somewhat differently by Boodin, "What does human nature contribute to nature?" Clearly the uniformities of nature are what they are, they are not made by our perceiving them, but they do gain significance when we become cognizant of them. Human experience contributes this significance, it does not contribute existence. Moreover, we must not conflate the motive for seeking the truth with the test for its validity; often the truth-seeking motive and the validity of the truth discovered are two different things, as when Oscar Minkowski and Josef von Mering removed a dog's pancreas in 1889 to see its role in digestion only to discover its role in diabetes as an insulin-producing organ. All our truth-seeking, all our truth finding, comes from ourselves; even our intuition resides in ourselves.

Finally, Boodin comes to his core concept—*pragmatic realism*. He unpacks this idea by first explaining that realism refers to the cognitive moment; idealism takes this to mean that realism is related to thought, in fact nothing exists apart from thought. Idealism, however, fails for the same reason that materialism fails, "both buy simplicity at the expense of facts," and mysticism is nothing but an "ineffable noetic intoxication."[72] Avoiding these cul-de-sacs, he defines pragmatic realism as follows: "Instead of the dogmatic method pursued by the old idealism and materialism alike, we must substitute the critical method. This method has been christened within recent years by C. S. Peirce and William James and called pragmatism. As I understand this method, it means, simply, to carry the scientific spirit into metaphysics. It means the willingness to acknowledge reality for what it is; what it is always meaning for us, what difference it makes to our reflective purposes."[73] This reality is more than mere perception, though that is part of it. Nor can we reduce reality to mere sensation, were that true, Helen Keller—a great communicator—would be inaccessible to us and we to her. The "multitudinous thing-contexts" and "our will attitudes" may be, at times, more important than sensation. For example, it may be more important that water satisfies our thirst than that it makes us wet. Things do have "inner meanings" such as the element gold has value or that the cinchona tree protects from fever and heals. Additionally, there are both "stuff" and "non-stuff" dimensions; the latter do not appear as immediate phenomena. Time, for example, "creeps into our equations and makes revision necessary" with waiting our only option for making those revisions clear. We must assume a direction by which to measure our finite standards. Our attitudes and purposes—these "non-stuff" dimensions—are in this sense simpler, more direct, and more immediately felt and, therefore, more knowable than the almost limitless things ("stuff") of this world. Both the "stuff" and the "non-stuff" things are real. Truth is found in things experienced, in things felt and recognized.

This discussion has paved the way for Boodin's introductory comments on metaphysics, a sort of prelude to *A Realistic Universe*. As quoted above, metaphysics is not, for Boodin, a pragmatic extra or special add-on, but a method to apply a scientific approach to metaphysics. Metaphysics, for all its historical missteps, presumptions, and attempts "to spin its spider-web of logic from its own demands," comprises that "persistent effort to see the various contexts of the world of objects as one pattern, the divine love for the wholeness of things."[74] It must employ the methods of science by testing provisional hypotheses for verification rather than by dogmatic a priori convictions. But its intentions are unique: to discover or develop a final interpretation of knowledge and other "overlapping problems of experience." Here Boodin has not been deceived as the positivists have: *everyone* has a metaphysic. "Common sense, with its implied dualism or materialism; the agnostic with his hide-and-seek game with the unknowable; the professed scientist with his fundamental assumptions—they all have it [a metaphysic] as truly as the systematic idealist or realist."[75]

Boodin makes numerous statements preparatory for his thorough examination of metaphysics in his forthcoming companion volume, and in the interest of moving on, we needn't recount them all here except to say we can see its contours coming. It is an opponent of materialism, which Boodin sees as an attempt to make a partial character of the world do the heavy lifting of the whole and as such plays fast and loose with facts. It is equally dismissive of mysticism, the unknowable or the occult. There are no "hidden essences of things." But metaphysics is also—perhaps preeminently—constructive. Because metaphysics is "the oldest of the sciences" (meant in the broadest historical sense), it laid the foundations for scientific procedure in the "laws of logic"; it set forth the general postulates of the physical sciences; it established its crude chemistry on the basis of Empedocles's four elements and understood the basis of proportional variation in chemistry; it found a teleological evolution based upon selective adaptation with Anaxagoras; it discovered psychological laws of association and founded the principles of ethics and politics with Plato and Aristotle. "Metaphysics does not transform the observed facts and values," argues Boodin, "[it] gives them a larger setting, and thus allows us better to appreciate their significance."[76]

Here, then, will also be found religion. Utterly rejecting Humean empiricism as "logical bankruptcy," Boodin affirms "the reality of religious ideals" in his final chapter. We want to know if religious ideals have any real basis. We know that humans since ancient times have acted as if they did. But they must be put to the test. Here it is argued that there is no difference between religious tests for truth as for scientific ones. After all, science too proceeds by faith, sometimes a slender faith. The real question is, Does the history of humanity give us reasons to regard religion *as if* it were real in order to attain to our highest levels of moral, ethical, and social behavior? Boodin answers in the affirmative because, in short, "life would be poorer without it." Its evidence may be seen in its historical persistence and in its inspirational character; these may be taken as evidence of its truth. A popular philosopher has correctly observed, "One lesson from history is that religion has many lives, and a habit of resurrection. How often in the past have God and religion died and been reborn!"[77] Moreover, he adds, "There is no significant example in history, before our time, of a society successfully maintaining moral life without the aid of religion."[78] Echoing the historian, Boodin declares, "If it is a fact that the religious ideal is thus essential to the highest unity and development of life [personal and social], then the religious ideal can be no mere shadow projected by the imagination of man; but it becomes objective; it thickens into being. It is the ultimate constitution of the cosmos."[79] This view may be objected to as dated—even benighted perhaps—but those of our own generation who would agree with Boodin are not hard to find; one thinks of Arthur Peacocke, John Polkinghorne, Alvin Plantinga, Alister McGrath, Rupert Sheldrake, Freeman Dyson, William Dembski, and others.

Boodin ends *Truth and Reality* on this religious note. And it is religion not given as a vague phrase but religion with real content; it must connect itself to our values and norms of life, it cannot collapse into a hazy pantheism because pantheism has no intrinsic worth. "I do not see how any one can love or worship things in general," says Boodin, "this medley of comedy and tragedy, of harmony and discord, which we call a world. . . . However satisfying such a view may be esthetically, it is not ethical. Pantheism is as unethical as materialism. A God that is identical with the totality of existence is helpless to redeem the world, as he is equally responsible for its sins and virtues."[80] Religion cannot just be a concept, an idea about God thrown in to fill in gaps in our knowledge; it must, as James pointed out in *Varieties of Religious Experience*, matter to our daily lives. It is not a metaphysic about the nature of God, an apologetic for God's existence, or a denominational position, it is a belief system energized by faith that gives genuine *meaning* to life. To use a Jamesian metaphor, Boodin's religion had to have cash-value for our lives.[81]

This ends Boodin's long road from time to truth and his theory of knowledge. It remains only to tie up some loose ends by way of general assessment and conclusions regarding Boodin's handling of time and its path toward his theory of knowledge.

Assessment and Conclusion

Despite the dedication of *Truth and Reality* to William James, there is always something more of Royce than James in Boodin because it is from the former (as a teacher and advisor) rather than the latter (as a friend and sifter of ideas) that the necessary tools were gained to lead him in the metaphysical directions that James simply could not follow; the clear aim of this theory of knowledge is to get to a broader, more encompassing metaphysic. Boodin was starstruck by James but it was Royce who was his guiding North Star, the compass though not the content of his philosophy. Boodin learned from Royce that the only way to untie the "worldknot" of ideas and their relationship to reality was through epistemology where the true ontology would reveal itself.[82] He also learned the basics of sound metaphysics from Royce's "Philosophy 9" course given at Harvard from 1894 until 1916, since Boodin was there from 1897 until receiving his doctorate under Royce in 1899.[83] Although Royce's grand metaphysical synthesis, the two-volume *World and the Individual* (1899–1901) was a work-in-progress at that time, Royce must have inundated the young Boodin with his metaphysical ideas, which were in full percolation. It should come as little surprise then that Boodin's initial road to his own grand metaphysic was through epistemology, concentrating, as Royce had, on the meaning of ideas, the nature of being, and the temporal. But it *is* possible to make too much of the influences bearing upon Boodin; in the end, he took the best of what he learned from Royce and James and refashioned it into something truly new and, in many ways, improved.

Boodin received mixed reviews. H. W. Wright happily reported that *Time and Reality* handled "many of the vexed problems of epistemology" and gave them "a detailed and thorough treatment. Although the reader is sometimes led along paths where it is difficult to keep one's footing, still the vigor and earnestness of the author's thought together with his fair and tolerant spirit will tempt him to continue until he emerges in the light of the concluding chapters."[84] Others had issues with various positions, definitions, and arguments proposed by Boodin, most of which appear based upon misreadings or misunderstandings. Typical is Charles B. Vibbert's objection to Boodin's "appeal to the immediate" as vague and unclear; at one time the immediate is identified with the perpetual, at others it is equated with conceptual construction, and later on the immediate is associated with the social agreement of sharing. What is unclear here is not Boodin's development of the immediate but Vibbert's insistence that each of these cannot coexist coherently. Why the immediate cannot be a perpetual conceptual construction that can be shared Vibbert fails to explain.[85]

Of all the critiques of *Truth and Reality*, that presented by Radoslave Tsanoff was the most sustained, eliciting published exchanges between the Bulgarian-born American immigrant idealist and his Swedish-born pragmatic counterpart. First of all, there is a sense in which all of Boodin's critics talk not *at* but *past* him. As an idealist, Tsanoff places his epistemic and metaphysical foundations upon thought while Boodin is anchored in pragmatic experience. Hence, Tsanoff speaks of "the impossibility of spanning the chasm between experience and reality," which represents if nothing else the essential thesis of all of Boodin's published work to date. Tsanoff, therefore, rejects Boodin's extensive exposition of the coordination of the cognitive and noncognitive facets of reality as "doomed to failure." But, other than Tsanoff's declaration, there is no reason to think Boodin has in any sense "failed." The distance between these two philosophers is perhaps best captured in Tsanoff's description of faith as "too filmy and irresponsible to afford any solid warp for the epistemological woof of the philosophic fabric," but this is only because Tsanoff cannot see faith—certainly as James and Boodin do—as the context for belief and action upon belief. Boodin responded by restating his four-fold theory of truth: the law of consistency, the law of totality, the subject-object law (or law of reference), and the law of finitude. Boodin is careful, however, to agree with the idealist's absolute and a higher consciousness, but he views them as epistemological ideals rather than ontological assumptions.[86]

In a similar fashion Paul Carus talks past Boodin's five truths (time, space, being, consciousness, and form) only to berate pragmatism for what he regards as its claim to be the only true philosophy, "a consummation of philosophic development." Equating pragmatism with agnosticism (a strange association to say the least), Carus argues that pragmatism is not properly part of philosophy at all but "an outburst of literary enthusiasm." Carus also insists that "philosophy . . . as the science of the sciences is not a phantom of the human mind" (what pragmatist ever said that it was) and that it would be wrong to say that "philosophy should be restricted limited to strictly scientific works" (apparently reading Boodin's interest and, at times, emphasis upon science incorrectly as a form of scientism). Boodin simply replied that pragmatism can both appreciate "the softer muses of literature" and at the same time not ignore "the severer muses of science." In answer to Carus's charge of arrogance, Boodin modestly referred to "the overlappings" of history that, in the end, would judge all philosophical systems.[87]

There are two final issues of a more positive nature that need to be addressed. First is that Boodin's emphasis on the negative aspect of time (the suggestion that so bothered Boyd Henry Bode and Percy Hughes) is an innovative application of Boolean logic already mentioned, and perhaps even an anticipation of cybernetic feedback theory. It is likely that Boodin got some of this directly from Royce who was an especially adept logician, and had made a thorough study of the discipline in the summer of 1879 when he was only twenty-four years old.[88] Royce undoubtedly read George Boole's masterpiece, *Laws of Thought* (1854), but he was unpersuaded by Boole's reductionism proposing that all truth could be stated as a mathematical formula. That didn't mean that Boole couldn't be put to good use. Although Boodin never mentions Boole in his writings, his use of negation as a positive operator for Boodin's temporalism is compelling, especially since Boole related his ideas concerning truth to his mathematical system and famed pragmatist Charles Sanders Peirce, who was generally considered Boole's equal in the field, had featured Boole in his Harvard lectures on the British logicians in December through January 1870. This was surely known by Royce and passed on to Boodin. Boodin's interesting use of time as a "not" operator in this sense and in a pragmatic context rescued Boole from what one critic called the "cold aloofness from the warmth of life."[89]

The second issue is, Did Boodin avoid the old trap of misplaced concreteness in his handling of time? Royce got caught in its jaws by conflating types of time—clock or mechanical time with physics time with the deep time of geology—as all relating living beings to nature when, in fact, these can just as easily alienate us from nature in ways that perhaps only organic time can truly resolve. James avoided this by focusing on time experientially. Boodin seems to recognize in clock time not real time but merely the expression of "succession or fleetingness of values" in which the "*fleetingness of process* . . . furnishes us with the real time content."[90] Avoiding Royce's jumble of times, Boodin states, "All quantitative standards are relative and arbitrary. They are the result of social convenience and agreement, but have no absolute basis. The time process continually eats into our standards of measurement. This is nowhere more obvious than in our measure of the time-process itself. The earth-clock [i.e. organic time] is necessarily our standard clock."[91] Because Boodin understood the specific differences in the various manifestations of time, he did not let mechanical or clock time (much less geological time!) become concrete in ways Royce had.

Here ends our discussion of Boodin's epistemology. He brought truly new and constructive perspectives to bear on some very old questions. In both senses, his first two book-length studies demonstrate that as the twentieth century opened, this was truly Boodin's time even if it was destined not to last. But he was only half done; his next task was to develop a new metaphysic built upon the foundation his epistemology had started.

Chapter 2

Pragmatic Realism
Boodin's Metaphysics

Although Boodin titled his metaphysics *A Realistic Universe*, it is perhaps best summarized in his phrase "pragmatic realism." As we have seen this was introduced and defined in *Truth and Reality*, but it is presented again here as an elaboration upon his summary article that appeared in the October issue of *Mind* in 1913 as "Pragmatic Realism—The Five Attributes" (being energy, time, space, consciousness, and form, all of which are irreducible to each other). We would do well to recall Boodin's own definition of pragmatic realism was explicitly based upon the pragmatism of Peirce and James, tested empirically in the sciences, and evaluated for its "cash-value" for our intellectual life. Boodin's pragmatic realism, therefore, rests on the twin pillars of practical experience and scientific verification so much a part of the pragmatic program of its leading representatives, not just Peirce and James but Dewey as well. It might also be synonymously designated, as Boodin does in chapter 3 of *A Realistic Universe*, "Pragmatic Energism." By energy Boodin means a discernable and knowable attribute of nature that in many ways serves as the drive train of a vibrant and evolving universe. In order to unpack this, some additional background on exactly what metaphysics is and its checkered fortunes should be helpful in providing some context for this aspect of Boodin's philosophy.

A Brief History of Metaphysics

The word stems from the Greek *meta ta physika* ("after the things of nature"). This broadly sweeping designation was used as a synopsis to cover a group of untitled texts by Hellenistic commentators succeeded in later generations studying Aristotle's writing designated today as the *Metaphysics*. Andronicus of Rhodes, who classified and cataloged Aristotle's works in the later first century BC, called metaphysics "the books after the books on nature." As such, Aristotle by way of Andronicus first gave us the term, although it can more accurately be said that of Aristotle's predecessors, Plato was surely more metaphysical than his teacher Socrates, and Aristotle, in turn, revised his teacher Plato's dualistic metaphysic rules by dualistic forms in favor of a plurality of substances.[1] Later scholars, especially in the middle ages, took this to mean that these "metaphysical" subjects were in some sense removed from the empirically real world and given to include things of a more esoteric and thereby difficult realm of inquiry. But how exactly did this happen?

It began with the second-century (CE) Greek philosopher Plotinus, the last representative of the classical period, who would carry Plato's idealism into the Middle Ages with a striving for union with the One or the Absolute (God). For Plotinus, God does not create, he emanates. He is reflected onto the lower planes, which are imitations of a perfect Absolute; with matter, soul, and intellect emanations of the Absolute, we have a metaphysics bordering on pantheism. Plotinus's neo-Platonism exerted a tremendous influence over medieval Christianity. Donald Palmer has said that Plotinus's flirtations with pantheism persisted "not to haunt the death scene of classicism, but the birth scene of Christian philosophy."[2] Thus many Latin Fathers such as Augustine were greatly influenced by Plotinus.[3] Especially influenced by Plotinus was Ambrose, who encouraged a spiritual path to God through personal mysticism, the knowledge of an all-encompassing God through transports of ecstasy.[4] But Copleston reminds us that care must be taken in making the emanations of God in Plotinus pantheistic. While it is true enough that Plotinus's God proceeds according to the necessities of nature and he rejects free creation ex nihilo, the primary principle remains in its place, undiminished and unimpaired, indeed forever transcendent over all subordinate beings. This cannot support true pantheism.[5] From Plotinus on, Platonism and neo-Platonism would weave its way in and out of Western metaphysics. Descartes would change this with his dualism and build his metaphysics upon the foundation of epistemology, *Cogito ergo sum*, set forth in his *Meditations* (1641). But philosophers could continue their infatuations with substance monism and pantheism under Spinoza, especially his *Tractatus Theologico-Politicus* (1670) and his *Ethics* (1677).

As this very brief summary suggests, medieval and modern metaphysics has been taken to mean things transcending nature, in a sense "above" and more "significant" and therefore of more intrinsic value than the immediate natural world, a new use of "meta" over the early classical Greek. Since Kant, metaphysics has come to mean a priori speculation on those things beyond scientific experimentation and empirical observation. The eighteenth-century skeptic David Hume sought to denigrate metaphysics by often using it synonymously with "excessively subtle" ruminations, even "sophistry and illusion." Although metaphysics is often today associated with the spiritual and the occult, less fancifully it is really just the effort to examine what kinds of things there are and their modes of being all of which include the following: the distinction between particulars and universals, individuals, and classes; the nature of relations, change, and causation; the nature of mind, matter, space, and time; the existence of things, properties, and events. Although the eighteenth and nineteenth centuries used metaphysics to include the nature of reality itself to the external world, the existence of other minds, the possibility of a priori knowledge, and the nature and relations of sensation, memory, abstraction, and so on, today these issues are usually taken up under epistemology.[6]

The British empiricists—Locke, Berkeley, Hume—seem mostly concerned with epistemology, but Locke and Berkeley had distinctly metaphysical interests; Locke, similar to Robert Boyle, in the nature of substance and his study of abstract ideas in order to conceptualize a theory of universals, and Berkeley in his version of "empirical" idealism as distinguished from Leibniz's "rational" idealism. Both are metaphysicians.[7] Less so is Hume, who will be mentioned at bit later.

In Germany the idealists Kant and Hegel made their own contribution to metaphysics. For Kant, there is a shift away from traditional metaphysics toward what might be called a metaphysics of experience. Kant believed that metaphysics devoted to speculations about God, freedom, and immortality went beyond the bounds of legitimate reason, and his *Critique of Pure Reason* (1781) was aimed, at least in part, to show that such "illegitimate" uses of reason ended up in hopeless contradictions. Fichte, however, proposed the first instance of absolute idealism with his *Wissenschaftslehre* ("Doctrine of Scientific Knowledge"), which he based upon the "pure I" or thorough subjectivity. Hegel, then, developed his metaphysic by proposing precisely what Kant denied, that through his system of categories "Spirit knows itself as Spirit," what Kant would certainly have regarded as an unwarranted conclusion on the basis of pure reason.[8] There were, of course, objections to Hegel's emphasis upon essences and its starry gaze at a vast impersonal, all-encompassing system, which he filled in with an esoteric occult mysticism.[9]

Existentialists like Kierkegaard have been especially outspoken on this point, while Marx tried to salvage out of Hegel a theory of economic determinism based upon a historical dialectic. Hegelian influences continued into the late nineteenth and early twentieth centuries under Thomas Hill Green, F. H. Bradley, Bernard Bosanquet, and John McTaggart. Throughout the twentieth century metaphysics captured the interests of philosophers, some of the most influential and important examples of which can be found in Edith Stein, Bertrand Russell, Teilhard de Chardin, Ludwig Wittgenstein, and Alfred North Whitehead to name a few. More recently, metaphysical philosophy has been pursued by John-Paul Sartre, Alvin Plantinga, Hilary Putnam, and Thomas Nagel. But overall they are a scarcer breed.

Even in Boodin's day such grand systems were on the decline. David Hume had already dismissed metaphysics as a smokescreen for popular superstitions, speculative ruminations that should be "committed to the flames." But he was not the only skeptical prophet of things to come for metaphysics. Some of the most vocal opponents of metaphysics came years later from logical positivists, heirs to the ideas of Auguste Comte, and more specifically to the Vienna Circle that rose to prominence in the 1920s and 1930s under Moritz Schlick, championing a strong verification principle. Although reactions against their extreme reductionism saw their influence wane, the general distaste for metaphysics can be found more recently in the growth of scientism under the so-called New Atheists (e.g., Richard Dawkins, Daniel Dennett, Sam Harris, and the late Christopher Hitchens) and an apparently increasing number of scientists (e.g., Francisco Ayala, Lawrence Krauss, Taner Edis, E. O. Wilson, Neil deGrasse Tyson, and the late Stephen Hawking) who have expanded their turf well beyond its empirical horizons. Quite different bedfellows are the postmodernists and deconstructionists typified in so much of Michel Foucault's work with its dismissal of eternal (or at least consistent) truths and endless word parsing and syntactical dissections. Here the rejection of metaphysics is based, of course, not on scientism but on thoroughgoing relativism that forces a rejection of coherence, system, and certainty.

"With enemies so divided amongst themselves," observes E. J. Lowe, "metaphysics may comfort itself with the thought that so many people can't be right. The very fact of such widespread disagreement over fundamentals demonstrates the need for critical and reflective metaphysical inquiry, pursued not dogmatically but in the spirit of Kant."[10] If so, the final verdict may have already been issued by Bernard Phillips long ago when he noted that "one of the lessons to be derived from the study of the history of philosophy is that metaphysics always buries its undertakers."[11] And there appears to be a growing awareness of this with a general call to seriously revisit metaphysical inquiry.[12] It is in that spirit we propose to resurrect Boodin's metaphysics.

A Realistic Universe: A New Metaphysics

Boodin's general metaphysical work came out at a less than opportune time for two reasons. First of all 1916 saw Europe locked in World War I. The Battle of the Somme was sufficient distraction to draw attention from intellectual matters, especially a book on metaphysics. Although true as a purely practical matter, that did not mean that Boodin was wrong to bill *A Realistic Universe* as "the first systematic effort in the English language to create a metaphysics in the intellectual climate of the twentieth century."[13] While he declared this in the preface to the revised edition, he clearly meant it to apply to his original effort in 1916. But that raises the second reason why 1916 was a less than ideal time to release a metaphysic deeply reliant not just on science but on physics in particular. Of course Boodin couldn't have known this when he first published *A Realistic Universe* and he wouldn't be the first author ever to get caught in the changing landscape and fortunes of history, but 1916 stood on the precipice of nothing short of a revolution in physics. He might have suspected that great changes were afoot, however. His close reading of French mathematician Henri Poincaré, an "overlapping genius" with the ability to "make the generalities of science available for the trained philosopher," was one of the reasons the second edition published fifteen years later needed so little revision.[14]

That reading of Poincaré was especially referring to Henri's *The Value of Science* (English translation 1907). There this world-renowned scientist-turned-philosopher warned that new discoveries particularly regarding radium and radiation were undermining classical Newtonian physics and other ideas such as Lavoisier's principle of the conservation of mass, not to mention the conservation of energy. A revolution in the physical sciences was afoot and over the years between Boodin's original metaphysical work and its revision in 1931 that would unfold in a number of momentous discoveries, ultimately establishing modern physics through two theories, the theory of general relativity and quantum theory. The first of these—relativity—was born from Einstein's "special theory" in 1905 dealing with time, space, and simultaneity memorialized in the famous equation $E = mc^2$. Although this was well before the first edition of *A Realistic Universe*, its stunning confirmation would not take place until 1919 when general (proposed in 1915) and special relativity could be demonstrated conclusively by the bending of starlight by the sun's gravity during a solar eclipse. This made Einstein a household name and synonymous with being inordinately brainy.[15] His image and his equation now sport everything from neckties to t-shirts, including the "nerdkini," a new style of swimwear. Rarely has a scientific figure become such an integral part of popular culture.

This was just part of the unfolding revolution. In 1913 the Danish physicist Neils Bohr published a paper in which he introduced Planck's constant (the measure of how much energy a photon increases when the frequency of an electromagnetic wave is increased by 1, which by definition is $h = 6.62607015 \times 10^{-34}$ J·Hz^{-1}) and referred to Einstein's work on the photoelectric effect; it was, by Bohr's own admission, an imperfect "atomic model." When Arnold Sommerfeld noted that "much of the wave theory still remains, even in spectroscopic processes of a decidedly quantum character" in his influential *Atomic Structure and Spectral Lines* (1922), Boodin stated that his observation made nature not merely a "mathematical genius" but an "aesthetic genius" with the most successful theories being those that are most beautiful.[16] By 1931, the date of Boodin's revised version of *A Realistic Universe*, Erwin Schrödinger and Werner Heisenberg were about to receive Nobel Prizes for establishing quantum mechanics on a firm mathematical footing. Work in the forthcoming years would be devoted to extending and interpreting this strange new world, all made stranger when twenty-nine-year-old Paul Dirac suggested that the electron had an antiparticle associated with it and one year later Caltech student Carl Anderson demonstrated the existence of the positron. Also, in 1931 Ernest Lawrence, not far from Boodin at the University of California at Berkeley, had completed a small prototype cyclotron that he hoped would accelerate protons sufficiently to break into the nucleus of atoms. Spencer Weart points out that 1931 was "a rich smorgasbord of opportunities" in physics, a fact worth remembering in assessing Boodin's metaphysics.[17] Indeed the state of physics during this period is an essential context for understanding *A Realistic Universe*, for without it Boodin's talk of electromagnetism, electrons, particles, waves, fields, and energy scarcely makes sense. The point is Boodin's effort to establish his metaphysics on a firm scientific footing based in the physical sciences was sound. He began his revised edition appropriately enough by declaring, "The philosopher must be a poet who makes use, so far as he is able, of the material of science."[18]

And poetic is precisely how Boodin launches his metaphysic. His opening chapter is labeled a "perspective" on the "five-fold truth," the core principles upon which his entire system would be based: being (i.e., energy), time, space, consciousness, and form. Almost from the very beginning Boodin presents his thesis statement: "Looked at from the side of process, nature is a lavish creator, and some of its gifts also have form as read or appreciated by human nature. This is not mere chance. It is part of the selective evolution of reality, for human nature is part of nature. Beauty is but nature become conscious of its formal character through its more developed organs of human nature. Thus do nature and human nature conspire to produce the sunset and the symphony."[19] Here we have a statement close to panentheism, the features of process and creativity through a participative interaction of humanity and nature itself, ideas reminiscent of Whitehead and Hartshorne. Although this is recapitulated in more prosaic form in the final chapter, it has been noted that Boodin's "grand purpose" is a *new* synthesis in the Platonist tradition.[20] Boodin agrees with Whitehead that philosophy is really a series of footnotes to Plato, and Aristotle's teacher is never far from his mind; Plato is mentioned forty-one times in *A Realistic Universe*.

In the first substantive chapter "being" (which Boodin likes to put in quotes), in the sense used here, needs to be qualified as energy to avoid conflating being with substance. As he states, "The substances, in other words, must be known through their activity; and, therefore energy, and not substance, becomes the fundamental thing; substances so-called are mere abstractions of the relative uniformities and constancies, physical and psychological, which we observe in the stream of processes."[21] There is, therefore, a sense in which being can also be associated with mind, all of which remains cast within the larger framework of *process* and, not insignificantly, *divinity*, by which I take Boodin to mean some transcendent teleological force (God, if you will). Here will be found the only real changes from the 1916 version (pages 22–23) where Boodin has added the revolutionary contributions of physics by Einstein, Bohr, Heisenberg, and Schrödinger already discussed.

Although Boodin claims to have made changes to chapter 3, "Pragmatic Energism," a careful perusal will, in fact, indicate no changes whatsoever. In any case, he opens by remaining true to pragmatism by proposing that "energy is what it does" and he adds that the "thin concept" of energy as a universal is all we have for this "pluralistic world," a universal that requires supplemental definitions of things like matter, spirit, electricity, and so on. While Boodin consistently rejects any reductionist formulas for nature, he also dismisses vitalist explanations as too simplistic and lacking explanatory value. What use is there, he suggests, to "introducing a water impulse to account for the unique properties of water"?[22] But this should not signal an abandonment of any idealist tendency. Interestingly, he invokes Maxwell's demon to make the point. Boodin simply states that the eminent physicist James Clerk Maxwell demonstrated that "new available energy" could be created by intelligent manipulation, but this is an underestimation—or perhaps better an understatement—of its significance; Boodin would have only served to bolster his case by delving into it in greater detail and so I shall do so here.

In an 1867 letter to mathematical physicist Peter Guthrie Tate and a few years later to Lord Rayleigh (destined for a Nobel Prize in physics in 1904), finally published in his book *The Theory of Heat* (1871), Maxwell proposed a thought experiment. It featured a gas-filled box containing two compartments connected by a trapdoor over which a "finite being" had control.[23] William Thomson, later known as Lord Kelvin, had more descriptively referred to this as "Maxwell's intelligent demon" (meaning *dæmon* as δαίμων in the Greek sense of "power" or "spirit"[24] rather than some malevolent entity) signifying the obvious implication of purposeful agency. In any case, this theoretical demon constantly intercepted and sorted the fast moving ("hot") molecules from the slow moving ("cold") molecules so as to completely divide the hot from the cold on either side. Maxwell thereby demonstrated that such complete separation could only occur through outside intervention. Maxwell's immediate point was to show that the second law of thermodynamics—entropy as a physical property of any thermodynamic system in which heat flows spontaneously from hotter to colder in an irreversible process—could not be naturally violated or reversed. In other words, any state of spontaneous evolution would always move toward equilibrium. While the immediate application is physics, it is fraught with analogous teleological implications in biology. Over a century later, astrophysicist Fred Hoyle pointed out, "Darwinists imagine groups of organisms with slight differences being sorted into distinct species. The problem is that outside intervention has no part in the Darwinian theory. How then does the sorting in the natural world occur?"[25] This question remains one of the most vexing of modern science and indeed of philosophy as well. Boodin provides a philosopher's answer, an answer that unpacks itself in the remaining chapters. And it is worth adding here that answers in philosophy are always more interesting and potentially significant than scientific ones (not that science should be ignored or that it cannot play an important role in mapping out those answers) because philosophy's epistemological and ontological reach is always much deeper—and *legitimately* so—than science. If there has been one overall mistake of scientists it has been to engage in the professional hubris of declaring themselves the arbiters of ultimate truth. More will be said on this in the next chapter and the epilogue.

As if to prepare his reader for his next book, *Cosmic Evolution*, Boodin repeatedly states that chance (often preceded by the adjective "mere") is inadequate to explain the diversity of life or many of its complex features.[26] He points out that even if life evolved from primitive organisms, it does not follow that more complex life can be explained by "a mere mechanical combination" of these comparatively simple forms; evolution is not just a material explanation with everything that follows just more of the same. Boodin points out that this is especially true in dealing with the mind and mind-like processes. While mind and matter cannot be simply conflated with matter, it should, according to Boodin, be clear that mind and matter have developed through an evolving creative synthesis only glimpsed in science "and in those instinctive demands for harmony and beauty which make themselves but feebly felt in us."[27]

In chapter 4, "Do Things Exist?," Boodin rejects the "dreamy confluence" of mysticism and questions Henri Bergson's convictions based solely upon intuition. Boodin prefers the empirical realities of the real world to Bergson's murky and question-begging *élan vital*. Boodin instead insists that we must have an instrumentalist means of reality, we must take it insofar as it realizes "*our* purposes." This seems akin to James's "cash-value" idea. The "our" is more important than the "my" here. For Boodin, social purposes are more important than individual ones; they are less prone to illusion, self-deception, and the idiosyncrasies of personal preference and indeed personal prejudice. When social agreement and *sensible* continuity are in agreement we find "that the *substance* of things is precisely what we must take it as in experience. . . . The only key we have to reality is what reality must be taken to be as in the progressive realization of the purposes of human nature."[28]

The next two chapters deal with epistemological concerns. These round out part 1 on "Energy and Things." Since we have already dealt with Boodin's theory of knowledge in the preceding chapter, it shall not be elaborated upon here. It should be said, however, that Boodin ends part 1 on a note of partial idealism. Although we cannot take things as having values in and of themselves, they have no "halo of value on their own account." Though sympathetic to Plato, Boodin pointed out that he turned the real world into "a world of normative ideals of which the concrete is only a poor imitation." Boodin couldn't live in a world of such shadows. Nevertheless, his cosmic order is no mindless machine; it must have a real existence fashioned in idealism. "If there is a conscious power that exercises selection in the larger universe," he declares, "then survival in the whole existence as well as in the world of human control, may depend upon its value fitness—its harmony with an ideal constitution."[29]

Chapters 7 and 8 deal with consciousness. Boodin's introduction of energy into his metaphysical system has already been noted, but here he adds that "mind-stuff is a distinctive type of energy" although mind and consciousness "do not necessarily coincide." Unlike mind, consciousness does not vary; it is more a state or condition, and, therefore, should be "treated as an independent dimension of reality." To define consciousness as a form of energy requires that the former must be smuggled into energy. Instead Boodin regards consciousness as an existent field. Like space, relations are not conditioned by their connection to space but rather relations to each other as conditioned by space; with consciousness, things or energies are not related to consciousness but related to each other "within the *field* [emphasis added] of consciousness." In other words, consciousness adds *awareness* over and above energetic relations. In the end, "mind-stuff" may indeed be regarded as energy but expressed within the field of consciousness.

Boodin makes his schematic of mind/consciousness separation clearer within the larger context of his process-based philosophy as follows:

> Any theory, whatever it calls itself, which strives to derive consciousness, will have the difficulty of materialism—in losing the quantitative and energetic in what is not energetic. This involves an unintelligible *saltus*; and we shall always, therefore, look for a smoother transition between consciousness, on the one hand, and the world of processes, with their qualitative variations, on the other. This is furnished in the theory of consciousness as a constant in the universe, though depending upon certain conditions for its manifestation, as electricity is now regarded as an original fact (by some the most original), though dependent upon certain conditions. This brings it into the realm of the familiar.[30]

With regard to attempting to "derive consciousness" it could be that Boodin has in mind here the newly proposed theory of behaviorism announced and promoted by John B. Watson, whose protégé, B. F. Skinner, would popularize in his utopian novel (some would say *dystopian* novel), *Walden Two* (1948). Watson rejected the concept of consciousness as something that could be clearly defined or even fruitfully studied and instead sought to understand human behavior, viewed as collections of reflexive environmental stimuli or developed from past historical experience, and manage it through operant conditioning. Even if not, the description here would seem to obtain. After all, Watson's whole project was to reduce psychology to empirical data points, the height of materialism that no longer even distinguishes between the quantitative and the energetic because the complexities of the human mind are reduced to a dump of empirical measures for the sole purpose of behavioral manipulation. Boodin's follow-up comment would seem directed at the kind of theory being promulgated by Watson: "To make psychological processes parallel to mechanical rearrangements can only convey sense to a man who does not think about it."[31] One man who *did* think about it years later was the biologist/polymath Ludwig von Bertalanffy, founder of systems theory, who utterly rejected this "robot model of man."[32]

In ways that Bertalanffy surely would have appreciated, Boodin went on to note important distinctions between machine logic and human rationality, marking the difference between automatic and significant activity. Here the process is from mechanical causality to teleological causality. Consciousness makes a difference by transforming cumulative habit into significant activity through the illumination of immediate value emergent from the other independent variable of energy. It is, after all, only through the "energetic structure" of the mind that meaning is accorded to awareness. For Boodin, there can be only two ways to explain consciousness: it must either be an ontological constant or it must be created over and over again after each occurrence as materialism must assume. "Such a heaping up of miracles is hardly consistent with the modern scientific spirit,"[33] he complained. But Boodin sorely underestimated the power of materialism to reduce everything—even miracle—in its path. Years later, in our own generation, Terrance McKenna would wryly observe that "modern science is based on the principle: 'Give us one free miracle and we'll explain the rest.' "[34]

Boodin rejects the notion that thought is simply a product of brain activity, an epiphenomenon of our neural synapse. Boodin even leaves open the possibility that "the energy underlying our personal continuity" may be more nuanced and accessible than science at present can prove, indicating that "the spontaneous trance" of "Mrs. Piper" may be suggestive of a "new state" of spiritual existence. It is easy to scoff at such an idea, but William James was seriously interested in spiritual phenomena, and famed medium

Leonora Piper, to whom Boodin refers, left the Harvard professor utterly baffled at her abilities.[35] James told the Society for Psychical Research in 1896, "If you will let me use the language of the professional logic-shop, a universal proposition can be made untrue by a particular instance. If you wish to upset the law that all crows are black, you mustn't seek to show that no crows are; it is enough if you prove one single crow to be white. My own white crow is Mrs. Piper."[36] Like James, late in life Boodin reportedly told his longtime friend Bernhard Mollenhauer, of San Diego, that he was interested in psychic phenomena and felt that great minds of the past may somehow influence thinkers at present.[37] This is in keeping with a man who said, "It is a matter of surprise how much of what seems original is foreshadowed by the geniuses of the past."[38]

The next two chapters cover "Knowing Minds." The starting point, Boodin indicates, is with the acknowledgment that the properties of mind are different from the properties of matter, despite the efforts of materialists to conflate the two. How then do we know ourselves? Boodin answers simply enough, through others. This becomes a social adjustment whereby we know our own significance through social situations. Mental acts reveal themselves as will and in the dual actions of attention and affection. What Boodin calls "the conative constitution" is exposed in various levels of complexity from "blind instinct, through impulse, desire, and wish, to organized character." This is not Schopenhauer's noumenal "Will" but a real will revealed in physical and social continuities, a will that is really free. If the mental is will, then it is expressed in the living activity found in tones of agreeableness or disagreeableness with the rest being physical in nature. How then do we know ourselves? Boodin says we basically determine this through language. But he answers as a pragmatist not as an analytic linguist:

> Knowing our own meaning, past and present, as knowing those of others, is largely an interpretation of language. But language after all is only the symbol of the contents of experience. Language could not convey the same meaning, unless we owned the meaning. We can recall blue sky when we perceive the words, because we have the actual meaning of blue sky, however fragmentary its concrete content. And failing this, as in an unknown tongue, we would simply have words staring us in the face, conveying nothing. Language, moreover, is discrete and stereotyped and fails to give an equivalent for the quivering transitions that

persist indefinitely within the systematic meaning. It is not fair
to substitute the tool, however important, for the living reality.[39]

Forgoing Boodin's discussion of social minds in his next chapter for
full treatment in chapter 6, he next addresses the attribute of space. The
leading standoff between Newton's concept of space as a real void and Leibniz's concept of space as geometric and relational is as old as Parmenides.
It was Parmenides of Elea, widely regarded as the most important of the
pre-Socratics, who argued that no void is possible because non-being is
unthinkable, and the atomists (Democritus and Leucippus) who insisted
that since there is motion then the void must be real, atoms and the void
being their two ultimate principles. But Boodin wants to affirm the reality
of both serial space and pure space as revealed in physics and astronomy.
He spends a fair amount of time discussing psychological space (especially
Kant's arguments for a priori space intuition) only to finally declare that all
theories of space must be "rewritten largely in biological terms" in which
"the principal coordinates to our space world are a motor affair."[40] He concludes with an interesting reference to Johann Gottlieb Fichte, whom he
calls "the most brilliant of modern idealists," by explaining that for Fichte
space becomes a "self-intuitive eternal system of truth" that is not simply
a type of eternal but *the* eternal. This goes too far, for then space loses all
real meaning for us, reducing it to a mere eternal nothingness.

So what then is the nature of space? There are two ways of discerning
this. First, geometrically through the predetermined properties of its postulates; second, through the empirical sciences that reveal real space. Both
contain dimensionality and both have homogeneity in common. Continuity
is also a feature of space, but care must be taken not to confuse a metaphysical continuum with the mathematical continuity of order. Thinking
universally then, what does science, as Boodin sees it, suggest about space
as an attribute? In a rather prescient statement, Boodin acknowledges the
infinity of space as a real possibility and if so, "there may be other worlds
quite unknown to us, and incapable of making any pragmatic difference to
the world as we know it."[41] Now this sounds like multiverse theory, currently
quite popular among practicing physicists such as David Deutsch, Leonard
Susskind, Don Page, Michio Kaku, and Sean Carroll. But if Boodin's pragmatism prevents him from assigning much significance to the multiverse
speculation, it should be encouraging to know that at least one important
physicist agrees, Freeman Dyson:

> Opinions vary widely concerning the proper limits of science. For me, the multiverse is philosophy and not science. Science is about facts that can be tested and mysteries that can be explored, and I see no way of testing hypotheses of the multiverse. Philosophy is about ideas that can be imagined and stories that can be told. I put narrow limits on science, but I recognize other sources of human wisdom going beyond science. Other sources of wisdom are literature, art, history, religion, and philosophy. The multiverse has its place in philosophy and literature.[42]

Dyson's view is right insofar as he demonstrates a genuinely pluralist epistemology, but philosophy is more than imagination and storytelling, it is testably forged in the fires of reason and discussion. What Dyson apparently means is that multiverse theory cannot be associated with any hard empirical data; it is a matter of hypothesis borne out by mathematical equations. In that sense it is truly speculative, but of course some speculations are always better than others. So Boodin is surely free to appropriately speculate upon the multiverse even if his philosophy agrees with science that little can come of it.

Boodin ends his discussion of space by pointing out the central error of "absolute idealism," perhaps alluding to his teacher Josiah Royce. He chides the absolute idealist for conceiving reality as ultimately a system of logic, and although logic is a very real part of our world and is an important means for our appreciating and understanding it, it cannot be reified into a fundamental reality. We live in a pluralistic world whose real unit can be found in the energy system in which real space becomes an essential condition. This is true of personal relations, of our social life, and of physical things as well. Boodin prefers these realities over the absolute idealist's abstractions.

Since the attribute of time has already been discussed at length, we can move on from chapters 14 and 15 to Boodin's fifth and final part, "Form and Reality." Chapter 16, "The Identity of Ideals," addresses the question of life's ideals—truth, beauty, virtue—by arguing their expression as forms, demands made upon what Boodin calls "the concrete will." Efforts to unify these ideals have been confused and approached biologically as aspects of adjustment or satisfaction yielding results that are too vague. For Boodin, ideals are matters of selective interest aimed at making experience significant and meaningful. More importantly, Boodin is no moral relativist; he sees morality related to truth insofar as it "gives dignity and calm to the soul" and can be expressed in otherworldly "noble and inspiring" imperatives such as "love your enemies" and "mystical unions" that form the basis for higher principles lifting out of the mundane world.[43] Although we cannot know an absolute truth, we create it though our collective experience, and despite its plasticity, it is real. Boodin plainly states, "There can be no private truth"; it is construed and verified socially.[44]

If the ideals are related to our "striving for evaluation," the delineations of those ideals are found in harmony, simplicity or economy, and universality. Recurring to James, Boodin argues that ideal activity toward beauty, truth, and morality must be determined by their "cash-value" in our practical lives. These must have distinct expression in unity, factual content, organization, no extraneous details, and social objectivity (perhaps better described as communal consent). In a very real sense then Boodin is an idealist whose pragmatism always keeps him tethered to the earth. This is important in his rendering of science, a subject never far from his philosophy. Significantly Boodin declares, "We cannot legislate arbitrarily to nature. We must try to discover clearness and distinctness *within* the relations of nature. That success here is possible is due to the fact that reason is not an arbitrary addition to nature," he adds, "but that reason grows up in the soil of nature, is nature's reflection upon itself."[45] Little could Boodin have suspected that right around the corner a host of postmodernists would attempt to transplant their assorted musings as the air plants of nature (i.e., science), chief offenders being Jacques Derrida, Jacques Lacan, Julia Kristeva, Bruno Latour, Jean Baudrillard, Giles Deleuze, and Félix Guattari. Science, relying upon more substantial ground, did not find these intellectual air plants useful or meaningful despite the academy's enthusiasms for them. Fortunately, most of these postmodernists' impositions upon our credulity have been exposed.[46]

In the next chapter Boodin takes his cue from Hume's is/ought problem to address not "is" but rather "Form and the Ought." He observes that we are so constructed that we seek to find order in the universe, and these are more than just mental constructs, they form the real architecture of the world we seek to know. Pointing out that Aristotle was clear that form is not a product of process but rather "legislates to" process, he explains that Herbert Spencer was wrong in seeing evolution as a one-way progression from formlessness to form but, anticipating his next book, a dynamic cosmic process. Form, not being physical or made up of "mind stuff," is eternal, a fact that requires us to contrast the physical world and its processes from the world of form. This comment of course assumes the reality of form and a world of forms. Flux and directional selection seem necessary to explain our world. "In the protean guises of the process, form legislates. And while it cannot stop the process," he writes, "it determines what can have meaning and existence in the process." Form creates as the artist creates, by selection and the elimination of the extraneous. Form is a very real ontological feature of our world, at once the heavenly pattern and the artist. Thus the universe and all that is in it is imbued with a transcendent teleology; form (the very embodiment of God) bestows upon it not just an ethical system but an ethical *process*. And this is Boodin's essential argument against chance: "If the process of the universe is merely a chance affair, no ideals can be enforced, or be binding whether mechanical or ethical. Science in such a world would have no guaranty for its ideals of simplicity and unity any more than ethics for its ideals of worth."

Here Boodin seems drawn to critique Darwin's theory of evolution, ending this chapter with a section titled "Evolution and Direction." This will be dealt with in great detail in the next chapter, but a few preliminary observations are in order. First of all, Boodin correctly understands that the operative principle of this theory rests on the explanatory power of chance and in this regard his opposition to any substantive formulations of chance has already been noted. However, even where chance is given over to natural selection (in and of itself *not* a blind process) we are left with little more than a tautology, a statement that is true but tells us nothing. As Boodin explains, "The criterion of the survival of the fittest [a phrase borrowed from Herbert Spencer, suggested by Alfred Russel Wallace, and accepted by Darwin] has meaning only when we define what environment is intended. Otherwise, it amounts to merely saying that what does survive is fit."[47] He is not the first to have pointed this out and this will be carefully unpacked in the forthcoming pages. Boodin also rather poetically ascribes the anthropological development of humans over time with meaning more than it knows by a "blindness" that "comes to fit a larger pattern," a pattern dictated by a directional evolution subsumed by teleological forms. And while we may think that Boodin is just waxing eloquent here, when he states that the specially endowed specie of being (*Homo sapiens*) awoke "from the long slumber of the ages, pregnant with the tendencies which ages of selection have forced him," he is not simply stating some reductionist ape-to-man scenario hidden in poetic language. He is stating what Raymond Tallis, after long and careful analysis, has said more directly, that "our bigger brains did not merely give us more of the same but helped to facilitate something qualitatively different; why we are not just very bright chimps."[48] Humankind is not, however, *utterly* distinct from the animal world; interesting (often important) affinities exist between the two but not to qualitative indistinguishability.

In summary, then, the "is" of this world amounts to conditions of survival that must be selective *and directional*. That this could be the result of mere chance is unlikely. Therefore, if we define directionality in terms of the teleology it implies and the qualitative difference it seems to make for humanity, then no doctrine of "survival of the fittest" gives us warrant for the kinds of excesses to which certain kinds of evolutionary theory are prone. In fact, this tendency within Darwinian thought to reduce everything in biological terms is dangerous. Max Horkheimer is worth quoting at length on this if only because I believe Boodin (at heart always a Swedish farm boy at one with nature) would heartily approve:

> Darwin was essentially a physical scientist, not a philosopher. Despite his own religious feeling, the philosophy underlying his ideas was plainly positivist. Thus his name has come to represent the idea of man's domination of nature in terms of common sense. One may even go so far as to say that the concept of survival of the fittest is merely the translation of the concepts of formalized reason into the vernacular of natural history. In popular Darwinism, reason is purely an organ; spirit or mind, a thing of nature. According to a current interpretation [written in 1947, presumably a reference to the neo-Darwinian synthesis], the struggle for life must necessarily, step by step, through natural selection, produce the reasonable out of the unreasonable. In other words, reason, while serving the function of dominating nature, is whittled down to being a part of nature; it is not an independent faculty but something organic, like tentacles or hands, developed through adaptation to natural conditions and surviving because it proves to be an adequate means of mastering them, especially in relation to acquiring food and averting danger. As a part of nature, reason is at the same time set against nature—the competitor and enemy of all life that is not its own.[49]

This brings us finally to "the Ought" of this world. This Boodin defines as "the consciousness of the formal character of the universe," which he likens to chords of music that form each movement of a symphony and ultimately the symphony as a whole. The directional unity and harmony we find in music give some sense of this "Ought" in cosmic evolution. This analogy is especially engaging and suggests that William Paley was wrong, nature is not like a watch, a contrivance of man, but something much more profound. One is reminded of the fourth-century cleric Gregory of Nazianzus, who preached in his Second Oration:

> Natural Law teaches us the same truth because, through these visible things and their order, it reasons back to their Author. For how could this Universe have come into being or been put together, unless God had called it into existence, and held it together? For every one who sees a beautifully made lute, and considers the skill with which it has been fitted together and arranged, or who hears its melody, would think of none but the lute-maker or the lute-player and would let their thoughts run back to him, although they might not know him by sight.[50]

Moreover, this cosmic music has its corollary in science. Nobel laureate Frank Wilczek points out that the equations for atoms and light in physics are "almost literally, the same equations that govern musical instruments and sound."[51]

The penultimate chapter, simply titled "Teleological Idealism," was originally published as "The Reinstatement of Teleology" in a 1913 issue of the *Harvard Theological Review*. It is a shame that Boodin did not use this title to conclude his grand metaphysical volume since the idea of "reinstatement" correctly suggests the recovery of something historically lost. The philosophizing of this as a form of idealism may be true as far as it goes but something is lost or at least obfuscated in such a move. At any rate, Boodin here seeks to capture "the drift in our cosmic weather" by outlining three types of theory: mechanism, finalism, and vitalism. In some ways this chapter serves as a preface to his *Cosmic Evolution* and a prelude to his *Three Interpretations of the Universe*, but since this must serve as an essential part of Boodin's metaphysic it should be outlined as presented here.

He begins with the mechanical view. This is one of chemical analysis and relationships, a workshop model of quantitative examination and experimental controls. Boodin has no issue with the tremendously fruitful methods of the mechanistic approach, but this is not necessarily the end of the story. Here, for example, "the Darwinian hypothesis of natural selection" may be useful in its own way and yet only a "partial account of the facts."

Natural selection, Boodin argues again and again, is at heart a negative factor, the elimination of the unfit. Finalism places the importance of evolution in its outcomes. Rather than discovering things sequentially by their natural causative factors, finalism looks to the future or *end in view*—the prospective view of the process. In vitalism there is the search for some added feature, some force that forms a common denominator for the whole. Although vitalism assumes a rather idealist posture, Boodin is critical of it. For example, he rejects Hans Driesch's "entelechies" because "they seem to have no meaning at all"; they are mere duplicates of the process they seek to explain and describe. Boodin concludes, "It is hard to see what we have gained by hypostasizing such tendencies and giving them a Greek name."[52] He is equally uncomplimentary of Bergson's "vital impulse," calling it, in the end, "no less blind than the elements of chemistry," that in its effort to explain everything, explains nothing. Diversity of the entire process remains unaccounted for. In this sense Maxwell's demon makes more sense because it suggests that "the natural order must be thought of as interpenetrated by an intelligent order."[53] Instead of these attempted solutions, Boodin calls for a new teleology, one in which nature is "leading in the direction of human nature" (finalism) by means of certain physical processes (mechanisms) not blind but teleologically directed. For Boodin, this is masterminded by God "interpenetrating our mind and nature and, in a manner which we can but faintly grasp, guiding to a meaningful issue." All of this occurs in a pluralistic universe.

The final chapter, as suggested earlier, is a restatement of Boodin's five attributes. It is really just a summary epilogue to the entire presentation that has preceded and need not be restated here. It behooves us only to now give a final appraisal of the work.

Assessment and Conclusion

The reviews of both editions of *A Realistic Universe* were generally favorable. Hoernlé's review of the 1916 publication was one that Boodin himself noted with satisfaction in the introduction to his revised edition fifteen years later, so we might choose this one to begin. Interestingly, Hoernlé begins on a somewhat sour note by questioning Boodin's attempt to "apply scientific method to philosophical problems" because he does not use Bertrand Russell's analytic method of mathematics. This is from today's perspective a dated complaint; few would argue today that science must be couched in the formulas of mathematics. This is a rather blinkered view of science and scientific method. And besides, Boodin virtually answers Hoernlé when he states, "Form manifests itself in the concrete process of history, in the real flux, which is not merely a rearranging of bits of substance or mathematical models."[54] Hoernlé seems not to think that Boodin's approach is scientific because he is too interested in the whole of things and, "Scientists never think in terms of the whole, nor do they ever use the whole range of experience."[55] Hoernlé is probably alluding to the inductive method, but it seems too restrictive to reject Boodin's science-based approach on this basis and quite a bibliography could be compiled of scientists who have held forth on "the whole" of things. He also thinks Boodin takes too much at face value but gives no real examples of his doing so. Nevertheless, Hoernlé ends on a very positive note calling his objections a few "critical growls." "Altogether," he concludes, "professor Boodin has written a book of exceptional interest and value, accurate and ample in scholarship, rich and varied in range, original in its total vision of the world," an assessment worthy of any dust jacket and one in which Boodin was justly proud.

Radisloff A. Tsanoff, the idealist with whom Boodin had spared earlier on his *Time and Reality*, reviewed both editions.[56] In a masterful sentence of summary economy Tsanoff describes *A Realistic Universe* as "a process of energy systems, involving change and transformation (time), translation and free mobility (space), and awareness (consciousness)." Form, Tsanoff concludes, provides direction, standards, criteria, and order. While Tsanoff does offer some modest criticisms, Tsanoff admits a truism, namely, that "his [Boodin's] approach is not invariably as incompatible with that of the objective idealist as his words might indicate." But in Boodin's defense it should be pointed out that he never rejects idealism per se, after all, why would he end on a note of "teleological idealism"? Boodin's objections are directed at *absolute* idealism, not idealism in general. In the end, Tsanoff admits he is not entirely convinced but acknowledges it is the product of "honest genuine thinking" and worthy of any general introduction to metaphysics. In the second edition, Tsanoff is more unrestrained in his praise, calling it "an outstanding contribution to living modern philosophy."

E. C. Wilm of Boston University, an expert on Bergson and an authority on Friedrich Schiller, reviewed the first edition and was apparently not too bothered by Boodin's jabs at Bergson. His summation is as follows: "In the style of the book, as in the thought, there are here and there loose ends, and rough places, as indeed there should be in a realistic universe like this. The book as a whole reveals an original, flexible and erudite intelligence, it abounds in shrewd and homely comments, and, with all that may be said in criticism of either conception or workmanship in the details, will stand as one of the substantial additions to the American literature of philosophy."[57] One could hardly do better than this.

Among idealists in general, Boodin received a favorable reading. But in the hands of any reductionist like socialist/activist Carl Haessler, he was savaged. Haessler dismissed Boodin's teleological idealism as demanding "psycho-analytic treatment" and denigrated the entire work as ending in the "stagnant tepidity of such doughy optimism."[58] Really this is more of a rant than a review and is mentioned here only to show the degree with which an otherwise valuable and thoughtful work can be despised.

One of the more interesting reviews came from Tulane University's M. T. McClure, who gave a detailed and thoroughly descriptive review of Boodin's book. McClure asks provocatively, "May a pragmatist be a metaphysician? And if so, what sort of metaphysics would it be?" These are not unreasonable questions. James never produced a metaphysical work and seemed to have no facility for it. Peirce was equally challenged. Though he long dreamed of completing a grand evolutionary metaphysics, his biographer admits that "he spent the rest of his life unpacking and classifying its elements, increasingly as a doctrine of signs. He never achieved his fantastic—and intoxicated—purpose of explaining the universe, material and immaterial, etc."[59] Here, McClure explains, Boodin has provided "an affirmative and constructive answer."[60]

So perhaps this is where we should end Boodin's metaphysics. He accomplished what his forebearers could not, and that in and of itself is a significant accomplishment. Although his mentor Royce did offer a complete metaphysical system, he was as is *not* principally known as a pragmatist but as an idealist. Well written and engaging, this is probably the best book for an overview of Boodin's philosophical thought. It is all here in one cover at least in summary form: his treatment of truth, being, time, space, consciousness, form, pragmatic energism, cosmic evolution, religion, and sociology. We will want to delve deeper into each of these aspects, but for anyone who wants a metaphysic from a pragmatic perspective, *A Realistic Universe* is the one place to go.

According to the Library of Congress *Report of the Copyright Entries for the Fiscal Year 1915–1916*, Boodin's *A Realistic Universe* is recorded as entry no. 7090 on December 20, 1916. By that time Josiah Royce had passed on the month before, to join the Absolute. Boodin knew perfectly well that his achievement—and it *is* an achievement—in metaphysics was part of a legacy bestowed upon him not by Peirce or James but by his beloved teacher to whom he appropriately dedicated the book. Royce's unmistakable spirit will reappear later as he turns his attentions to religion and the community, but for now Boodin was looking to present his evolutionary cosmology.

Of course being written during a very transformative—even revolutionary—period in science, the question naturally arises: Did Boodin's close reading of science upon which he built his metaphysic hold up over time? Twenty-seven years later Boodin reexamined his position in light of new developments and happily reported that his "formulation of cosmic attributes . . . remains relevant . . . though in some respects the advance of science has enabled me to make my meaning clearer."[61]

Chapter 3

Evolution, 1925

There probably is nothing more central to the current "culture wars" than the modern theory of evolution. For most, that means the theory proposed by Charles Darwin as he laid it out in *Origin of Species* (1859), and more controversially in that theory's applications to human beings in *Descent of Man* (1871). As a backdrop to our main investigation of Boodin's *Cosmic Evolution* we need to understand the general context in which it was written and its influences and impacts upon subsequent thought. That requires a thorough understanding of Darwin and his theory. Once accomplished, three remarkable events in the history of modern evolution can be examined in turn. All, interestingly enough, occurred in 1925: the famous (some say infamous) Scopes "monkey" trial, where allegedly religious fundamentalism came face-to-face with modern science; the discovery of the first hominid ancestor in South Africa, supposedly confirming the ape-to-man common descent that Darwin had predicted over fifty years before; and Boodin's publication of *Cosmic Evolution*, which proposes a unified theory of the organic and physical worlds on very different principles premised upon idealist pragmatic process philosophy. In this sense, *Cosmic Evolution* might well be considered Boodin's fullest metaphysical work because the epistemological and ontological frameworks are all there. To see how different this is, the positivist perspectives of Darwinism must be seen in contrast.

The Man and His Theory

Before examining Darwin the "scientist" we should at least review briefly those influences bearing upon his *Weltanschauung*, and when we do, it can be readily seen why Janet Browne, this iconic figure's leading biographer, has stated unequivocally that "Darwin was profoundly conditioned to become the author of a doctrine inimical to religion."[1] His grandfather, Erasmus Darwin, had some years before written a rather scandalous evolutionary treatise *Zoonomia* (1794–1796) that pushed God back into a hazy deistic existence far removed from humankind. His father Robert was cut from the same cloth. Among those who knew them, Adrian Desmond and James Moore probably do not overstate the case when they say that even in Charles's childhood the "name Darwin was already associated with subversive atheism" and that "Dr. Robert was himself a closet freethinker."[2]

Darwin derived his *Weltanschauung* from sources other than his father and grandfather. That would come when his physician father, worried that the young lad was interested only in "shooting, dogs, and rat-catching," sent him off at age sixteen to the University of Edinburgh to study medicine. He hated the classes but delighted in the loose body of freethinkers he found at the Plinian Society he joined on November 21, 1826, where he attended all but one of the ensuing nineteen meetings. There he was exposed to some of the most "scandalous" ideas of the day—reductionist accounts of the mind/body problem and all sorts of materialist notions considered too dangerous for polite company. More importantly perhaps, he got from long walks with Robert Edmond Grant, sixteen years his senior and an expert on aquatic invertebrates, the views of a hardened and thorough materialist who easily brought the young boy under his spell. Through Grant, Darwin was exposed to radical transmutationist ideas, a nature bound only by laws of contingency, necessity, and chance. Grant was a confirmed atheist. As Desmond and Moore put it, "Nothing was sacred for Grant. As a freethinker, he saw no spiritual power behind nature's throne."[3]

Darwin dismissed this early period in his life and claimed that Grant left little impression on him; Grant, reflecting back to that time in his *Tabular View of the Primary Divisions of the Animal Kingdom* (1861), reminisced in a dedicatory letter reminding his protégé of their common labors nearly forty years earlier "in the same rich field of philosophic inquiry." Grant was clearly not referring to aquatic invertebrates. Considering all this, Janet Browne seems quite justified in claiming, "Young Darwin, it now turns out, was well aware of evolutionary views and perfectly capable of grasping the full implications of what Grant had to say."[4] So whether he carried it in the back or in the forefront of his mind, Darwin long had a template of what materialistic, positivistic science looked like.[5]

It is, therefore, hard to take comments from certain Darwin apologists at face value. For example, Karl W. Giberson insists, "He [Darwin] was, in fact, a sincere religious believer who began his career with a strong faith in the Bible and plans to become an Anglican clergyman. He did eventually lose his childhood faith," Giberson adds, "but it was reluctantly and not until middle age; long after his voyage on the *Beagle*."[6] Similarly, Alister E. McGrath argues that Darwin's *Origin* contains not even "the whiff of a personal atheism."[7] Michael Ruse tries to finesse Darwin's beliefs a bit by saying, "Darwin was never an atheist; at most an agnostic."[8] All of these comments stem from one principal source: Darwin's *Autobiography*. Although originally intended only for family and close friends, it was always destined for publication. Darwin "was constructing himself in the shape in which he wished to know himself and to be known by."[9] It must, therefore, be handled with caution. Browne admits that the *Autobiography* seems to have "thrown up a smoke screen almost as effective as if no records had been left at all."[10] Perhaps it is even worse. Darwin conveniently forgets his introduction to the materialism of the Plinians and to Grant, and he further confuses matters by suggesting that when first coming aboard the *Beagle* he was fond of quoting from the Bible and engaging in impromptu homiletic, which made him the butt of many jokes among the crew. He further claims to have found the natural theology of William Paley persuasive and even delightful. "It is difficult, however," notes historian of science Phillip R. Sloan, "to find compelling evidence from Darwin's early writings that warrants this reading. It is also difficult to find evidence from this documentary base that he ever adhered strongly to the 'watchmaker-designer' model of a creator-nature relationship integral to the tradition of British natural theology."[11]

It seems clear from all that has been said here that Darwin—ever the promotor and rhetorician—developed a certain narrative that presented himself as a pious youth whose passion for science led him to the evidence-based truths of a chance-driven world guided by impersonal forces wholly indifferent to the plight of humankind or any other organism. He came to this with no preconception or a priori assumptions except perhaps for some benevolent God now made superfluous by the inescapable "facts" of science. Of course the analysis above indicates otherwise; that more accurate account would hardly establish Darwin as a wholly unbiased, careful observer of nature whose evolutionary theory was confirmed by the sheer weight of the evidence.[12]

So where in all this is the *real* Charles Darwin? It seems obvious that Darwin had some view of nature more akin to Robert Edmond Grant's than William Paley's. His interest in Auguste Comte has already been noted and is confirmed in his many mentions of the French positivist in his notebooks.[13] This is not meant to suggest that Darwin deliberately went out to craft a theory of evolution inherently reductionist and inclined toward a positivistic verificationism, only that with Grant's ideas and those of the Plinians as a template—a schematic of how data fit into the larger picture of things—it shouldn't be especially surprising to find Darwin's theory of evolution bereft of any guidance or guide; chance and necessity rules supreme. While Darwin late in life claimed to be an agnostic, it seems that Johns Hopkins historian of science Maurice Mandelbaum more accurately defined his theistic leanings thus: "In the end his Agnosticism was not one brought about by an equal balance of arguments too abstruse for the human mind; it was an Agnosticism based on an incapacity to deny what there was no good reason for affirming. Thus, those who, at the time, regarded Agnosticism as merely an undogmatic form of atheism would, in my opinion, be correct in so characterizing Darwin's own personal position."[14] This is why Darwin could largely agree with the atheists Edward Aveling and Ludwig Büchner when they came to visit him just months before his death. His only real qualm was in adopting what he felt was an unnecessarily aggressive tone. "Is anything gained," he asked, "by trying to force these new ideas upon the mass of mankind? It is all very well for educated, cultured, thoughtful people; but are the masses yet ripe for it?"[15] Ever the strategist, Darwin just couldn't see the value of such a frontal assault in promotion of a controversial idea. But that did not mean he thought the idea itself to be false. This is precisely what one would expect of an *undogmatic atheist* like Darwin.

Now none of this speaks to his theory. In fact, to try to argue for or against *any* theory on the basis of the personal beliefs of its proponent is clearly an example of the genetic fallacy. But, as historian John C. Greene has shown, worldviews do matter and they help us sort through what kind of theory we may be dealing with. I've brought up Darwin's own beliefs to show that the theory he eventually developed and argued for throughout his life was precisely the kind of theory he was conditioned by family and personal experience to produce—*undogmatic atheism*. This would not be necessary if it wasn't proclaimed otherwise by so many "experts," some of which I've already mentioned. The reason they protest against the association of Darwin with any form of atheism is usually based on one of two equally problematic positions. One is based upon the idea that Darwin produced a theory that was science pure and simple. As such, Darwinian evolution does not and should not demand any kind of hard atheistic conclusion. Any metaphysical position derived therefrom (including the error of turning Darwinian evolution into a religion) is in the eye of the beholder, not theory itself. This is Michael Ruse's position. Ruse is himself a professed atheist but not of the Richard Dawkins variety. Ruse is too adept at philosophy not to recognize the pitfalls of the hard fundamentalist brand of atheism espoused by Dawkins, Peter Atkins, and some others. So Ruse adopts a superficial reading of Darwin and takes his *Autobiography* as it portrays him. The other position is held by Darwinian theists like Giberson and McGrath (other famous examples would include Brown University biologist Ken Miller and former NIH director Francis Collins). For them, Darwin left room for God in his theory and there is nothing, in their view, atheistic about it. They argue that God could use chance as a means of injecting novelty and diversity into the processes of life. But as explained earlier, Curtis Johnson has ably demonstrated that Darwin himself never opened the door to such an interpretation. Darwinian theism is simply a false reading of the theory itself; it amounts (to borrow an apt cliché) to fitting a square peg into a round hole. In short, it is incoherent. In this sense, from a Darwinian standpoint, Richard Dawkins is right when he declared: "The universe we observe has precisely the properties we should expect if there is, at bottom, no design, no purpose, no evil and no good, nothing but blind, pitiless indifference."[16] Whether one agrees with Dawkins or not, it *is* consistent with the theory itself.

This brings up two issues worth noting. First is Darwin's complete rejection of teleology in his system. Thomas Henry Huxley, Darwin's so-called bulldog defender, observed, "That which struck the present writer most forcibly on his first perusal of the 'Origin of Species' was the conviction that Teleology, as commonly understood, had received its deathblow at Mr. Darwin's hand."[17] The second issue relates to what Dawkins alludes to in his description, namely the "good." If we find any "good" in life, for Darwin, it is found in ourselves, not in nature "red in tooth and claw." But what are the meanings and consequences of any ethical "good" in a universe bereft of purpose? It ended up being a problem that vexed Thomas Henry Huxley. Historian Jacques Barzun has pointed out that Huxley's "rejection of everything untested by him was radical, revolutionary, heedless of consequences. And it left him and his world naked before moral adversity. Europe became more and more like the vaunted jungle of the evolutionary books, and Huxley died heavyhearted with forebodings of the kind of future he had helped to prepare."[18] These are certainly the metaphysical issues that must be addressed in any discussion of Darwinian evolution. These points cannot be sloughed over as not germane to the theory; the question of an ethical good in a purposeless world rests at the heart of Darwin's theory and all the isms that stem from it.

Boodin avoided the pitfalls of Darwinian evolution by rejecting it. But he did not reject it solely on the basis of its metaphysical implications; he rejected it also because he thought it was bad science, mostly because he thought the explanatory power of chance was extremely limited.[19] But there are two other more specific reasons Boodin found to reject it. One is the phrase by which Darwinian evolution became best known, "survival of the fittest." Actually, as mentioned earlier, this originally came from Herbert Spencer. But in the face of repeated misunderstandings regarding the principle of natural selection, Wallace suggested that Darwin use it. Darwin agreed and he adopted the phrase that following year in his *Variation of Animals and Plants Under Domestication* (1868) and in the fifth edition of *Origin* published one year later. While some historians and biologists bewail the popular catch phrase for natural selection as inaccurate if not misleading, we are left with the fact that Spencer invented it, Wallace recommended it, and Darwin adopted it. Today the phrase is generally regarded as synonymous with natural selection.

Be that as it may, survival of the fittest is a tautology, a statement that is technically true but that tells us nothing. As Boodin explains, "Organisms must, of course, be able to maintain themselves in their life environment and in the physical environment, in order to leave descendants and determine the character of the race"; or stated somewhat differently to the same effect, "Life forms which cannot maintain themselves in the struggle of forces, inorganic and organic, are doomed. This is the significance of natural selection."[20] This is tantamount to saying, "That which *cannot* survive the competition for life *won't* survive the competition for life." This is true enough but uninformative. Additionally, Boodin emphasized the fact that natural selection is a negative or eliminative force in nature. It winnows out the unfit only. Wallace had pointed this out years before and so he had to find the creative power in natural selection elsewhere, which he did in *Darwinism* (1889), *Man's Place in the Universe* (1903), and *The World of Life* (1910).[21] Again, Boodin points out the circular nature of describing natural selection in negative terms, saying, "Natural selection is a negative principle at best. It can produce nothing. It is merely stating that what cannot survive perishes, which is a truism."[22]

Boodin is not the only one to notice this. Survival of the fittest as a tautology has its own interesting historiography worth recounting here if only to point out the fact that Boodin raised an issue that has persisted in the history of biology. He wasn't the first. The idea began with Samuel Butler's *Evolution Old and New* (1879) in which Butler admitted that any organism born "fitter for its conditions, will commonly life longer" is a "truism" that tells us nothing. It is a sin que non for modification but no better or worse than any other sin que non for modification. Other scientists such as geologist/zoologist James Dwight Dana and the early geneticist and plant physiologist Hugo de Vries agreed that survival of the fittest is an uninteresting truism without using the word tautology but conveying, as had Butler, its essential meaning. Then in 1972 the philosopher of science Karl Popper shook the field of evolutionary biology by arguing that the phrase was a tautology and, therefore, untestable. Nevertheless, he thought Darwinian evolution could serve as an important *metaphysical* research program. He recanted this a few years later, although Elgin and Sober argue, for a variety of reasons, that Popper retained his old view in a new disguise because he never really regarded it as anything other than a metaphysical statement leading to research.[23]

Popper is right on this point and Boodin is confused. In *Time and Reality* Boodin appears to rest the metaphysical nature of Darwinian evolution on its ability to satisfy its truth claims, saying, "If this theory really holds, if we can satisfactorily meet the facts of life that way, then it is a metaphysical theory," but "if . . . the hypothesis fails to meet the intended facts even for its biological purpose, then the hypothesis is unmetaphysical."[24] This is a strange mistake for Boodin to make, especially since he clearly rejects Darwinian mechanisms as valid, for surely he knows that metaphysical claims can be thoroughly mistaken. In fact, the positivists rejected all metaphysical statements but only by ignoring the metaphysical plank in their own eye. As James K. Feibleman has pointed out, the statement that all metaphysics is nonsense is itself a metaphysical statement, "A warning carried too far in this field leads to the very error against which the warning was directed."[25] Boodin never revisits this and at almost every turn he argues against Darwin's mechanistic reductionism as if, in fact, it *was* a metaphysic, which, of course, it really is. It is just a bad one.

In any case, the most sustained attack on the circularity of Darwinian evolution has come from biologist Robert Henry Peters, who indicates that the theory of evolution generally, including several ecological concepts derived therefrom, "are actually tautologies and, as such, cannot make empirically testable predictions. They are not scientific theories at all."[26] In answer to numerous responses challenging his central argument, Peters reiterated that "the general theory of evolution through natural selection is tautological because it defines fitness in terms of differential survival, although other problems, such as the historical dimension, do exist. Specific theories may be predictive because they define fitness independently (e.g., the capacity to burrow), but their falsification cannot affect the general argument, since in all cases the fittest are, always and only, those which survive."[27]

Edward S. Reed sees the matter differently. According to him, natural selection "is not a tautology by any means," it is "a hypothetical law of nature," and although Henry C. Byerly disagrees with Reed that it is in any sense a "law of nature," he does believe that natural selection can be formulated into a hypothesis that is ultimately testable.[28] Ronald H. Brady reframes the entire question and in a carefully argued paper examines the tautology question at length. Responding to the claims of Haldane and others that tautology is perfectly acceptable in science because mathematics is inherently so, Brady cautions not to get our analytic/synthetic distinctions conflated. "Well yes, it is," he answers, "but mathematical theorems follow from axioms, by definition. Mathematical truths are analytic in nature. Causal explanation is synthetic." Tautologies are indeed fatal to causal arguments, but survival of the fittest aims at explaining differential reproduction (or more accurately differential mortality), which "has no explanatory power, but remains a datum to be explained. If natural selection means anything in a causal framework, it means that a causal factor exists, independent of differential reproduction (or mortality), the discovery of which could explain the differential." In this explanation, the tautology is removed. However, Brady argues that the organism as defined is too indeterminate—too incomplete—to be tested. For a Darwinian evolutionist this is a little like being rescued from the burning building only to be run over by the fire truck. In the end, Brady is so unimpressed with the potential of Darwinian evolution for fruitful inquiry that he suggests "until such time as a complete theory is put forward, the effort expended on applications of natural selection would be better spent elsewhere" even to the point of replacing "Darwin's speculation" altogether.[29]

Still, the problem of tautology refuses to die for natural selection. Tam Hunt has more recently explained that generally speaking evolutionary theory is expressed in a circular manner. Instead it is suggested that natural selection be defined "as the theory that attempts to predict and retrodict evolutionary change through environmental forces acting upon organisms." This comes at the cost of admitting that natural selection as currently construed is "not predictive in the way that mature scientific theories should be."[30] And even more recently Doulcier, Takacs, and Bourrat readily acknowledge the tautology problem that still plagues natural selection. Their solution seems to be to redefine fitness as something other than reproductive output, which opens up other issues.³¹ The point is that the tautological nature of natural selection as commonly described is not likely to go away anytime soon.

In order to finish this point, it is best to return to historian John C. Greene, who reminds us that when John Herschel, who Darwin highly respected and who met the young voyager in Cape Town years earlier, first read *Origin* he referred to natural selection as the "law of higgledy-piggledy." This was no doubt Herschel's way of pointing out the vacuousness of Darwin's leading idea. "True," writes Greene, "the fittest always survived. But who were the fittest? By definition they were those that survived. Survival was both the test and the outcome of fitness. There was no escaping the circularity of the argument, no way of proving any substantive excellence in what survived. Yet," he adds, "strange to say, the competition for survival seemed to have produced a trend toward higher forms of life, something that looked like progress, however sporadic and piecemeal."³²

Julian Huxley, Thomas Henry's grandson and later to become one of the leading architects of the neo-Darwinian synthesis with his popular *Evolution: The Modern Synthesis* (1942), noticed this too. But he needed to account for this progress without invoking some larger teleological force, especially since he was convinced that Darwinian evolution was not goal-directed in any way. He first began working on the problem in 1923 through several essays just two years before Boodin would publish *Cosmic Evolution*. Huxley argued that history could be seen as an evolutionary process "towards a realization of the things judged by the human mind to have value," those values being "increase of power, of knowledge, of purpose, of emotion, of harmony, of independence" leading to "the embracing of ever larger syntheses by the organism possessing them."³³ These ideas were further developed in *The Science of Life* (1931), coauthored with G. P. Wells, son of famed science fiction writer and historian H. G. Wells. But Huxley's problem was that he was constantly invoking vitalistic or teleological language to account for progress in a system that by definition was blind and directionless. To handle this difficulty Huxley adopted the term "anagenesis" in 1957. Animals, he argued should be classed in terms of their morphology (in *clades*) and *also* in grades based upon the biological improvement—that is, "detailed adaptation to a restricted niche, specializing for a particular way of life, increased efficiency of a given structure or function, greater differentiation of functions, improvement of structural and physiological plan, and higher general organization"—they exhibited. Although Huxley insisted the argument wasn't circular, he was unable to explain his criterion of fitness as other than Dobzhansky's criterion of increased population that formed the basis of his objection in the first place. How could a newly evolved parasite, for example, be considered an "improvement" upon its ancestor? Huxley never had a good answer. "Progress" is what "progress" was—survival—and survival was still those that survived. As Greene has noted, Huxley's efforts to "graft the concepts of improvement and progress onto evolutionary theory have borne little fruit despite the sanction provided by Darwin's own writings [which were largely speculative]." Today biologists usually set off words like "higher," "lower," "improvement," and "progress" with scare quotes. "Evolutionary biologists," Greene concludes, "can neither live with nor live without the idea of progress."

Beyond the problem of tautology in survival of the fittest, a second problem lurks. It was a second scientific (or at least logical) reason Boodin objected to Darwin's argument for evolution; this was his reliance upon the domestic breeding analogy for natural selection. Wallace's original letter to Darwin sent in the spring of 1858, "On the Tendency of Varieties to Depart Indefinitely from the Original Type," prompting Darwin to complete and publish *Origin* in November of 1859, was, in part, written to make precisely the point that "domestic varieties, when turned wild, *must* return to something near the type of the original wild stock, *or become altogether extinct*." In other words, "they will vary, and the variations which tend to adapt them to the wild state, and therefore approximate them to wild animals, will be preserved. Those animals which do not vary sufficiently will perish."[34] Darwin was so swept away by Wallace's discussion of common descent and speciation that he ignored this aspect of the letter. But Wallace never backed away from his position; he always argued that Darwin's artificial selection as an analogy for natural selection was false because the selection of breeders (Darwin was particularly fond of pigeon breeding examples and even for a time raised his own pigeons for experimental purposes) was intentional and directed in ways that *natural* selection was not. Boodin agreed. In fact, Boodin explained that Mendelian genetics, which neither Darwin nor Wallace could have known about at the time, only exacerbated the problem of Darwin's analogy. Echoing Wallace, Boodin argued: "Mendel's law shows how jealous nature is of the preservation of the type. Cross-breeding of types either leads to sterility, where the types constitute remote species; or where they are less variant, leads to segregation of the original characters thus blended and the reestablishing of the original types. Thus nature enforces a certain simplicity of form where otherwise would exist endless chaos."[35]

Indeed long after this (well after the neo-Darwinian synthesis had been completed in the 1940s), outside of the Anglo-American scientific community (particularly among the French), doubts concerning Darwin's analogy persisted.[36] Rémy Collin, professor of medicine and director of the *Institut d'Histologie*, insisted, "Natural selection . . . cannot properly be compared with artificial selection, since the latter, by deliberate and intelligent choice of crosses, by preventing some unions taking place, and by manipulating the environment, introduces precisely the purposive factor that Darwinism denies in natural processes."[37] And well over a century after Wallace, one of France's most eminent biologists, Pierre-Paul Grassé, echoed precisely the British naturalist's point. Far from imitating natural selection as Darwin claimed, Grassé pointed out, "The products of domestication deviate more or less from those found in free nature and sometimes border on monstrosity. It is hard to visualize how lapdogs, Yorkshire terriers, or Pekinese could survive in the wild. Certainly they would not last long in the woodlands or pastures of our temperate zone. The albino rat, albino angora rabbit, and how many other domesticated animals would, if set free in the countryside, perish in only a few days."[38]

This is more serious for Darwin's theory than it might at first appear and it goes directly to the heart of the theory itself. Darwin gave pride of place to his artificial selection or domestic breeding analogy in *Origin*. Using it in his lead chapter, "Under Domestication," it runs for forty-three pages in the first edition and is not diminished in the sixth and final edition. Bert Theunissen has provided careful contemporary analysis of Darwin's analogy and found problems with it.[39] For one thing, Theunissen explains that breeders like John Saunders Sebright, whom Darwin cited in *Variation*, did not provide data from which Darwin could build his analogy. Sebright's crucial ingredients were inbreeding and outcrossing; the parallel between artificial and natural selection failed to account for species change but instead supported species constancy. Darwin's claim in *Origin* that small, imperceptible differences through persistent selection represented the main method breeders used to create new varieties belies the fact that most domestic breeding practices were more diverse and complex than he suggested. Another problem is that Darwin's experience with pigeon breeding was extremely limited. Purchasing his first bird in May 1855, he had dispensed with them all by September 1858, too short a time to know whether or not artificial selection could establish new and stable varieties. Darwin hastily worked through a mass of contradictory literature and failed to consider crossing in the creation of Shorthorn cattle, Flemish and Friesian horses, and New Lancaster and Southdown sheep. Minister-naturalist John Duns could see the amateurishness of Darwin's analogy and challenged his "favourite pigeon argument" on the basis of twenty years' observation of pigeon breeding. "Has man's intelligence gone out in seeking variation by selection only?" he asked. Crossbreeding was used in pigeon breeding "to an extent which is destructive of the conclusions from Mr. Darwin's 'facts.' " He concluded that "cross-breeding, and breeding in-and-in, *under man's watchful care and discriminating intelligence* [emphasis added], can alone give the key to variation." Darwin's only reply was to call Duns's remarks "clever."

Darwin was not ignorant of Wallace's objection; they had corresponded on the subject many times. But rather than answer Wallace, Darwin amazingly enlisted Wallace on behalf of his artificial selection analogy, saying, "It has often been argued that no light is thrown on the changes which natural species are believed to undergo from the admitted changes of domestic races, as the latter are said to be mere temporary productions, always reverting, as soon as they become feral, to their pristine form. This argument has been well combated by Mr. Wallace."[40] This is patently false. One can only assume that Darwin's promotional zeal for his own ideas got the better of him.

Even supporters of Darwinian evolution pose a problem for the artificial selection analogy. Although Jean Gayon defends it, he is very clear about the significant role it played in his theory. This was more than a mere pedagogical device. Gayon states, "It was methodologically essential; without it, the subtle interrelationship between variation, heredity and modification, so characteristic of the Darwinian hypothesis of selection, would have been nothing more than empty selection without any empirical content."[41] So the sheer importance of the analogy itself speaks to the heightened significance of the issues above. One point is clear: Wallace's assessment and Darwin's analogy cannot both be right. Ironically, current opinion seems to favor Wallace, even among those who would defend Darwinian evolution. Biologist David N. Reznick has clearly said, "Domestic animals rapidly evolve, or die out, when they are released back into the wild because they have been released from the shielding influences of humans' care and are once again subject to natural selection."[42] This was Wallace's point precisely. Thus nature itself is the best arbiter *against* the artificial selection analogy. On this basis the argument seems thoroughly defeated. Arthur Koestler observed that Darwin's greatest defenders seem "disarmingly naïve concerning the implications" of their own statements. They seem guided "by the maxim that a bad theory is better than no theory, [and] they are unable or unwilling to realize that the citadel they are defending lies in ruins."[43] He was writing more than forty years ago, and since then the situation has only worsened, as we shall see.

While there is much more to Darwinian evolution than the inherent construction of the "survival of the fittest" argument and the analogy between domestic breeders and natural selection (this will become much clearer in the discussion of the neo-Darwinian synthesis), it can be said that the Darwinian paradigm appears to be much more tenuous than is often proclaimed. But the endurance of the theory at least in America has much to do with its unique history. The reason, in this chapter, for locating 1925 as a special time for evolutionary theory will be examined in the next three sections.

Darwin in Dayton, Tennessee

The previous section critiqued Darwinian evolution on the basis two serious flaws: (1) the tautological nature of natural selection generally and "survival of the fittest" specifically and (2) the failure of Darwin's domestic breeding (artificial selection) analogy. Neither of those objections have any theological or religious basis. Not so when the theory was brought into the courtroom of Dayton, Tennessee, on a hot and humid July 10, 1925. Tucked away in the southeastern corner of the state, the jurisdictional seat of Rhea County had a population under fourteen thousand, and here, supposedly, religion would take on science in a most dramatic fashion.

However the whole thing was one big publicity stunt orchestrated by local businessman George W. Rappleyea. Rappleyea persuaded civic leaders that a high-profile court case challenging the state's recently enacted Butler Act, forbidding the teaching in public schools that humans descended from lower animals (specifically apes), would boost the town's flagging economy.[44] Unlike other state laws in Florida and Oklahoma forbidding the teaching of evolution, this one carried a fine of $100 to $500. This punishment allowed a very young American Civil Liberties Union (ACLU) to have a test case, and it publicly proclaimed its willingness to defend any teacher violating the law. It was a perfect setup for Rappleyea to persuade John Scopes, a substitute biology teacher, to "admit" he had done precisely what the law forbade. The trial could not have had better combatants: three-time Democratic presidential candidate William Jennings Bryan led the prosecution with famed trial lawyer Clarence Darrow, who had made his reputation one year earlier in the sensational thrill killer case of Lerner and Loeb, defending the young teacher. It was the dream of every business owner in Dayton: social, political, and economic theater bringing visitors and money into town. The trial lasted through July 21, ending up with a guilty verdict and a $100 fine that was later overturned on a technicality. But the so-called monkey trial had its intended effect, as a circus-like atmosphere pervaded the entire proceedings with live radio coverage (the first of its kind) and H. L. Mencken's punchy, satirical blow-by-blow account for readers of the *Baltimore Evening Sun*, with other papers picking up the report nationwide.[45]

These details don't even begin to convey the importance of this otherwise trivial event as it lodged itself into the cultural memory of America. Nothing has played a greater role in establishing that cultural memory than Frederick Lewis Allen's classic portrayal of life in America's so-called roaring twenties, *Only Yesterday*. First published in 1931, this modestly sized but engagingly written history has been in continuous print ever since. *Only Yesterday* proved to be a cash cow for Harper & Row, selling more than one million copies and running through twenty-two printings. Today it is available in over three thousand libraries worldwide and is frequently assigned reading for undergraduate history classes. Allen presents caricatures of Bryan and Darrow; the former "the perfect embodiment of old-fashioned American idealism—friendly, naïve, provincial," the latter "a radical, a friend of the underdog, an agnostic."[46] In this telling, Bryan displayed the "sort of religious faith . . . [that] could not take the witness stand and face reason as a prosecutor." Although a technical victory for the religious right, the case signaled a shift that pushed "civilized opinion" away from "Fundamentalist certainty."[47]

The fact of the matter is that Bryan was no fool. He has been described by serious scholars as well-read and informed on the issues of evolution (*not* a proponent of young earth creationism) whose "actual religious beliefs were far from orthodox fundamentalism."[48] For Bryan the issue was nothing short of "the democratic control of public education" protecting itself against the private opinions of school employees and defending public morals, which he was convinced Darwinian evolution undermined; for Darrow it was a matter of teaching the truth, no matter what the public might think of it, and making sure that the science classroom remained *scientific*. Yet in the retelling these less compelling aspects were lost in the flashy contrast of urbane science-based evolutionists versus Bible thumping creationists, the former representing the rising urban elites personified in F. Scott Fitzgerald's "smart set" and the latter, the shrinking untouchable caste of country yokels whose heroic figure was a seventeenth-century prelate, Bishop Ussher, who thought the earth was created on Sunday, October 23, 4004 BC.

If Allen helped establish this mythology, nothing solidified it more than the movie *Inherit the Wind* (1960). Set in the fictional southern town of Hillsboro, this movie was made memorable with brilliant portrayals of Darrow by Spencer Tracey and Bryan by Frederic March. Another theatrical team might have taken the same themes and cast worldly-wise evolutionists as wine-sipping, jazz-loving, sexually liberated, bilingual Parisian expatriates pitted against teetotaling, hymn-singing, inbred ignorant rubes speaking bad English. All the stereotypes find their way into the film that is based on a far less notable play by the same title. Historian Ronald Numbers said it best when he justifiably complained: "I have no quarrel with the playwrights' invention of dialog to heighten the drama of their fictional encounter. I do, however, question the judgment of the historians who drafted the National Standards for United States History, which recommends the movie as an aid to understanding Bryan's mind and 'fundamentalist thinking' generally. This strikes me as being a little like recommending *Gone with the Wind* as a historically reliable account of the Civil War."[49] And yet it is Hollywood-as-historian that has made this fiction a reality in the minds of many Americans, including—rather egregiously—high school teachers. The careful Internet meta-analysis of Shala Barczewska has clearly demonstrated its telling effects: "Education websites identified in this study . . . tend to favor a simplified understanding of *ITW*. This is a pity as the movie, the trial, the prejudices, and the dichotomies all provide fertile ground for lessons involving critical thinking on a variety of issues."[50] Unfortunately, whether in Hollywood or the classroom, critical thinking appears to be in short supply. In Hollywood it is forgivable—perhaps even expected—but the naive acceptance of it in the classroom is disappointing to say the least.

With the false impressions of the monkey trial now lodged firmly in the public's mind, there was a real-life manifestation of it on the American sociocultural scene and it was a direct consequence of the Darrow/Bryan exchange in Dayton. To understand what happened, a little backtracking is in order. From 1910 to 1915, Christian businessman Lyman Stewart and his brother, Milton, financed the publication of twelve apologetic tracts titled *The Fundamentals: A Testimony to the Truth*. The books essentially defended five religious principles seen as crucial to the Christian faith: (1) the inspiration and infallibility of the Bible, (2) the deity of Christ and his virgin birth, (3) the atonement of Christ as a substitutionary sacrifice for human sins, (4) the literal resurrection of Christ, and (5) the literal return of Christ in a second coming. The fundamentalist movement as it took shape in America was the product of a conference held in Philadelphia in 1919 where the World's Christian Fundamentals Association (WCFA) was established. Hugh Ross, an astrophysicist and progressive creationist, explains it thus: "Shocked by Bryan's apparent disgrace, many fundamentalists became convinced that bolting the door on the geological time scale was their only hope for upholding the Bible as true. They decided to lock belief in a universe only thousands of years old into their doctrinal position on creation. So the five fundamentals of the faith became six."[51] Prominent in the WCFA was Harry Rimmer. The same year as the Dayton trial, Rimmer published a pamphlet, *Modern Science and the Deluge*, affirming the young earth position. One year after that, Rimmer's "Science Research Bureau" published *Monkeyshines: Fakes, Fables, Facts Concerning Evolution*. Others soon followed: *Facts of Biology and the Theories of Evolution* (1929), *Modern Science and the First Day of Genesis* (1929), *Embryology and the Recapitulation Theory* (1935), and book-length works such as *The Theory of Evolution and the Facts of Science* (1935), *The Harmony of Science and Scripture* (1936), *Modern Science and the Genesis Record* (1941), and *Lot's Wife and the Science of Physics* (1947).[52] The net effect of the WCFA and Rimmer's efforts was a polarizing public relations campaign that did nothing but confirm the public's suspicions that the monkey trial happened exactly as reported.

Today this sort of *sola scriptura* "science" can be found in creationist theme parks like the Creationist Museum located in Petersburg, Kentucky, and The Ark Encounter in Williamstown, Kentucky. They are run through Ken Ham's Answers in Genesis organization, a lucrative business that reported revenue in 2018 of $34.7 million. Its stated goals are to promote its views of Christian apologetics, evangelism, biblical inerrancy, and young earth creationism. Its opponents are many, including not just Darwinists but also progressive creationists and proponents of intelligent design. Nevertheless, as its revenue indicates, it is highly successful, and on any given day the expansive parking lot is full of wide-eyed believers, arriving in church vans and buses, eager for their excursion into the land of make-believe. So in the end, the myth of *Inherit the Wind* has become real, only now attended not by people seeking entertainment but by those seeking a religious experience, the reward of their unbridled fideism. One wonders what William James would have made of it all.

This is one aspect of evolution that presented itself in 1925. It is in many ways the ugliest in that it arose from an episode in American history that was concocted, then misreported, and then memorialized in a film that took root in the nation's popular culture. It mixes elements of faith with science in a way that makes them oppositional, and it created one side of the destructive culture wars currently raging. But like all wars, there is always another side to a conflict. It is that "other side" to which the next section is devoted.

Darwin "Vindicated": The Case of Raymond Dart and the Taung Child

Half a world away from Dayton, Tennessee, in the small town of Taung, which is eighty-six miles north of Kimberley, South Africa, an intact fossilized skull was excavated from a limestone cliff at a depth of about fifty feet by the Northern Lime Company. Suspecting that the area might hold remains of prehistoric humans like those found at Broken Hill, Rhodesia, the general manager of the Northern Lime Company permitted geologists to inspect the site. A colleague at the University of Witwatersrand in Johannesburg informed anthropologist Raymond Dart of the finding. Examining the remains, Dart was convinced this was a true humanoid ancestor, naming the almost perfectly preserved skull of a three- to four-year-old child *Australopithecus africanus*. Dart reported his findings on February 7, 1925, declaring, "Apart from the evidential completeness, the specimen is of importance because it exhibits an extinct race of ape *intermediate between living anthropoids and man*." Dart was ecstatic; along with other findings in the region, the "Taung child" could help construct an early picture of human evolution, "thus vindicating the Darwinian claim that Africa would prove to be the cradle of mankind."[53] Although many within the anthropological community initially protested that it was really an early ape, paleontologist Robert Broom, a fellow at the Royal Society and later curator of the Transvaal Museum, agreed with Dart, thus splitting the experts on the issue. Today the anthropological community confirms Dart's discovery, making it the type specimen for other *A. africanus* findings.[54]

This kind of assessment in anthropology helped establish the Darwinian model of common descent more firmly than ever. But the advent of Mendelian genetics offered special challenges to other aspects of the theory. The problem of heredity had vexed Darwin, whose only answer was to fall back on a form of neo-Lamarckian inheritance of acquired characteristics he called pangenesis. Some early geneticists like Hugo de Vries doubted Darwin's gradualist scenario of evolution and argued for a more saltationist mutation theory. De Vries famously said, "Natural selection may explain the survival of the fittest, but it cannot explain the arrival of the fittest." In fact, despite the great confirmatory discoveries in anthropology occurring at this time, some have claimed that other aspects of Darwinism were in retreat.

In fact, the alleged "eclipse" of Darwinism is really part of a triumphalist narrative told largely by Julian Huxley of "anagenesis" fame and Harvard biologist Ernst Mayr, and it was given credence by historian Peter J. Bowler. According to this tale, Darwinism was in serious decline until certain researchers combined modern genetics with Darwin's theory of common descent by means of natural selection, rescuing it from an almost certain demise—the cavalry riding to the rescue of the beleaguered pioneers (in this case biological pioneers) under attack. Some of the leading players were Ronald Fisher and his 1930 book *The Genetical Theory of Natural Selection*, Theodosius Dobzhansky's *Genetics of the Origin of Species* (1937), Julian Huxley's *Evolution: The Modern Synthesis* (1942), and Ernst Mayr's *Systematics and the Origin of Species* (1942). It was all capped off by David Lack's 1947 book, *Darwin's Finches*.[55] The biological sciences seemed at long last unified under one theory. But historian Betty Smocovitis thinks something else was going on. In fact, the neo-Darwinian synthesis is less about science and much more a "mechanistic and materialistic" manifesto for positivism (positivism, and all its variant forms, is mentioned forty-six times in the article).[56]

Nevertheless, with all their genetic and empirical field observations under their belt, by 1959 nothing seemed more secure than Darwin's theory. It was being portrayed as irrefutable as gravity! In that year, of course, the centennial of Darwin's *Origin of Species* was celebrated at a huge conference held at the University of Chicago, November 24–28, 1959. It drew 2,500 participants with almost 250 delegates from 189 colleges. Many of the synthesizers of modern genetics with Darwinian evolution were there. This rather curious event did two things. First, it trumpeted the striking victory the neo-Darwinian synthesizers had achieved in a remarkably short period of time by proclaiming with loud fanfare "the agreement over the centrality of natural selection as *the* mechanism of evolution" thereby enhancing the "sense of unity and consensus," but at the same time it also gave "rise to a constricting new orthodoxy—as the panel discussions and somewhat exclusive invitation list indicate. This 'hardening' of the synthesis around a selectionist orthodoxy" established the clear contours of the neo-Darwinian paradigm.[57] This conference put *everyone* on notice. More interestingly, the Darwin Centennial Cerebration—and the three-volume published proceedings that emanated from it—had all the attributes of a religious encyclical. From here on everyone working in the life sciences would be expected to show some fealty to the neo-Darwinian paradigm. As Smocovitis observes, the event "had little to do with the historical Darwin or the development of his work; instead it revealed much about postwar American culture and its embrace of a new synthetic science of evolutionary biology, a science that could potentially redirect the future of 'modern man.' "[58] This was scientism exerted with religious force, the *mentalité* defining a new socio-scientific order.

Only recently has this triumphalist story been challenged. Ron Amundson has brought considerable expertise to bear upon what he calls "Synthesis Historiography, or SH"—a largely mythical creation of Ernst Mayr.[59] According to Amundson, Mayr made essentialism "the central pillar" of SH by equating essentialism with typology, or typological thinking, as the source of species fixism before Darwin, an assertion that is simply historically false.[60] In fact, despite Mayr's claim suggesting rampant pre-Darwinian fixism rooted in Platonist essentialism, only Louis Agassiz could be considered a true Platonic species fixist.[61] Others have agreed with Amundson's analysis.[62] Amundson's critique is damning, with Günter Wagner calling Mayr's SH a "cartoon version of the history of biology."[63]

More recently Mark B. Adams has reviewed the history of the evolutionary/neo-Darwinian synthesis and found it highly politicized, more aimed at professional turf building and personal aggrandizement than science. The key to a *real* evolutionary synthesis rests in its ability to explain *macroevolutionary* processes, genuine demonstrable and empirically confirmable examples of speciation or organic transformation. Adams asks, "When . . . was the problem of 'big evolution' (whether called 'macroevolution,' 'megaevolution,' 'transpecific evolution,' or 'evolution in depth') finally settled? The answer is, quite simply: It never has been."[64] Geneticist Yuri Filipchenko, leader of experimental zoology in Russia and mentor of one of the key founders of the neo-Darwinian synthesis, Theodosius Dobzhansky, was a firm believer in evolution but doubted the capacity of natural selection to explain in any significant way evolutionary change; he also did not think that specific categories above the species level were the result of Mendelian genetics. Highly impressed with Filipchenko as a geneticist, breeder, biometrician, and evolutionary theorist, Dobzhansky knew full well that the neologisms his mentor had used to distinguish varieties from species and other taxonomic categories in his 1937 book were intended to show that genetics could *not* be enlisted to support evolution, and yet a decade later at Columbia Dobzhansky co-opted the same terms to argue the opposite. When in a 1973 interview Adams asked him to account for the change, Dobzhansky simply said, "He [Filipchenko] bet on the wrong horse!" Adams was astonished. He realized that Dobzhansky "was not a scientist who had been certain of his own approach, but someone who had realized it might have gone either way, and chose the option that, if it turned out to be right, would both justify and empower his newly coined specialty, 'population genetics.'"[65]

Adams is making an important point. Despite its claims, the neo-Darwinian synthesis has been unable to demonstrate one instance of macroevolution. The confirmatory evidence usually presented are the Galapagos finches. Initially reported by David Lack, finches are the iconic exemplars of natural selection "in action." This is especially true since the longitudinal studies of Peter and Rosemary Grant beginning in the 1970s and popularized in Jonathan Weiner's Pulitzer Prize–winning account, *The Beak of the Finch* (1994). The "discovery" of the evolutionary significance of the finches is that during droughts, the ground finches with larger beaks survived because they could crack open the hard spiny fruits more readily than the smaller-beaked birds. More importantly, according to the Grants, they produced offspring with larger beaks on average, thus "confirming" evolution by means of natural selection in a stunningly brief period of time. Then in 1983 when torrential rains dramatically changed the vegetation, ground finch beak size decreased to take advantage of greater food source availability. So evolution was deemed to respond in reverse order due to changed circumstances. The problem with all this is that *none* of it demonstrates sustained morphologic transformation. It had always been known that temporary adaptations can and do occur in response to environmental change, but the extrapolation to evolutionary change seems questionable. As McKay and Zink have explained: "We argue that instead of providing an icon of insular speciation and adaptive radiation, which is featured in nearly every textbook on evolutionary biology, Darwin's ground finch represents a potentially more interesting phenomenon, one of transient morphs trapped in an unpredictable cycle of Sisyphean evolution. Instead of revealing details of the origin of species, the mechanisms underlying the transient occurrence of ecomorphs provide one of the best illustrations of the antagonistic effects of natural selection and introgression."[66] Oscillating change is not exactly evolutionary transformation.

So this brings us to the end of what must seem to the reader as a long digression from our main subject. But it is necessary to let Boodin's own evolutionary philosophy stand out in bold relief against these two very different but very prominent explanations of life's origins and diversity—biblical creationism and Darwinism. The issue is not creationism versus evolution, for as we are about to see there are many kinds of evolution. In most cases, the so-called culture war that has raged between creationists of various stripes and neo-Darwinians reveals more rancor than reason. The former argue from the position of their faith-based convictions, the latter defend a speculation based more upon professional turf protection and academic expectation than actual empirical evidence allows. Both challenge each other on the basis of deep-seated anxieties; the one concerned for the social, moral, and ethical implications of a Christian world they already see slipping away, the other convinced that the abdication of science in favor of religious superstition will cause the gains of the Enlightenment to collapse into an irretrievable Dark Age. There really is a third way in this, shown, in many cases, by some interesting and adept process philosophers such as Henri Bergson, Alfred North Whitehead, and Teilhard de Chardin. Although there are important differences in detail, Boodin can be considered one of these, and his *Cosmic Evolution* takes us refreshingly away from the dogmatism of the creationists and the positivistic scientism of the Darwinians.

Cosmic Evolution

When William Werkmeister ("Werkie") published his widely used college textbook one year before Boodin's death in 1949, he memorialized (albeit ever so briefly) for many American philosophers and many more students how he would be remembered, if at all.[67] Interestingly, Werkmeister regards "cosmic evolution" as a "specifically defined thesis" that forms the culmination or epitome of Boodin's metaphysical system, a conclusion that is hard to disagree with. That thesis is fairly summarized by Werkmeister and may be said to be a "functional realism" of dynamic interrelationships united in an intimate embrace of "empirical realism" (the method) and "cosmic idealism" (the conclusion or culmination). Here science has come to the aid by revealing a plurality of energy systems (of which humans are a part) that operate through fields. As such, it is impossible to view the universe as anything other than an organic whole.

The structure of *Cosmic Evolution* is likened by its author to a three-movement symphony representing variations on the theme of cosmic interaction. As in all Boodin's work, science guides but does not determine the way. He invokes the fine-tuning argument early on by suggesting that the conditions of life require such well-coordinated and tight tolerances that chance is immediately ruled out. Tightening his perspective to biology, Boodin cites Henry Fairfield Osborn approvingly at length. Osborn, a paleontologist and geologist of considerable stature, was a proponent of orthogenesis, the idea that evolution was not a willy-nilly affair but directional, from simpler, lower life forms to higher, more complex forms. To Boodin's pleasure, Osborn rejected Bergson's *élan vital* (life force) as incapable of causing phenomena to spring forth autonomously. In other words, there must be some teleological force to "call them forth." Although physics occupies an important place in *Cosmic Evolution*, especially in part 3, Darwinian evolution is mentioned (never approvingly) twenty-two times. In explaining how directionality might occur by evolutionary means, Boodin suggests a pluralistic universe that functions through "a system of compensating rhythms where worlds grow up and die as parts of a self-sustaining whole. The life cycles of the earth no more happen by chance than those of the individual organism which is a part of its history. It is absurd to suppose that the cosmic system as a whole emerges from chaos."[68]

Thus, the biological and the cosmic worlds are interrelated. While it is true that the biological and larger physical worlds evolve in different contexts, they are, for Boodin, clearly intrinsically part of a larger whole. Both are teleological from top to bottom, and at times our poet-philosopher waxes eloquent on the subject, as when he analogizes efficient and first causes by saying, "If Aphrodite did rise in her full-formed beauty from the mists of the sea, it was not altogether owing to the potencies of salt water. There was also the cosmic genius of Zeus."[69] He further emphasized the social and cooperative nature of the universe, saying, "As mind or intelligence with us is fundamentally social, a focus on mental interactions, so there may be in the cosmos a continuum of spiritual interactions of various levels of which we are ignorant or at best catch a glimpse in the intuitions of genius, in mystical communion, in the intimations of beauty and immortality."[70] Lest we dismiss this as the musings of a hopeless idealist, it is worth noting that the eminent naturalist Alfred Russel Wallace and paleontologist Robert Broom (who confirmed the Taung child's hominid ancestry), agreed. That was the whole point of Wallace's *The World of Life* (1910), and Broom could endorse an "ape-to-man" paleontology without succumbing to an "ape-to-man" metaphysic. In the same work in which Broom supported Dart's conclusions regarding the Taung skull, Broom also concluded the following:

> The great religions of the world—the Jewish, the Christian, and the Mohammedan—all believe that a supreme Spiritual Power has created and rules the Universe. But they believe also in other

spiritual beings—archangels, angels, devils, and a variety of less clearly defined types. How they have arrived at these conclusions we need not consider. And further, they believe that in man at least there is a spiritual element, the soul.

Now it seems to the writer that the facts of Science do not contradict the main conclusions of those religions, but are possibly in considerable harmony with them.

Much of evolution looks as if it had been planned to result in man, and in other animals and plants to make the world a suitable place for him to dwell in. It is hard to believe that the huge-brained thinking ape was an accident. But if we become convinced that man is the result of the working out of millions of years of planning, we seem forced to the further conclusion that the aim has not been merely the production of a large-brained erect walking ape, but that the aim has been the production of human personalities, and the personality is evidently a new spiritual being that will probably survive the death of the body.[71]

This is in substantial agreement with Boodin. There is no indication that they knew one another, but a science/philosophy collaboration would have been fraught with possibilities. Could Boodin have written this? He practically did. He wrote,

If we take the long view of human history that biology and palaeontology are opening up to us, can we deny that there has been advance since the first apelike ancestors of man appeared? For while man did not descend from the present apes, the evidence shows that man and ape were closely related in their origin. They had a nearly equal start. The distance from Pithecanthropus to Newton is vastly greater than that which separates Pithecanthropus from his ape kin; and we need not suppose that Pithecanthropus is the first in the human series. If we view man as the palaeontologist views him, we can hardly deny advance. And if we view man as the anthropologist views him, the advance is even more striking."[72]

In other words, he viewed the paleontological evidence as not only showing hominid kinship to modern humankind but also showing (in true orthogenetic fashion) progress, a progress explicable by a large teleological force or guidance. This is the leitmotif of Boodin's entire corpus. Although it is cast in much more detailed and sophisticated epistemological and metaphysical terms than in Broom, the essence is the same. It is no wonder that Boodin's symphony ends on a grand finale titled *Cosmic Religion*, not surprising for a book that mentions *soul* 151 times.

But it would be incomplete to hear any symphony's beginning and ending without hearing the development of all the harmonies and cadences in between. One of those important "harmonies" appears in chapter 2, "Evolution as Creative Adaptation." Citing Leucippus, "Nothing happens without a reason," Boodin rarely misses an opportunity to attack materialism, probably because of its rising prominence during his professional life. "Materialism has substituted magic for sober thought," he complained. "The whole process of evolution becomes a succession of miracles without intelligible ground in the process. The appearance of a world of chance of any order at all, the emergence of life, with its series of forms and organs, the final appearance of intelligence and beauty—all are miracles. The materialist scientist has a truly marvellous appetite for miracles."[73] Sixty years later ethnobotanist Terrence McKenna expressed this as reductionist science's "one free miracle."[74]

For Boodin, the problem with much of current philosophy and science comes from what he calls "vicious bifurcation," where man is abstracted from nature, mind from body, qualities from things, the individual from his environment, and the earth from the cosmos, thus creating absurdities. While there are many sources, Darwin is singled out in this section. The tautological nature of the Darwinian argument has already been covered, but Boodin also noted another important criticism: "Many of the variations are merely negative, i.e., they consist in subtraction of characters rather than addition."[75] So if natural selection is eliminative, as discussed earlier, and operates by forces of subtraction, where is the building capacity of any of Darwin's mechanisms? Interestingly, in 2010 Lehigh University biochemist Michael J. Behe published a controversial but well documented article that suggested the "first rule of adaptive evolution" is "loss-of-function" mutations.[76] Behe, a vocal proponent of intelligent design, in his *Edge of Evolution* (2007), subsequently expanded this to claim that the dearth of evidence for new mutations in nature suggests a tightening of his earlier estimate of natural selection's boundary problem (i.e., its ability to explain biological change) from class and above to family and above.[77] Stated another way, he believes that the continued findings of molecular biology over the past decade or so, including the ongoing project by Michigan State University's Richard Lenski examining some 60,000 generations of *E. coli*, demonstrate a constriction of neo-Darwinian evolution even more damning than before. If Behe is right, Boodin's comment, at least as far as it goes, is not only accurate but prescient. However, as far as intelligent design is concerned, there are more issues involved, as we shall see in chapter 6.

Along the way, Boodin refuses to revert to mysticism. He rejects vitalism as failing to explain key organizational features of nature and says of Bergson that "any age should take seriously such an incoherent mixture of mysticism and science is evidence of nothing so much as a want of logical thinking."[78] For him, "inherent potencies" moved by a "vital impulse" is so much verbiage lacking in empirical content or explanatory value.

In chapter 3 the evolution of cosmic interaction is explained. He wishes to distinguish his view of evolution from purely mechanical causes in which chance governs change and all final causes are dismissed out of hand. Here "creative synthesis" comes in to play with the issue being to explain "the impetus to organization." Citing Samuel Alexander's *Space, Time, and Deity* (1920), Boodin rejected his version of emergent evolution for its pantheistic tone as a work of "magic, not logic." Despite the ability of this "magnificent magician" to get wondrous things out of space and time, Boodin wryly observed that for the rest of us, space and time yields only space and time. Instead, Boodin emphasizes the importance of intelligence as an integral part of the process of creative adaptation to the whole cosmic structure. And here rests the central thesis, namely that the entire cosmos moves from chaos to order by a "plurality of histories of different levels" all interacting at quantitative and qualitative levels toward teleological goals through dynamic fields—*energy fields*—controlled by the total field woven together "into a cosmic plot."[79] What ties this together is Boodin's conception of an active, dynamic divinity. Here is not an anthropomorphic God waving his magic wand, but a more complex, more nuanced force or power. It seems panenthic in form. "The organization of our earth into a system fitted for life," he explains, "the appearance of levels of life, intelligence, creative imagination, the intuition of divinity—these are possible because there are levels of life, intelligence, creative imagination, divinity in the cosmos." Avoiding Gnosticism, he writes (again rather panenthically), "Life, thought, beauty, God must be incarnate in matter to be effective in the cosmos.... There is nothing degraded or evil about matter, as the mystics have always maintained. Nor is matter non-being in the sense of being unreal."[80]

Part 2 covers cosmic evolution and its relationship to human beings. Here Boodin starts by differentiating different forms of realism. Naive realism—the idea that our sense perceptions are direct with no explanation for how those perceptions may change despite the permanency of things being perceived; how illusions come about; how colors, sounds, smells can be independent of us—is seen as the false bifurcation of nature and human nature. Only when epistemological realism and idealism are united against such bifurcation so that sense qualities become functions of "the creative interaction of the energies of physical nature with the organization of human nature, its sense-organs, and nervous system" will the problem be resolved. A significant feature of this cosmic relationship is found teleologically, through "the realization of needs and interests in social relations."[81]

Boodin then turns his attention to the mind/body problem. Here he views the mind as an "incarnation of matter" interacting with a cosmos possessed of mental levels. Again, he emphasizes energy levels. "When mind does appear in the fullness of time," he explains, it is a unique level of control interpenetrating the lower levels and giving new quality and tone to the functioning of all levels. . . . For mind, as we know it, is not an abstract entity, but an energy system."[82] As such, Cartesian dualism is rejected as is Bergson's compartmentalization of mind/body as abstractions. Furthermore, all forms of materialistic or physicalist reductionism are dismissed as inadequate to the task of explaining mental functioning. Rather, minds are emergent in nature, forming into a new pattern of *social* organism, a "new step in creative synthesis." Ultimately, this is culminated in the creative adaptation to a new psychic nature of social relations where the divine will "help build a new city of God, a new equilibrium of spirits."[83] There is in this the implication of process theology. Here temporality and creativity become realities in an evolutionary world and universe in which deity itself is in cooperative process with the free choices of his creatures. The word "help" in the above quote cannot be sloughed over or ignored. Charles Hartshorne seems compelling in this. Like Whitehead, Boodin sees this panenthic God as One (in contradistinction to Hartshorne's series of entities), but the panentheism remains: "God is not a passive spectator of nature. He does not live in blissful and indifferent isolation, as Aristotle conceived Him. Rather He interpenetrates nature, becomes progressively incarnate in nature, and is responsive to the striving of nature. There is nothing foreign or indifferent to him."[84]

In any case, Boodin's handling of the mind/body problem is complementary of the distinguished Nobel laureate and neurophysiologist John C. Eccles. For Eccles,

The anthropic principle [already referenced by Boodin] achieves a new dimension in the coming-to-be of each of us as unique self-conscious beings. It is this transcendence that has been the motive of my life's work, culminating in the effort to understand the brain in order to present the mind-brain problem in scientific terms. I maintain that the human mystery is incredibly demeaned by scientific reductionism with its claim in promissory materialism to account eventually for all the spiritual world in terms of neural activity. This belief must be classed as superstition.[85]

Boodin's new "city of God" is perhaps simply the social expression of theology explained by Eccles as an act of divine creativity: "It is the certainty of the inner core of unique individuality that necessitates the 'Divine creation.' I submit that no other explanation is tenable; neither the genetic uniqueness with its fantastically impossible lottery, nor the environmental differentiations which do not *determine* one's uniqueness, but merely modify it."[86] Thus, the human soul or psyche is more than a collection of genes, more than organic physiological functions, more than a mere epiphenomenon of the brain. It is, as Boodin once paraphrased Plotinus, "the generosity of God" expressing itself in "individual souls" divinely participatory in the biological act of conception itself. It is not *super*natural but a part of nature itself, a nature that cannot, as far as we currently know, be discerned solely through empirical means.

For Eccles as for Boodin, Darwinian materialism leaves consciousness utterly inexplicable. Their solutions are largely commensurate with each other even to the point of Boodin's frequent mention of energy fields coinciding with Eccles's suggestion that the immaterial mind and its probabilistic synaptic events are analogous to probabilistic quantum fields, thus linking modern physics and neuroscience.[87] Boodin writes, "The unity, duration, and forward-looking co-ordination implied in the simplest type of minded behaviour are not resolvable into the external relations of a mechanical system. Mind must be recognized as a new type of adaptation, a striving after a new equilibrium with the cosmos."[88] Eccles agreed.

In the final part, Boodin relates his concept of cosmic evolution to relativity and the new theories in physics then emerging from the European continent. Here Boodin gives his appraisal of Whitehead. It should be stated at the outset that Boodin and Whitehead share much in common: both are avowedly process philosophers; both are thoroughly non-reductionist; both have comprehensive metaphysical systems (although Boodin's analysis here preceded Whitehead's *Process and Reality* by four years); both had little use for Cartesian dualism; both talked a great deal about the bifurcation of nature as seen in Descartes, Boyle, and Locke; and both had a distinctly poetic nature. While science held the close attention of both philosophers, Boodin's biographer is probably right in saying that for him science was used "not as a means of discovery but as evidence for beliefs which had emanated elsewhere."[89]

With this as a backdrop, Boodin's analysis of Whitehead can only be given in summary form. He begins by pointing out Whitehead's continuous duration of nature extending in space and time, the extension of which is duration. There are things within this duration that have extension in space and duration Whitehead calls events. These extended "events" (spatial or temporal) are made up of smaller and smaller units until a zero of spatial extension (a point) is reached and a zero of duration (an instant) is reached. Motion consists of space and time and so we have point-instants. The problem, as Boodin sees it, is that a point is not an infinitesimal extension and an instant is not an infinitesimal duration. So in satisfaction of Whitehead's requirements of extensive abstraction, "matter must be treated as infinitely divisible and we arrive at a physical point—an event-particle."

Sorting this out is important for Boodin because, as he says, "metaphysics as I understand it deals with the constitution of the real world as revealed in human experience. And recent discoveries in science affect our general conception of events in nature, including ourselves as parts of nature." Whitehead agreed, and so engaging with his ideas had the utmost relevance for Boodin. But Boodin has problems with Whitehead's conception of nature because the eminent mathematician represents it as "extensive abstractions" that are impossible to conceive in the manner described. In other words, "it is hard to see how points, instants, and physical point-particles . . . [are] events which possess spatial extension, temporal duration, and physical thickness." These do not, as Boodin sees it, "possess the character of extension and inclusion." Whitehead leaves no room for mind in nature. "Nature," Boodin says of Whitehead,

> is to be self-contained to a theory of nature. He did not leave the least hole through which he could chuck embarrassing entities. The percipient organism is indeed part of nature, but it is to be conceived as a strictly physical thing. If he had admitted mental perspectives as a part of nature, then he might have had a locus for such creative additions as points and Newtonian uniformities and other abstract and sometimes misplaced limits. In any

case, I cannot see how, in any real sense, points, instants, and point-particles can be said to be contained in nature, or how such entities can be said to have a real complexity. There is, it seems to me, an unbridgeable chasm in the argument.[90]

Since Whitehead's magnum opus in metaphysics, *Process and Reality* (1929), lay ahead of Boodin's comments in *Cosmic Evolution* by four years, it seems wise to let them stand where they will. I suspect that Boodin felt that Whitehead failed to account for mind within nature, as when Whitehead suggested that sensations were "qualities of the mind alone. These sensations are projected by the mind so as to clothe appropriate bodies in external nature. Thus the bodies are perceived as with qualities which in reality do not belong to them, qualities which in fact are purely the offspring of the mind. . . . The poets are entirely mistaken. They should address their lyrics to themselves, and should turn them into odes of self-congratulation on the excellency of the mind. Nature is a dull affair, soundless, endlessly, meaninglessly."[91] As *Science and the Modern World* was written at the same time as *Cosmic Evolution*, Boodin couldn't have been commenting on it here, but our poet-philosopher would unquestionably have balked at writing odes in this manner to himself.

Irrespective of Whitehead, Boodin was certain that the new revelations of the theory of relativity "must be understood as part of fields of energy with their space-time structure."[92] For him, then, mind is a "reflex of nature" in which the cosmic process "furnishes a sufficient cause for the ordering genius of man." The universe is too wholistic for Boodin to see it otherwise, and order then becomes "a co-operative enterprise between the finite mind and the cosmos." This becomes somewhat clearer in consciousness, which is related to the mind/body problem—the aspect of sensitivity, "of taking note" of things implied in all action. Boodin resolves this by treating consciousness as a universal class concept arising through energy fields. Through these energy fields comes consciousness through creative synthesis and ultimately social organization. This, he believes, removes the separation between the conscious and unconscious. This is Boodin's way of escaping the difficulty of naive realism: "There is no more consciousness in general than there is colour in general. And the similarities of consciousness must be determined in the same way as we determine the groupings of colours. They are, as a matter of fact, not a color, blue, and consciousness, but a selective response which we call blue, and so with every other quality of reality."[93] As we proceed to understand mental perspectives more generally,

Boodin explains them more in terms of energy types that are pluralistic and cause energy exchanges between each other while remaining distinct. This kind of activity is a universal constant where it exists.

The culmination of Boodin's cosmic scheme is really two-fold. It is (1) a dynamic and interactive wholism at all physical, biological, and social levels and (2) a material and spiritual world united under panenthic theistic care and control that requires mutual participation. Whatever the details, it is clear that in Boodin's framework it is pluralistic even to the point of positing multiverses. There might be, he argues, "an infinite number of closed island universes." Indeed, he echoes some current physicists in suggesting that "the cosmic field may be closed and yet be infinite, i.e., if there is an infinite number of universes." It should be pointed out that Boodin suggested that "the cosmos is self-sustaining, there must be a running up process to compensate for the running down process; and the levels within the cosmos as a whole must be eternal."[94] Belgian priest Georges Lemaître proposed what would become the "big bang" theory in 1927, which Boodin dismissed as mere "fireworks." Although big bang cosmology is the reigning paradigm, there is a strong body of current evidence suggesting that Boodin may again be prescient. Paul J. Steinhardt in the physics department at Princeton and Neil Turok in mathematical physics at Cambridge have proposed an eternal "bouncing" universe.[95] Indeed its rival, steady-state cosmology, is anything but dead. Its computations have been confirmed by Plamena Marcheva and Stoil Ivanov.[96]

Above all, the cosmos is a *creative* enterprise, and thus Boodin may be considered, strictly speaking, a *creationist*. But his is surely *not* the creationism of Christian fundamentalism that thinks it now has "science" on its side with the big bang theory. As João Barbosa has pointed out, creation is really a conceptual apparatus—a *themata*—that historian of science Gerald Holton argues, consciously or not, pervades the conceptual frameworks of scientists in the experimental and theoretical areas of their work. This thematic dimension comprises concepts, methodologies, and hypotheses with metaphysical, aesthetic, logical, or epistemological implications associated with the cultural context and the individual psychology of scientists. When Alexandre Friedman used "creation of the universe" in a scientific article published in 1922, he introduced creation as a scientific *thema* then picked up by all manner of cosmologists, physics-based scientists looking to explain cosmological origins and theologians looking for metaphysical coherence. What is important is that as a *thema* creation has had a polarizing effect because "creation" can serve several masters:

This arrival of creation to modern cosmology revealed a very interesting aspect of the concept as a thema, which is that *creation* takes several specific forms in the same subject area. In fact, *creation* manifested itself in cosmology under three derived forms: *creation from nothing*, although it is only implicit; *creation from shapeless material*; and *continuous creation*. Applied to the big bang cosmology and the steady-state cosmology, these three forms of the fundamental *thema* of *creation* are specific *themata* of cosmology that have equivalent specific *themata* in other areas. For example, we can see the creation from nothing in theology (where is usually named as *creation ex nihilo*); the creation from Ylem, in the Gamow hypothesis, has equivalents in areas such as art (for example, the creation of a sculpture from shapeless materials); the continuous creation of matter, specific of the steady-state cosmology, has in spontaneous generation of life, an outdated idea of spontaneous appearance of life from non-living matter, its equivalent in biology. In other words: the fundamental *thema* of *creation* has disciplinary and cultural transversality, which is a characteristic of any *thema*, that is also expressed through its derived *themata*, each of them assuming a specific form in a certain area.[97]

For Boodin, creation can occur through preexistent matter in a steady-state universe. Therefore like nearly all panentheists he believes that creation occurs *ex materia* in *ex deo*. The issue of Barbosa's claim of an "outdated" form of spontaneous generation of life implied by the steady-state view is obviously negated by a non-spontaneous or intentional theistic effect inherent in any panenthic theology. So the term *creation* applied to Boodin cannot be construed as the overly simplistic media-and-culture-war debates depict it. In this sense Boodin is no more a "creationist" than Whitehead or Hartshorne. But *creation* is a useful descriptive term invoking action leading to process, and so it shouldn't be denied to any of them, even if its current connotations are extremely misleading. Here the concept of creation *as creativeness* is crucial to rescue Boodin from merely proposing a theology of nature. No, he says, "This is not a religion of nature in the sense of levelling all to the less-developed stages of nature—brute and matter. It is in the upper creative reaches that the meaning and goal of the universe, the genius of divine creativeness, is foreshadowed."[98] Here is creation made manifest. In this sense, Boodin could be nothing but a creationist.

Boodin ends it all (as indeed any good pragmatist should) by asking how all this impacts us. What is its cash-value? Here our insightful philosopher turns into the knowing sage and expresses his *creationist* theology in the most poetic fashion:

> The finding of the cosmic path, the straight and narrow way that leads to salvation, must be one of trial-and-error experimentation, of creative discovery through the ages. Salvation is a co-operative undertaking. Divine genius furnishes the impulse of higher orientation, but the finite individual must respond creatively to this impulse. He who fails to find the past, merely oscillates and beats out his life in vain. And oh, the tragedy of it! He who finds the way, finds life, development, divinity. And oh, the glory of it! But this insight must be bought with anguish and blood. No advance is made without paying the price. Sometimes in a moment of beauty, sometimes in a fleeting mystic sense of unity, we may feel the order, the way, which our intellect is so inadequate to trace. And be the moment ever so brief, it will shed its radiance over a life—a radiance from the divine source of beauty, gentle and constructive in its influence, enveloping like love, creative like genius, tender as the soft kiss of a child.[99]

Assessment and Conclusion

The chapter began by taking 1925 as a signal year in three conceptions of evolution. One is the reductionist positivism of Darwinism; the second is the equally reductionist *sola scriptura* dogmatism of religious fundamentalism; the third is the one presented by Boodin, an evolution enriched by cooperative creativity and process that is in harmony with our highest ideals and the hardest empiricism of science. Of course, there are many others. All cannot be covered here, nor should they be in an examination of one man's ideas. But two are worth mentioning because they bring out certain aspects of Boodin's cosmic evolution.

The first is the work of a philosopher who could not be further from Boodin but was his contemporary. In an especially and uncharacteristically cantankerous moment our poet-philosopher railed against those who reduced the creative processes of the universe to chance and relegated what is real to animal instincts and mere behavioral processes.

What fanciful dogmatism to regard the former activities as accidental and foreign to nature and to limit the scope of nature to what is real to a pig. No doubt to a pig it is thus limited. But we do not necessarily have to take the pig's satisfaction as the ultimate criterion of value. There is rank and dignity the reality of which the pig doesn't dream . . . in spite of the pig-trough philosophy of our fashionable materialism. There evidently are pigs enough to make a market for it and to enthrone it in high places.[100]

One wallowing in the "pig-trough" of materialism is Roy Wood Sellars. Sellars was a University of Michigan philosopher known for his theory of critical realism. A thoroughgoing materialist, Sellars's epistemology is based upon disclosing objects that exist as external things that are mediated by subjective meanings by the perceiving organism. Content and object are ontologically different for the critical realist. In short, "knowing . . . is found to be interpretive of objects."[101] His book *Evolutionary Naturalism* (1922) is a direct challenge in almost every respect to Boodin's *Cosmic Evolution*, though the latter never mentions it directly. Sellars proposed what he considers to be a non-reductive naturalism; it is non-reductive because nature itself undergoes constant change with newly emergent patterns.[102] He therefore rejects the mechanical view of nature for a more dynamic form of materialism. He sees this as at one with his metaphysical scheme of critical realism. He reveals his approach early on:

But instead of bringing mind down to the brain as kinetically conceived, why may we not bring the brain up to the mind as empirically analyzed? Such is the endeavor of evolutionary naturalism. We shall hold that even psychophysiological parallelism does not do justice to the empirical facts. If naturalism is the view of the world which founds itself upon the results of science, it follows—does it not?—that the texture and breadth of naturalism will alter as the sciences alter and as science is enlarged by the frank admission of new sciences into the commonwealth of tested knowledge.[103]

For Sellars, mind is a physical category best described and studied through behaviorism already mentioned in chapter 2.

As we shall see in chapter 6, Sellars's evolutionary naturalism has had considerable staying power, especially in this rising age of reductionism. But

Sellars's critical realism, which his evolutionary naturalism seeks to defend, was by no means seen as unproblematic by his colleagues. For example, George H. Sabine of the University of Missouri concluded, "It can not be said that he [Sellars] has any theory of the general structure of knowledge. And yet without it critical realism can not vindicate its claim to be an independent theory of knowledge." And Maurice Picard noted that Sellars's heavy reliance upon biology reduces the knowledge process to simply a matter of biological adaptation. To attempt an answer to the question of a reality beyond that of the knower by relying on the "special sciences" of "enlightened common sense" Picard finds wanting: "That he [Sellars] is not entirely successful from the metaphysical standpoint is partly due to the limits which bound his [materialist] faith."[104]

Where does that objectively leave Sellars? For one thing, Sellars admitted that Darwinian evolution was incomplete in many regards, but it was clearly foundational for him. From what has already been said about Darwinian theory here and in previous chapters, this puts Sellars on an extremely shaky foundation. In his effort to escape naive realism it seems he has fallen prey to a very naive view of science. He speaks of science in general and Darwinian theory in particular as unquestionable "facts" that proceed simplistically through history as building blocks one upon another. But, as also pointed out earlier, science is open to constant revision. The science that Sellars so confidently relies upon in biology is subject to a host of objections in terms of both evolutionary biology and psychology. Darwinian evolution is surrounded by well-credentialed science-based nay-sayers at present.[105] And as for behaviorism, although Sellars was critical of its denial of consciousness, his *Evolutionary Naturalism* mentions it approvingly and indeed uncritically twenty-three times in ways that few today would follow as enthusiastically as had John B. Watson and B. F. Skinner in the early to mid-twentieth century. The complexity of human psychology cannot be reduced to examples of salivating dogs, to response-on-command laboratory monkeys, or even to the operant conditioning of human beings under highly controlled conditions.

There are other problems. Sellars wants to create the numinous by euphonious phrases and assurances that he has only the highest aspirations of the human condition in mind in his evolutionary naturalism. It should not go unnoticed that he was the author of the first humanist manifesto in 1933, a document meant to establish a thoroughly non-transcendent religion seeking "the good life" for all humankind. (John Dewey's name on the list of signatories is revealing.) One is reminded of Auguste Comte, who tried to create a religion of scientism, an effort that Thomas Henry Huxley found off-putting.[106] While Sellars reminds us that he is no positivist, he certainly *acts* like one. But as one philosopher/historian of the period notes, "Humanity is not God, as Comte thought; no one who is familiar with humanity will care to worship it."[107] Robert J. Kreyche, after a careful analysis of Sellars's thought, argues that "despite the insistence which Sellars places upon the non-reductionist type of analysis which evolutionary naturalism employs, naturalism, whichever its form, whether it be mechanical or evolutionary, is but a caricature of what a true philosophy must be—a rational interpretation of the facts of experience. Naturalism is, whether regarded either from the standpoint of its methods or of its contents, by its very nature, reductionist."[108] By the same token, the effort to establish humanism upon a naturalist ontology is problematic because, as Kreyche observes,

> if science as such has nothing to say concerning the basis of our value-judgments, and if, moreover, philosophic knowledge is intrinsically dependent upon and essentially related to our scientific knowledge of things, then, clearly, philosophy is no more capable of discovering the valuableness-aspect of reality than science is. The only avenue of approach, therefore, to which one can resort, and to which Sellars himself actually does resort, is an empirical analysis of our value-judgments—an approach which . . . leaves unexplored the *fundamentum* upon which those judgments rest.[109]

John C. Greene, writing years after Sellars, points out the difficulty of wedding a scientistic naturalism to a humanistic impulse because such a view "strips nature of aim, purpose, and value, and hence of any meaning other than purely scientific intelligibility. Caught in this dilemma," he adds, "the advocates of evolutionary humanism are reduced to claiming the sanction of evolutionary biology for values that originated elsewhere and to introducing the forbidden elements of teleology and value into their science in figures of speech."[110] This disingenuous procedure is not likely to produce productive results for society or for science.

But perhaps the most telling criticism of Sellars's evolutionary naturalism resides in his own materialism. Whitehead observed that no complete evolutionary theory is truly compatible with materialism:

> The aboriginal stuff, or material, from which a materialistic philosophy starts is incapable of evolution. This material is in itself the ultimate substance. Evolution, on the materialistic theory, is reduced to the role of being another word for the description of the changes of the external relations between portions of matter. There is nothing to evolve, because one set of external relations is as good as any other set of external relations. There can merely be change, purposeless and unprogressive. But the whole point of the modern doctrine is the evolution of the complex organism from antecedent states of less complex organisms. The doctrine thus cries aloud for a conception of organism as fundamental for nature.[111]

All Sellars can offer in reply to this cry are reductionisms that amount not to genuine evolution but to trivialities of atomistic change. The fact that modern evolutionists of the Darwinian persuasion—including going back to Darwin himself—can't see this can only be ascribed to blindness brought on by an overweening scientism ill-suited to sound metaphysical thinking.

Much more could—and will—be said regarding Sellars, but suffice it to say that Boodin's work by comparison has much stronger scientific and philosophical legs to stand on. Part of this has to do with those Darwinian nay-sayers. Neither religiously motivated nor philosophically indisposed to evolutionary reductionisms of the kind just described, a growing number within the scientific community are calling for a replacement of Darwinian evolution for something that takes greater account of epigenetics and Lamarckian inheritance. This "third way" (see note 105) of evolution has many varied proponents, but Denis Noble, a British biologist and cardiovascular physiologist, expresses a view at least ostensibly similar to Boodin's reference that evolution must be regarded as "creative synthesis, as productive reorganization" but *not* "as a synthesis of chance or as a reorganization independent of the environment with which life must effect energy exchange."[112]

Noble believes that the neo-Darwinian synthesis places far too much emphasis upon DNA and has dismissed all aspects Lamarckianism out of hand. For him, the evolutionary synthesis is bankrupt. Instead he insists that all living systems are purposive in a non-transcendent teleological way (he, in fact, chides the biological community for its apparent fear or disregard of any and all forms of teleology). He calls his key thesis *Biological Relativity*—"the theory that all levels in organisms have causal efficacy. There is no privileged level from which all others may be derived."[113] This is the basis upon which these living systems use to establish emergent forms. Stochasticity (i.e., randomness) provides the extensive variation, leaving it "wide open to an organism to select what works in any given situation."[114] How Noble gets around the implicit intentionality of *selection* by non-intentional entities remains unexplained. But he gives a hint of his confusion when he says, "Natural purposiveness is to human purposiveness much as natural selection is to artificial selection (meaning human) selection."[115] We're back to Darwin's much beloved breeding analogy, but as already discussed this analogy cannot work without assuming the same kind of intentionality in nature as that exercised by intelligent human selectors. It was a problem for Darwin and it remains a problem for Noble. Although this is hardly his intention, the only avenue he has open is to invoke some form of panpsychism.

Panpsychism is hardly new. The sixteenth-century philosopher and Platonist Francesco Patrizi of Cherzo attempted in his major works to incorporate a systematic account of the natural world within an overall methodological and metaphysical context, anticipating Leibniz, Whitehead, Hartshorne, and many others. In doing so, he coined the term *panpsychism* in his *New Philosophy of the Universe* (1591). The long history of those arguing for a "world-soul," or as the German metaphysician Rudolf Hermann Lotz put it "the cosmos is but the veil of an infinite realm of mental life," has been a familiar stock-in-trade of some idealist and process philosophers.

But lately reductionist materialists have adopted panpsychism as a means of getting around the problem of how mind can be derived from non-mental things. Typical examples are Galen Strawson (son of Peter F. Strawson), Thomas Nagel, and Michal Ruse. All are atheists, but as Ruse admits, "this move [toward panpsychism] is today positively trendy in some very respectable circles."[116] Here, in broad outline, the proposal is that the universe consists of self-regulating and self-organizing non-transcendent teleological systems at interrelated multiple levels that function much as organisms do, not unlike the Gaia hypothesis. The degree to which spatio-temporal substances have some "mind" lurking within them remains unclear and is quite variable among its menagerie of proponents. In any case, it would seem that panpsychism has become a workaround for those seeking new ground for materialism. Not insignificantly, Boodin recognized this potential in panpsychism. He said,

> The world is neither better nor worse for our metaphysical conceptions. And if panpsychism is indifferent to the realization of ideals, if it reduces the higher to the lower categories, if it fails

to give us a preferential basis of values, if it offers no call to our creative capacities, it is teleologically indistinguishable from the crassest type of materialism. . . . We must not be misled by mere words. It is a mistake to suppose that by adopting mere euphonious terms for these situations, such as "vital impulse" or "panpsychism," we have either explained or dignified the process.[117]

Today, despite its lure for agnostics and atheists, completely non-reductionist panpsychist theists persist. One of the most prominent in the tradition of Whitehead and Hartshorne is biologist Rupert Sheldrake, who considers Strawson's adoption of panpsychism a confusing "revival of animism."[118] Others, like Joanna Leidenhag, have found comfortable moorings for panpsychism in thoroughly orthodox Christianity. For her, it is quite possible to "affirm panpsychism as a theory of consciousness, without inheriting either an event ontology or the Process doctrine of God."[119] Details follow. For our purposes, however, it is sufficient to state quite clearly that despite his strong theism, his "cosmic idealism," and his definite affinity for process philosophy, Boodin did not consider himself a panpsychist. As indicated above, he clearly declared against it. Now juxtaposing Boodin and Noble, the two seem to have less in common than one might initially think.

It remains only to examine the critical reception that *Cosmic Evolution* received. Reviews were mixed. Michael B. Foster was thoroughly unimpressed. Accusing Boodin of "more or less frivolous rhapsodizing," he dismisses all talk of the evolutionary environment's "higher levels" as unhelpful and not "wholly real." He rather unfairly and caustically charges Boodin with "naïve confusion scarcely creditable to a professed philosopher."[120] It can be said in reply that perhaps the reviewer did not understand Boodin's higher cosmic levels because they were simply over his head. A little less acerbic was A. K. Stout, who thought Boodin gave a "constructive metaphysics" but attempted to "beguile" the reader with "purple passages" that too often intruded into the argument. Stout also thought Boodin's frequent use of "energy" was a "defect." Although Stout finds the book "suggestive and stimulating," he remains unconvinced by Boodin's arguments. "His philosophy is ostensibly reached by interpretation of the results of physical science," he adds, "but the teaching of science does not demand this interpretation; Prof. Boodin appears simply to take it for granted that there is no alternative."[121] Now this is either unfair or presumptuous. Boodin is well aware of alternatives and, as we have seen, addresses most of them. Stout's assertion about what science does and does not demand seems arbitrary; many scientific phenomena are open to a wide range of interpretations and it often takes years to sort through them based upon other related evidence. Nowhere do I see in Boodin's arguments the kind of "demanding" dogmatism with which Stout accuses him.

Fortunately, there were friendlier voices on Boodin's behalf. Hoernlé, Boodin's collegial defender, praised the uniqueness of his overall argument. There is, Hoernlé concludes, "no doubt that he goes with the idealists in so conceiving the universe that the fundamental values are not subjective illusions, but grounded in the very nature of things. His cosmology is at once a philosophy of science and a philosophy of religion."[122] He further alludes encouragingly to Boodin's future exposition on the nature of God, which would be forthcoming in his *Three Interpretations of the Universe* and a more popular version, *God and Creation: God, a Cosmic Philosophy of Religion*, both published in 1934. Most enthusiastic of all is G. W. T. Whitney. The reviewer concludes that the book's

> discussion of relativity is remarkable for its comprehensiveness, clarity, and lack of all unnecessary technicality. An eminent authority on the subject has said that the chapter on "Theories of Relativity" is the best interpretation of relativity that he has seen. But what impresses me most is the author's breadth of vision, sanity of judgment, and wholesome idealism. He has shown how impossible it is to derive the concrete reality of our living experience from a few abstract categories, and especially he has shown how inadequate are all forms of materialism. One cannot read *Cosmic Evolution* without feeling that Professor Boodin possesses the intuitive insight of the poet as well as the profound knowledge of the philosophers.[123]

So even in his own day he was indeed recognized as the poet-philosopher—he was not entirely forgotten it seems.

The reader's perseverance through this long and winding but necessary chapter is now rewarded with the fact that Boodin's complete philosophy has largely been exposited. What lies ahead are fuller details of his theological and sociological views. It certainly should be no shock at this point to realize that suffused into all of our poetic philosopher's thought was a sense of transcendent direction, of Final End. After all, Boodin declared, "from the point of view of reason it is easier to read nature as striving to express certain types or ideals than to read ideals as chance. Nature seems to be, somehow, leading in the direction of human nature; striving for a type somehow to be determining the direction of the series; and freedom and significant expression of life to be all the same time the end to be realized." The nature of that "striving" will be explored in the next chapter of this text, in the three works that comprise Boodin's theology: *Three Views of the Universe*, *God and Creation*, and *The Religion of Tomorrow*.

Chapter 4

A Theological Trilogy

Boodin's theism is cast not in familiar religious categories but in the thought and language of philosophy. Thus, extended exegetical interpretations and "proof text" demonstrations are clearly not part of his modus operandi. And yet theism is there as much as in Alfred North Whitehead, Charles Hartshorne, David Ray Griffin, or John Cobb. However much each of the aforementioned might balk, or at least quibble, at some of Boodin's approaches and conclusions, they are of one panenthic process piece. Boodin presented his theological views in three broad formats comprising a trilogy beginning with cosmology; then a briefer and more accessible treatment of theism specifically, in Boodin's words, "consistent with the fundamental intuitions of Christianity"; and finally an extended "sermon" for the times aimed at cultivating the "sacramental communion" necessary for what his teacher, Josiah Royce, would have called "the Beloved Community" and ultimately the world. Unlike his first two works of this nature, the last one has a distinctly more social or communal emphasis. Each in its turn, *Three Interpretations of the Universe*, *God and Creation*, and *Religion of Tomorrow* present the theological implications of his entire corpus of religious thought.

Three Views of the Universe

This work is dedicated to the French philosopher André LaLande (the student of Émile Durkheim and proponent of *involution*, a movement toward homogeneity and unification), who Boodin regards as "one of the sanest and most constructive representatives of contemporary thought." Unlike Boodin's other works, this one is more generally historical and most specifically focused on tracing out three key concepts in the history of evolutionary cosmology: preformation, emergence, and creation. Here the reader is given a masterful tour of the leading proponents of each view, including Boodin's own position. Moreover, much of the book is devoted to detailed commentaries on the works of Plato, Aristotle, Plotinus, and the early church Fathers. Our poet-philosopher bids that his colleagues "leave their verbal game of technical quibbles" and direct their attentions to that which matters most, namely, "man's place in the universe."

Boodin sets out his task immediately, which is to distinguish and elucidate three interpretations of history (taken on their face to be evolutionary) that present the order of the world and the universe. As such, each takes on its own ontological character. The first is *preformation*: evolutionary development that is latent in a process that is unfolding or expressed through earlier stages of its history. The second is *emergence*: the appearance of wholly novel characteristics and structures that are apparently blind and without guidance internally or externally. The third is *creation*: the appearance of new forms, structures, characters, or stages controlled and/or animated by some actuality. In general, preformation and creation may be distinguished by their teleological natures.

The most thorough preformationists of the ancient world were the early Stoics, according to Boodin the only ones to take time seriously. While they were not panpsychists, they were pluralists whose earth, air, fire, and water ultimately reduced to a practical dualism of active factors (fire and air) and passive factors (earth and water). Stoics "spoke the language of cosmic immanence" with individuality as the universal law. Just as the Stoics are not panpsychists, neither are they pantheists. God is a pervasive force but is not present everywhere in the same way. Cleanthes presents his hymn to Zeus "as the omnipotent, king of kings, whose universal Word (*Logos*) pulsates through all" limited by the works of the sinner but compensated for without preventing.[1] A God of such cosmic proportions could seem distant and impersonal, but the Stoics did feel a sense of cosmic companionship, "a Friend behind the phenomena," to quote Edwyn Bevan's *Stoics and Sceptics*.[2]

Such ideas were clearly not inimical to Christianity, and Scotus Erigena (often spelled Eriugena) is its finest example. His *On the Division of Nature* is the first great philosophical system of the Middle Ages, and historian David C. Lindberg considers him "undoubtedly the ablest scholar of the ninth century in the Latin West."[3] His synthesis of Christian theology with neo-Platonism established a comprehensive natural philosophy and through his disciples exerted a continuous influence on Western thought. For Erigena, God concretes the world out of himself, making creation an unfoldment in an eternal process. The trinity is a relation within God and at the same time a relation of God to his world. According to Boodin, "in the divine Intellect or Word exist all the primal causes—the Platonic Ideas, the genera, the species, the individual forms. All the seeds are contained in the first seed. Erigena implied the extreme form of preformation we find later in Leibniz."[4]

Other examples, including some modern representatives like Georg Wilhelm Friedrich Hegel, Friedrich Wilhelm Joseph Schelling, Arthur Schopenhauer, and Eduard von Hartmann, are Boodin's preformationists. All of them exhibit in varying degrees and assorted nuances the basic definitional characteristics of the label. To belabor each one would become tedious and tax the reader's patience. Instead we will proceed with a discussion of emergence.

Before proceeding, some more general discussion of emergence may be helpful. Since emergence is defined generally as "the idea that novel, irreducible features (properties or substances) can arise from sufficiently complex systems of more fundamental elements,"[5] it has a very wide range of interpretations, from materialistic to teleological. As such, questions of emergence often filter into discussions of science and religion. Whether or not emergence is wholly naturalistic or involves a more transcendent process depends upon how this malleable term is used, and a review of the historiography of this vague and complicated term would include a host of disparate figures, from John Stuart Mill and C. Lloyd Morgan to Alfred North Whitehead and Teilhard de Chardin.[6] This is why Rupert Sheldrake, himself a teleological emergent evolutionist who will be discussed in detail in chapter 6, can also speak of Darwin and Marx as being emergent evolutionists of a very different stripe.[7] The issue is, how bounded by deterministic naturalistic laws is this so-called *emergence*? For Darwin and Marx, they are strictly bounded; for Sheldrake, Teilhard, or Whitehead, not so much if at all.

Indeed emergence can be used to expressly avoid reductionist or eliminative accounts of mind and human existence. In this sense emergence can stand as a direct challenge to physicalism. Sheldrake's emergent wholism is one example as is Arthur Koestler's emergent holons. Similarly, William Hasker's view of humans as ontologically emergent souls is supportive of an expressly theistic worldview. But this is not Boodin's emergence.

So how exactly is Boodin using emergence in *Three Interpretations of the Universe*? For him, emergence is "the appearance of new characteristics and structures with no apparent guidance from within or without."[8] It is the last half of this definition that matters here. Boodin is relying on C. D. Broad's idea of "materialistic emergence" found in *The Mind and Its Place in Nature* (1925). What Broad means by this is that "from simpler material situations in nature, there arise more and more complex situations which possess characteristic qualities which cannot be reduced to those of the simpler situations."[9] As we will see, Ernst Mayr thought of emergence in this way too—a probabilistic expression of purely material agents. It excludes any and all assumptions of a nonmaterial component or force such as Driesch's entelechy or Teilhard's Omega Point. Boodin uses Broad's materialistic emergence (admitting that Broad himself is not a strict emergentist) as part of his historical tool kit and as a heuristic device to distinguish various forms of evolution relevant to his book. Of course materialistic emergence existed long before Broad gave it a name, and in explicating this particular view of emergent evolution we can see that Boodin agrees with Sheldrake that Darwin is an emergent evolutionist.

Therefore, in contrast to either preformation or creation, emergence is usually prone to reductionism of one kind or another. In other words, one side views reality in terms of abiding structure with a plot-based experience leading toward and indeed part of a larger whole; the other sees only a concatenation of parts mechanically determined by ironclad laws knowing neither whence it came nor whither it is going. It is episodic, without structure or direction, formed only by the happenstance of chance and necessity. This does not mean that just anything can happen, but it does mean that what happens simply *is*—there is no goal or grief in any of it. With emergence the clear division between an idealistic interpretation of nature and naturalistic materialism becomes apparent. Materialistic emergence can be found among the ancient atomists, particularly Leucippus and Democritus.

But it found its most far-reaching expression in Lucretius. It shouldn't be surprising that our poet-philosopher found him fascinating, for it is said that this first-century Roman used "his poetic skill as honey to disguise the wormwood of [his] philosophy."[10] Although little has survived from Lucretius, Boodin says that his immortal poem [*De rerum natura, On the Nature of the Universe*] "furnishes the most complete and convenient statement of atomism."[11] Lucretius is praised for following empirical evidence instead of allowing his metaphysic to drive his theory. For Lucretius everything derives from the "restlessness" of nature and in it there can be found no final cause. Furthermore, the mind/body problem, while in some measure inscrutable, is nonetheless locked in an inseparable union. For Boodin,

Lucretius is the most thoroughgoing empirical emergentist who is all the more to be admired for his reliance upon experience as his guide. Having said this, Boodin admits that Lucretius's empirical generalizations do not necessarily follow—indeed are inconsistent with—his notion that empirical reality derives from blind combinations of atomic stuff. Lucretius is the focus of Boodin's attentions because in this Roman poet we have the beginnings of a species of empirical accounts of nature that will reappear throughout the centuries. In Lucretius's poetics Boodin predictably finds "the genuine scientific spirit" even if he objects to its conclusions.[12]

From here Boodin moves on to the later Stoics. Since little has survived of their work, Boodin views them through the eyes of the neo-Platonist Plotinus. These Stoics, Boodin explains, developed a theory of strict emergence of higher from lower through chance. Plotinus refers to them as materialists. Unlike their preformationist predecessors, the later Stoics produced the most complete theory of emergence. If we take Plotinus at his word, these Stoics were pantheistic monists, indistinguishable from evolutionary materialists. If the later Stoics are our guides, then as Erasmus Darwin once quipped, "Unitarianism is a featherbed to catch a falling Christian," and then surely pantheism is a featherbed to catch the falling theist of *any* particular stripe.

Boodin's next general summary of emergence is dialectical materialism, starting with Feuerbach's *The Essence of Christianity* (1841). Here Feuerbach develops a pantheism premised on human self-consciousness as the only absolute. According to him, "*Man feels* nothing towards God which he does not feel towards man. *Homo homini deus est.* Man is god to man."[13] In effect, in Feuerbach's hands theology is reduced to merely a form of anthropology. *The Essence of Christianity* exerted a powerful influence over Karl Marx and Friedrich Engels. Feuerbach later recanted his harsh eighteenth-century mechanistic materialism for a more idealist approach. The Marxists rejected this and incorporated Feuerbach's original formation in conjunction with their economic determinism as part of their doctrinal formula, although Engels did agree with Feuerbach's rejection of strictly mechanistic materialism for a more emergentist expression of the same idea and held up Feuerbach's scientific materialism as foundational to human knowledge. The idealistic faith in progress witnessed by Marx and Engels is conditional; it may, of course, prove itself false to the dialectic of history and need to be superseded. Nevertheless, Engels provided a certain strain of idealism within its own materialistic emergence theory, which might explain the perennial lure of Communism among impassioned youth.

Boodin provides detailed examinations of emergence in the sciences generally, specifically in physics, biology, and psychology (exemplified in the mind/body problem). They need not be belabored here except to say that each becomes a venue for Boodin's reiteration of his cosmic idealism in opposition to reductionist materialism for which a few examples will suffice. In physics, Boodin observes that all deterministic certainties have been demolished because at least since Heisenberg indeterminacy in conjunction with emergence is an established fact, and while emergence certainly highlights a significant aspect of our world, it by no means explains it. Indeed the old dogmatism of nineteenth-century science has been shattered by the discoveries of twentieth-century physics. Boodin concludes, "Truly science is relative."[14] This is a humbling dictum that "science"-obsessed materialists like Roy Wood Sellars fatally ignore; much in the same way that Darwin failed to be cognizant of the fact that, as Boodin notes, "A mere heaping up of observations will never give us science."[15]

Here we must be careful not to confuse our metaphysics with our science. Boodin knows very clearly that science is alive, it moves and grows; as such, metaphysics never plays journalist to science—it shows trends rather than news. What then are those trends? For Boodin, while determinism and indeterminism are simply postulates, the evidence favors the latter. Indeed mechanistic materialism is yielding to structure and wholeness with the so-called new physics leading the way. The rich forms of Plato are triumphing over the reductionist materialism of Leucippus; organic wholism is supplanting the merely mechanical universe. Indeed, for Boodin, physics is revealing what he has insisted upon all along, namely that metaphysics must make structure a fundamental aspect of nature enervated by indeterminate possibilities rather than reduced to a lockstep of pure mechanical actions and reactions.

If these are the suggestive directions of physics, Boodin argues that they are no less true for biology. All things that we know of the nature of life and its origins, speculative though they may be, suggest "the result of cosmic control, cosmic architecture."[16] In contrast, Boodin depicts biology as stuck in the old mechanistic emergent evolution based upon an outdated view of the universe as a machine. The result is the most "thoroughgoing mechanistic hypothesis in modern times."[17] Enough has already been said of Darwin's theory, but what of the science of life's origins about which Darwin said little? A number of fascinating experiments aimed at demonstrating the feasibility of life's chemistry generated from early conditions on the prebiotic earth might have given him pause. The most important of these was conducted by Stanley Miller and Harold Urey in 1953. Reproducing what they believed to be the earth's primordial reducing atmosphere, they showed that lightning under such primitive conditions could produce amino acids, the building blocks of proteins necessary for early life. Unfortunately, the assumptions regarding the nature of that atmosphere were called into question by geochemists in the 1980s. Thus, the era of optimism in the 1950s and 1960s, that the problem of the origin of life could be solved by invoking basic chemical principles involving chance and necessity, fell into as uncertain a position as before the Miller-Urey experiment.[18] More recent investigations, however, cast serious doubt on reductionist accounts of the origin of life.[19] Boodin's conclusions regarding the transcendent nature of cosmic control and cosmic architecture, therefore, remain as alive as ever.

Boodin ends his discussion of emergence in biology by calling on the discipline to abandon the "old atomic physics" in favor of the "new physics of cosmic structure." This has finally begun. For example, Johnjoe McFadden and Jim Al-Khalili have examined the possibilities of merging the "spooky reality" of quantum physics with biology. But their view of life seems as reductionist as the old biology. For example, they describe the essential principle of life by means of "enzymes, the engines of life, whose extraordinary catalytic power is provided by their ability to choreography the motions of fundamental particles and thereby dip into the quantum world to harness its strange laws."[20] Here adjectives substitute for evidence all guided by every materialist's confident faith in apotheosized laws directing everything as surely as Lucretius's atoms. And their hopeful explanation for the origin of life is little better. Here they speculate that a "quantum proto-replicator molecule [might] eventually collapse into a self-replicator state . . . and the first self-replicator will have been born into the classical world."[21] Now the machines of biology can take hold and explain the rest. Here we have the self-replicator-as-magic-wand theory; science has given way to mysticism. Other quantum biology scenarios appear to be old ideas simply dressed in new quantum clothes. Researchers looking for ways to incorporate quantum physics into the science of life seem incapable of avoiding old threadbare notions of selection effect and selective advantage in applying quantum concepts to the actual processes of life and evolutionary development.[22] Boodin (ever the optimist) was perhaps too sanguine in his hopes to rescue biology from itself. Old paradigms die hard.

The problem besetting biology of freeing itself from old reductionisms is no less pronounced in psychology. Here Boodin explains that the neurological model described by Huxley as epiphenomenal ("psychological events are concomitant with physiological changes" *and nothing else*) has "been invented in the service of an atomic theory of mental facts" for the self-serving purpose of satisfying materialistic prejudice.[23] And the behaviorists (under John Watson and later B. F. Skinner), whose problems we have already discussed, are little better. The behaviorist attempts to explain human mental action by means of physiology, forgetting that even the observer must have reference to their own introspective experience to interpret behavior. And as for Pavlov's dog conditioned to respond to the sound of a bell with a watering mouth, reducing thought to the action of the salivary glands "does not seem to make us any wiser about the process of thinking." Gestalt psychology similarly offers some new insights, but overall it fails as an explanation of human psychology.

Instead of these incomplete hypotheses Boodin asks for an abandonment of the old dualisms of Descartes and Spinoza in favor of Aristotle's more fruitful conception of the soul as a whole-form or whole-control.[24] For Boodin, everything has a particle and a field character. Then, rather presciently, he states, "The dynamic field has become grooved through habit."[25] This should be read in conjunction with his earlier statement regarding biology: "The organism is more than a collection of factors. It is a whole. It remembers, it intuits, it looks forward according to its own genius. The multicellular organism is more than interacting cells. It organizes these into an energetic unity with a pattern of its own."[26] Of course, this is precisely what mental activity as an epiphenomenon denies. More importantly, Boodin's comments here will need to be recalled when we discuss his relationship to the ideas of Rupert Sheldrake in chapter 6. For now, it is sufficient to note that Boodin sees biological life similarly to the physical world, that they are parts of energy fields that are coordinated and in various measures teleological. For this reason, emergence fails to capture the intricate details of physical, biological, and mental worlds (what Boodin calls groups). Viewed through any of these lenses, emergence is left wanting. In fact, we can see the progress of nature culminating in the psychological group.

Finally, we come to creation. This is Boodin's favored view of the universe. After developing his own cosmology in *Cosmic Evolution* (see previous chapter), he discovered much to his delight that in rereading *Timaeus*, "Plato's footprints [were] everywhere over the ground that I had traversed."[27] As Boodin sees it, creation values emergence and novelty in nature and affirms precisely the guidance denied by the strict emergentist. The key was Plato's discovery of teleology through Anaxagoras's *Nous* or mind. Moreover, it is not enough for Plato to have an aesthetically pleasing cosmology, it must pass the pragmatic test of experience. Boodin rejects the claim of British idealist philosopher A. E. Taylor that Plato implies along with Whitehead "the impersonal ingression of abstract eternal objects into the world of events"; Taylor, he insists misses the critical "dynamics in Plato which is always the activity of the soul endowed with mind, in short is personal."[28] This is the creative personality inherent in Plato's system. The heart of Plato, found in his great cosmological drama the *Timaeus*, is summarized as follows: "God's intelligence created the system of forms before he imaged them in a sensuous world. The word, eternal, does not for Plato mean everlasting but precisely the intelligible—that which is grasped by reason because the creation of reason. The forms of things, therefore, are eternal even if they were created with the creation of the world. The forms are timeless for Plato."[29]

Plato regarded the explanation of nature through chance as preposterous—even criminal—impiety. Furthermore, Plato's God is alive and dynamic, not like Aristotle and his God self-absorbed in his own perfection. Plato's God works *toward* perfection and encourages it. Neither is his God a vague impersonal deity like Fichte and Emerson propose: Plato's God takes part in the struggle in a panenthic embrace. Boodin's conclusion is both simple and profound: "Plato's cosmology is an honest attempt to give a cosmic explanation of human experience."[30]

From his commentary on Plato, Boodin moves on to his student Aristotle. Admitting that William Harvey was much keener than Darwin in seeing the conception of Aristotle's whole-form or entelechy, he says that the latter's discovery of this important feature of Aristotle could have spared the world all of his mechanistic—indeed positivistic—speculations up to our own day. To Harvey's teleological sagacity might also be added that of Robert Boyle and Maupertuis. Interestingly, in Maupertuis can be seen a strong French proponent of Newtonian physics and at the same time a vigorous defender of final cause.[31] In this sense it is simply wrong to conclude that the so-called scientific revolution naturally brought an end to teleology.[32] In any case, Aristotle might have let his biology turn into a cosmology of naturalistic pantheism had it not been that he was (fortunately) haunted by his teacher. Instead, Aristotle's "Unmoved Mover" turns him into a species fixist; however, there is emergence in Aristotle (the temporal process demands it) and time and motion are eternal features of the world. For Aristotle, God does not create matter, and although he is perfect, he is not omnipotent. The physical world is limited by empirical conditions and empirical individuality. God works with this world; it would exist without him but it would lack meaning and coherence. If Boodin is somewhat less enthusiastic about Aristotle than Plato, it is surely not that he fails to appreciate him. In particular, Aristotle saves us from the soul-deadening scientism plaguing the modern world since the Victorian era. Boodin ends his commentary on Aristotle thus:

> We need more than science if we are to become acquainted with the ultimate individuality of things. Science can reveal to us the aspects of nature. It can give us a concept of structure. But it cannot as science introduce us to individual reality; and ultimately reality is individual as Aristotle pointed out. To know the individual requires a different way of knowing from that of mere logic. It requires intuitive reason, as Aristotle says. It requires appreciation and communion. Purposive realization must include will and feeling, as well as ideas, and therefore real knowledge of purposive reality must include will and feeling as well as ideas.[33]

If this is the lesson learned from Aristotle, it is no small one indeed.

Like Lucretius, Boodin is drawn to the neo-Platonist Plotinus as more of a poet than a philosopher. Here Plato's rationalism has turned to mysticism in Plotinus. As we saw in chapter 2, both by timing and inclination Plotinus became a pivotal figure linking the ancient and Christian worlds together. Plotinus is concerned with personal salvation and Boodin regards him as an "extreme creationist" despite the fact that he regards Plotinus's view of the soul as an abstraction. Similarly, Plotinus's God remains real but vague, a distant entity that must be kept separate from matter, something taken as having a contaminating effect, this even though Plotinus attacked the Gnostics. Plotinus's emphasis on the ecstatic encounter with God was far too mystical for Boodin, but it is an important influence on Christian theology in general.[34] As Copleston points out, even though he never became a Christian (although he must have known of it), Plotinus "was a resolute witness to spiritual and moral ideals, not only in his own writings but also in his own life, and it was the spiritual idealism of his philosophy that enabled it to exercise such an influence on the great Latin doctor, St. Augustine of Hippo."[35]

As such, Plotinus easily takes us to Boodin's creation in the Hebrew and Christian traditions. At the outset, Boodin flaunts convention by rejecting ex nihilo creation. Citing the biblical scholar James Hastings, Boodin agrees that nowhere is creation out of nothing expressly taught in the Bible.[36] It is most nearly approached in the apocryphal/deuterocanonical second book of Macabees 7:28: "I beseech thee, my son, look upon the heaven and the earth and all that is therein, and consider that God made them of things that were not." Jerome's Latin translation rendered this critical passage, "God created them out of nothing." Now the original Hebrew was a phrase best translated "out of non-beings." This clearly Platonic reference does not mean nothing, but rather the world of flux instead of being (ideas, order, and form). Jerome's *Vulgate* had no equivalent for this "so a mistranslation finally determined Christian theology on this point." Boodin adds, "Why an ambiguous passage in an unimportant book—never recognized in the Hebrew canon of the Old Testament—should be given such weight in face of the clear dualism of 'Moses and the Prophets' can only be accounted for by the will-to-believe what suits our prejudices."[37]

Although Boodin obviously rejects the out-of-nothing conception of creation, he recognized a specific and very old creation of the earth. The common link between Christianity and the ancient world, for Boodin, comes from Plato, even though the early Church fathers interpreted Plato as having plagiarized Moses. More accurately, "The two great traditions—*Genesis* and Plato's *Timaeus*—furnish the true intuition which must be the basis of the philosophy of creation."[38] While there need not have been any philosophical or logical reason why Christianity had to turn to a monistic view of creation, the most important influence was historical. The power of the Roman state became the Christian model of a creative God. Here the world was brought forth by singular decree rather than teleological emergence. Boodin would agree with Whitehead that "the Church gave unto God the attributes which belonged exclusively to Caesar." Lost was "the Galilean vision of humility" that flickered dimly through the ages.[39] In short, creation matched the attributes artificially applied to deity.

Origen and Augustine were clearly neo-Platonists. Nevertheless, the Christian church eventually shifted its emphasis toward Aristotle under the influence of Thomas Aquinas, and by the time of Copernicus and Galileo it was the Aristotelian worldview against which they battled. Thus, we can discern three distinct forms of creation under Christianity: divine volition under Augustine, preformationist division under Erigena, and final cosmic emanation under Aquinas.

Boodin ends his *Three Interpretations of the Universe* by adopting a steady-state view. "The cosmos as a whole never began," he declares.

"Its constitution is constant. That is what we mean by the uniformity of nature."[40] Lest we think that this fails to accord with science as known today, it would be well to consider a growing body of evidence supporting some version of an eternal universe. Although the "big bang" theory supposedly supplanted Fred Hoyle's steady-state theory, it has recently been pointed out by physicist Francis M. Sanchez that far from a dead theory, "forgotten steady-state cosmology had correctly foreseen the acceleration of the galaxy recession and the critical flatness. The main argument that led to its discarding, the discovery of the Cosmic Microwave Background (CMB), was in fact not pertinent. Indeed, the steady-state cosmology is the only one that predicted correctly its temperature, from the Helium density."[41] And other well-respected physicists are questioning the viability of what Boodin called Lemaître's "fireworks" theory.[42]

Whatever the definitive answer might be, our poet-philosopher states an important truth when he says near the end of his volume, "Those who look for Spirit and God in the first stuff of things are looking in the wrong direction."[43] Applying sound process thinking, Boodin shows us where to look for spiritual things; these are found in the highest part of creation (those closest to divinity) not furthest from it. There is a hierarchy of being from "electron to man" and it is perceived most fully in humanity, especially the mind. In fact, beyond mind is spirit itself. "That is the reason," says Boodin, "no man can be a great scientist unless he is a poet." In any case, the goal is to find creativity, and this comes only "from our rapport with the genius of the whole."[44] And we end here on this fitting note.

Three Views: A Summation

Before departing this important work altogether, we might do well to present a summary view of the long and varied territory that we have just traversed. That can be done by means of table 4.1. Besides the few already discussed, only some of the more interesting inclusions in Boodin's categories will be highlighted.

The first one to catch the eye is Boodin's designation of Whitehead as a preformationist. This certainly surprised several reviewers. Charles F. Sawhill Virtue considers it an "astonishing interpretation of [Whitehead] the leading exponent of Platonic-Aristotelian creationist philosophy in our day."[45] This brings up Boodin's whole relationship to Whitehead. Their many similarities have already been mentioned in the previous chapter, but no

Table 4.1. Boodin's Three Interpretations of the Universe and Its Proponents

Preformation	Emergence	Creation
Greek Stoics (3rd century BCE)	Lucretius (99 BCE–55 BCE)	Genesis (ca. 1145 BCE/1405 BCE)
Erigena (ca. 800–877)	Feuerbach (1804–1872)	Plato (ca. 429?–347 BCE)
Cusanus (1401–1464)	Darwin (1809–1882)	Aristotle (384–322 BCE)
Bruno (1548–1600)	Marx (1818–1883)	Plotinus (204?–270 CE)
Böhme (1575–1624)	Spencer (1820–1903)	Origen (184–253 CE)
Schelling (1775–1854)	Morgan (1852–1936)	Augustine (354–430)
Schopenhauer (1788–1860)	Alexander (1859–1938)	Aquinas (1225–1274)
Von Hartmann (1842–1906)	Driesch (1867–1941)	Owen (1804–1892)
Whitehead (1861–1947)	Bergson (1859–1941)	Boodin (1869–1950)

person has followed Boodin's work more closely than Alfred Hoernlé. His assessment has particular value and therefore is worth quoting at length:

> When one surveys, in perspective, Boodin's philosophical development, from his doctoral thesis, *A Theory of Time* (1899), to these two most recent volumes, it becomes clear that the central problem of his thinking is the characteristically Platonic one of the temporal, the eternal, and their interpenetration. In general, his problem is thus the same as that which Whitehead states in terms of the "ingression" of "eternal objects" into the flux of "events." But, in spite of this common Platonism, the thought of Whitehead's *Process and Reality* is, obviously, not very congenial to Boodin. For, quite apart from the fact that he avoids completely the new and often obscure technical terminology employed by Whitehead, Boodin is a Platonist, not only in his conception of the problem, but also, unlike Whitehead, in his solution of it. One of the great experiences of his life was the discovery, made after he had developed for himself the main outlines of his cosmology, that these coincided in essentials with the cosmology of the *Timaeus*. Boodin's few references

to Whitehead, in *Three Interpretations*, are without exception critical; and it is interesting, if also somewhat startling, to find him assimilate Whitehead's concept of God, as given in *Process and Reality*, in the chapter on "Preformation in Ancient and Medieval Thought," to "the tradition of Erigena, Cusanus, Bruno and Bœhme." (*Three Interpretations*, 41)[46]

But exactly how "startling" is Boodin's categorization of Whitehead as a preformationist? In a dozen references to Whitehead's part 5, chapter 2, Boodin explains that in the manner of Erigena, Cusanus, Bruno, and Böhme, Whitehead divides God into primordial and consequent natures. In this first aspect God is the unlimited conceptual realization of all potentiality and as such is not *prior to* creation but *concomitant with* creation. He agrees with Erigena that God's nature is creative and to think of God apart from creation is a pure abstraction. Boodin quotes Whitehead on God as "the unconditional actuality of conceptual feeling at the base of things; so that, by reason of this primordial actuality, there is an order in the relevance of eternal objects to the process of creation." God cannot be opposed in achieving static completion in the world, for he is the principle of order. God and the world are "in the grip of the ultimate metaphysical ground," again quoting Whitehead, "the creative advance into novelty. Either of them, God and the World, is the instrument of novelty for the other." In reality neither the primordial nor the consequent sides of God can be separated. Whitehead's grounding of the primordial in the unconscious, with God becoming subjectively conscious in concreting himself by means of his wisdom into the world, Boodin regards as essentially that found in Böhme. Whitehead's union of God and the world, saying, "Each is all in all. Thus each temporal occasion embodies God, and is embodied in God," echoes Erigena, Cusanus, Bruno, and Böhme. Although Whitehead adds "a certain dualism . . . foreign to his predecessors," God becomes a multiplicity of actual parts in the process of creation but at the same time its "conceptual realization." Thus we have a multiplicity as well as a unity in God. But God is more than creation; he is also salvation. "He is," in Boodin's words, "the consummation of the striving of the individual to attain fullness of life or satisfaction through the urge that is in him by virtue of the incarnation of the eternal God in the temporal. God is the beginning, the urge, and the consummation of the process." The individual is the convergence of the various conceptual relations and the creative process of which they are a part; the individual is a microcosm or mirror of the whole. Cusanus and Bruno work out the implications of this, and Boodin sees this as a fundamental part of Whitehead's philosophy. Interestingly, Boodin offers no critique of this position, only to say, "we are merely pointing out that Whitehead is a part of a great tradition, the essential lineaments of which are to be found in Scotus Erigena."[47]

I will leave it to Whiteheadian scholars to make their own assessments of Boodin's position. Recalling his definition of preformation as latent evolutionary development with later forms or stages unfolding or expressing themselves explicitly in already preexistent forms or stages, then this preformation might also be found in Whitehead's *Religion in the Making*. For example, "the creative energy finds in the maintenance of that complex form a centre of experienced perceptivity focusing the universe into one unity. It survives because the universe is a process of attaining instances of definite experience out of its own elements. Each such instance embraces the whole, omitting nothing, whether it be ideal form or actual fact."[48] This certainly sounds preformationist. But we are cautioned against inferring too much from this earlier work since Whitehead was still developing his ideas. He also suggests that some see God as playing a less central role in *Process and Reality* than here, in which case it is nonetheless striking how much Boodin got out of his reading of Whitehead's chapter "God and the World."[49] If God's role is diminished here, then finding it in Whitehead's earlier work only indicates a consistent, and in some measure sustained, preformationist perspective.

In the emergence category (see table 4.1), the entries on Hans Driesch and Henri Bergson might warrant some comment, especially since Boodin associates emergence particularly with the mechanistic/reductionist conception of the universe and Driesch and Bergson are generally regarded as non-reductionist vitalists. But as pointed out in the previous chapter, Boodin is unimpressed with either one. He finds Driesch's entelechy so accountable for everything that it ends up accounting for nothing. As a concept it is simply too vague to be meaningful, and Driesch's complaints against mechanical explanations are equally mechanical themselves. Although it is more metaphysical than Driesch's vitalism, Bergson's *élan vital* "no more furnishes an account of the process of evolution in the particular environment than does the mechanical hypothesis."[50] More tellingly, Boodin noted Bergson's vitalistic reductionism early on. "Vital impulse, as pictured by Bergson," he noted, "is no less blind than the elements of chemistry."[51] Boodin was an idealist in many respects, but he was also (like all good pragmatists) an empiricist. He never let mere verbiage distract him from facts and evidence.

Finally, we come to creation. Here the inclusion of famed anatomist/biologist Richard Owen is my own. It should be instructive, however, for two important reasons: first, as a general consideration it follows Boodin's definitional formula so well that it turns his "three views" from a mere conceptual framework fixed in history into a dynamic analytic tool amenable to other applications of important figures; second, more specifically it highlights the clash of two very distinct evolutionary theories, the functionalist emergent evolutionary theory of Darwin versus the structuralist creationist evolutionary theory of Owen.

Richard Owen's career and writings are usually used as a positional foil to highlight the "brilliance" of his leading rival Charles Darwin. This, however, is inaccurate scientifically and historically. By all accounts Owen was the leading comparative anatomist and zoologist of his day, a worthy protégé of famed French naturalist and founder of modern paleontology Georges Cuvier, with whom he had studied briefly in Paris. Five years Darwin's senior, he lived a long and fruitful but rather frustrated life, dying in 1892. Although he was Darwin and Huxley's bitter rival, the view of him as an angry taciturn species fixist is only partly true; he *was* frequently angry and usually acerbic, but he was *not* a species fixist like his Swiss-born American counterpart Louis Agassiz. Owen was the first to accurately distinguish between the concepts of analogy and homology, and he designed the "terrible lizards" whose fossilized remains were being discovered and cataloged all over the world for the exhibition grounds at Sydenham in 1854. Dubbing them *dinosaurs*, Owen coined the word by which these prehistoric reptilians are known today.

What is of most interest is the fact that Owen was a Platonist and a religious idealist but not a fanatic. A staunch Anglican, Owen found Robert Edmond Grant (Darwin's walking companion), the professor of zoology at University College, deeply disturbing, and Adrian Desmond's superlative study of paleontology in the Victorian era considers much of Owen's science as a reaction against Grant's atheistic leanings. It is also obvious that Owen saw in Darwin's evolutionary theory precisely the kind of dangerous, godless metaphysic-as-science so much a part of Grant's ideology. But he was not a species fixist. When Robert Chamber's transmutationist *Vestiges of the Natural History of Creation* appeared in 1844, Owen withheld public support for lack of confirming evidence, but privately he wrote to Chambers in such congratulatory tones that he had his quiet sympathies if not his vocal support.[52] Owen's reputation as a species fixist came from no less than his rival Darwin, who claimed as much in his *Origin*. Darwin's retraction in the subsequent edition published a few months later did little to dampen the original impression left by the first. In fact, Darwin, and especially Huxley and his X Club cadre, made a habit of misrepresenting Owen on numerous issues.[53] Nevertheless, not long after the appearance of *Vestiges* Owen gave what is now considered his most important lecture. On February 9, 1849, he delivered "On the Nature of Limbs" to members of the Royal Institution of Great Britain. This launched Owen's earnest attempt at developing a structuralist version of biological evolution.[54]

But Owen's evolutionary views are only part of the story. It was William Daniel Conybeare, Christ Church clergyman and an active geologist and paleontologist, who suggested that the archetype could be reinterpreted in a way that could Christianize and Platonize the science of vertebrates. Others, such as John Duns, professor of Natural Science at New College (Edinburgh) and one-time editor of the *North British Review*, believed that in Owen's hands the archetype became "the true Ariadne thread by which he [God] is guided in the midst of what, without this thought, must have been a tangled maze."[55] And Owen used the archetype to its fullest purpose; the term appears fifty times in his comparatively short 119-page *On the Nature of Limbs*. Furthermore, he makes the Platonist connection almost immediately:

> "On the *idea* of limbs" might be understood only by those who knew the word was used in the sense it bears in the Platonic philosophy. . . .
>
> The "Bedeutung," or signification of a part in an animal body, may be explained as the essential nature of such part—as being that essentiality which it retains under every modification of size and form, and for whatever office such modifications may adapt it. I have used therefore the word "Nature" in the sense of the German "Bedeutung," as signifying that essential character of a part which belongs to it in its relation to a predetermined pattern, answering to the "idea" of the Archetypal World in the Platonic cosmogony, which archetype or primal pattern is the basis supporting all the modifications of such part for specific powers in all animals possessing it, and to which archetypical form we come, in the course of our comparison of those modifications, finally to reduce their subject.[56]

Despite Owen's clear and obvious connection of his structuralist theory to Platonist philosophy, others disagree. Michael Ruse thinks that Owen had no clear idea of exactly what the archetype was himself and used the term confusingly.[57] But here the "confusion" seems more in Ruse's mind than Owen's. At least the usage described above seems clear and unequivocal. Owen's biographer, Nicolaas Rupke, rejects Owen's Platonism because he believes the archetype to be an Aristotelian rather than a Platonic idea.[58] But Rupke is simply mistaken. The archetype is well represented in Plato's *Republic* and in *Timaeus* as Doherty has long ago demonstrated.[59] It might even be considered its hallmark. Whatever the historians may say, I see no reason not to follow a fairly straightforward reading of Owen on this issue. As Desmond points out, the archetype became Owen's sensible alternative to materialistic transmutation. Turning to the German idealists, it gave him "the 'primal pattern' on which all vertebrates were based. This was a kind of creative blueprint, what Plato would have called the 'Divine Idea.' "[60] In fact, Owen regarded the entire fossil record as "a progressive incarnation of the Ideal Archetype."[61] All in all, Owen stands as the best contemporaneous challenge to Darwin's materialistic emergent evolution after the manner of Boodin's own creation/Platonist process philosophy. We must leave behind our caricatured picture of Owen as an obstinate opponent of evolution in any and all forms and see him as he really was—as his modern biographer more constructively sees him—continuing the idealist tradition in science as a program (founded in German Romanticism) of autogenesis and homological research. He was turning the preformationism of Albrecht von Haller and Charles Bonnet via the epigenetic embryology of Caspar Friedrich Wolff into a feasible creationist theory.[62] Rupke fittingly portrays him thus:

> Owen can be given due credit for his contributions to the question of origin of species, without the need to ignore, belittle, dismiss, or distort him. After all, by the middle of the 1840s, he appreciated the importance of zoogeography in its bearing on the riddle of organic diversity, and he led the way in extending today's zoogeographical provinces to the fossil record. During the 1840s and 1850s his work in comparative anatomy and paleontology became the cornerstone of the evidence for evolution. With little more than a flick of the fingers, Owen's archetype could be turned into an ancestor.[63]

But, more importantly, Owen's Platonist archetype was *not* blind. Rather than evolving by what John Herschel called "the law of higgledy-piggledy," Owen's biological forms had an implicit direction, a teleological significance.

This type of structuralist evolutionary theory was carried on well after Owen. For example, it could also take on a neo-Aristotelian form as it did when morphologist D'Arcy Wentworth Thompson published his massive tome, *On Growth and Form*, in 1917. Heavily borrowing from Greek tradition, Thompson quite clearly invokes purpose in nature at the outset, saying, "Like warp and woof, mechanism and teleology are interwoven together, and we must not cleave to the one nor despise the other; for their union is rooted in the very nature of totality."[64] Today structuralist evolution is alive and well with physician/biochemist Michael Denton, who revised his initial Darwinian skepticism to include an Owenite form of structuralist evolution that he unashamedly admits "might be seen as a first step back to teleology and the notion that the laws of nature are 'intelligently' fine-tuned to generate a set of life forms on earth up to and including mankind."[65] The point is, structuralist evolution is thematically inclined toward a Boodian approach to biological change.

Quite clearly Boodin's *Three Views of the Universe* offers much to think about, and its applications using his framework appear rich and plentiful. But he knew a book with such a formidable title and an ever more formidable girth (over five hundred pages) would be forbidding for all but the most committed reader. Therefore, he wisely provided a more accessible companion volume as his "personal synthesis" to what was, for him, ultimately a religious question. The next section presents his message in a less detailed and technical format for a larger audience. It is in every way a logical outgrowth and expression of this much larger work.

God and Creation

Written in 1934, this companion volume sets forth Boodin's ambitious goal of offering a genuinely fresh theology for his time. If *Three Views of the Universe* has all the marks of a scholarly treatise written for a few, this book has the emphasis and tone of a public manifesto written for everyone. Boodin is, as we have seen, very sympathetic with the Platonic tradition that infused so much of early Christianity, and he regards Thomas Aquinas as the best theologian of Catholicism at its zenith. But Aquinas presented the best theology the thirteenth century had to offer; a new approach was overdue. Protestantism, according to Boodin, offered little to replace the Church with which it battled. It substituted the Bible for central authority but offered up no theology of its own, merely interpreting it essentially as

Catholicism had. If this assessment is as shocking—perhaps more—than Boodin's categorization of Whitehead as a preformationist, we should reflect on Luther himself as a telling example. As one historian has well remarked of this Protestant firebrand, "Though his theology was founded with trusting literalness on the Scriptures, his interpretation unconsciously retained late medieval traditions. His nationalism made him a modern, [but] his theology belonged to the Age of Faith. His rebellion was far more against Catholic organization and ritual than against Catholic doctrine; most of this remained with him to the end. . . . Theologically the line was anchored on Augustine's notions of predestination and grace, which in turn were rooted in the Epistles of Paul, who had never known Christ."[66]

Bereft of any new theology, Boodin argues that Protestantism served up a confusion of "sugar-coated" emergence, materialism dressed up in clerical garb, vague starstruck emotion, humanitarianism, secular envy, and a picture of God as Santa Claus. It is, for Boodin, at its best "a crazy quilt of new patches on an old garment." Seeking to save others, it found itself in need of its own salvation, something that can only be achieved "through a vital idealism which furnishes the inspiration for a new kingdom of man."[67] Instead it has chosen theatrics and circus spectacle to maintain a hollow relevance. Arriving on the scene more than a decade after his death, one can only speculate as to what Boodin would have thought of Vatican II, but it seems likely that he would have ascribed Catholicism's efforts at modernization as equally vacuous.

Boodin sees the situation as dire: "Materialistic naturalism has sapped the foundations of the old theology. There is a general confusion about moral standards."[68] The Great War brought only disillusionment, science and technology has replaced the economics of old, and atheistic communism triumphed in Russia with the threat of its globalization. It seems fair to say that nearly a century later, things have not improved. We have since witnessed a second World War, technological advance outstripping our human capacity to accommodate it, the replacement of Russian communist materialism with the calculated cynicism of an inner circle of geopolitical manipulators, and the widespread moral confusion that privileges individual and group rights over civic and communal responsibilities. The issues that prompted Boodin to write this book are only exacerbated today. Therefore, we might give heed to our poet-philosopher as he presents the nature of the problem, suggests the corrective cosmological view, and recommends a goal worthy of his cosmic vision. In other words, what kind of God should we seek?

The rest of *God and Creation* is an answer to that question. Since much of this book is a reiteration of material covered in its parent volume, a summary of its leading points should give a clear idea of Boodin's theology for the masses. These can be subsumed under three general headings: (1) the nature of God, (2) the nature of the universe, and (3) God's interaction with humankind. These three themes are, of course, interrelated, and it is difficult to discuss any single one as an abstraction wholly distinct from any of the others. It might be said that they ultimately combine to form a synergistic whole of divine creativity and love. Nonetheless, teasing out each of these theistic dimensions will elucidate that whole.

First, then, is the nature of God. For all of Boodin's talk of wholeness, it cannot be applied with a broad and indiscriminate brush. "The God of religion cannot be conceived as merely the whole of things. God must be conceived as an energizing spirit in the universe who furnished the inspiration for creating an ideal realm of values—a kingdom of heaven—in a distressed and struggling world." Put another way, "nature cannot become God though it adapts itself in a measure to God and though its order is the expression of the genius of God."[69] If some analogy for God is sought, music frequently serves the purpose with allusions to the "God-stream of energy" surging through cosmic space and time like music forming a harmonic symphonic whole. (One is reminded of Johannes Kepler's music of the spheres in his 1619 *Harmonices Mundi*.) We perceive this wholeness not from any *sola scriptura* dogmatism but "as science discloses."[70] God's action is imminent and transcendent in its extensive quality just as our soul is transcendent beyond our biological being. He is more akin to an incorporeal Mind than any kind of anthropomorphic image we can conceive, which is bound to be misleading.

The next aspect of Boodin's theology is the nature of the universe. God is a "dynamic structure" that operates through a "hierarchy of fields." Moreover, this dynamic structure expresses itself as a creative intelligence that possesses genuine—divinely prototypical—personality. But just as we cannot anthropomorphize God, neither can we biologize the universe like the Gaia hypothesis; too much of the universe is obviously inorganic, but the universe does appear toward an overall stasis.[71] Nevertheless, any reasonable cosmological theory must to some extent be animistic; this would be the nature of a cosmos overseen by life and mind. The key issue is to decide whether everything (to borrow from Whitehead) is just a footnote to Plato, in which case the universe and all that is in it is a creative reality of forms and structures held together in a transcendent teleology, or whether everything is just a footnote to Democritus (or if you prefer his Roman counterpart, Lucretius) with the universe "playing a blind game of marbles."[72] Rather than bouncing between extremes of materialist reductionism on the one hand and idealist—even mystical—panpsychism on the other, Boodin advises, "We must, with Plato, recognize two fundamental principles in the cosmos—spirit and matter."[73] And this Boodin has woven into a rich tapestry of dualistic pragmatism. In summing up the nature of the universe, it can be said that we live in an emergent world but not a *strictly* emergent world; there is correspondence among and between emergent histories and a cohering guidance for them all. They come together in a cosmic pluralism. Obviously Darwinian evolution thoroughly rejects this idea, but others who have balked at Darwin's positivism have not always done much better. One of the most popular Darwinian alternatives of Boodin's generation was Henri Bergson, whose vague ideas Boodin calls a kind of "theistic mysticism,"[74] which is not to his credit since he regards all mystics as merely reiterating mundane truths or uttering nonsense.

The third aspect of Boodin's theology is God's interaction with humankind. Like the good process philosopher that he is, Boodin never sees God as a one-directional Caesar dictating and decreeing to humanity. Instead there is always prevenient grace that stresses God and humanity as an interactive *becoming*. "Wherever you are, and whatever you are," he insists, "if you will assent and cooperate, He will transform your sordidness and create communion, harmony, and salvation."[75] Furthermore, the teleological nature of the universe is more than just a vague label; it relates directly to us, and in fact, in a sense, it *needs* us. "The divine Spirit always permeates everything," Boodin says, "but it is only where there is the will to respond that grace becomes efficient in the individual life. This is the mystery of creation and salvation. The goal of creation is not to make a perfect world but an effort of infinite patience to create spiritual value in the world."[76] In fact, for Boodin, religion is the creation of values and of building them into even nobler values. It is eminently human and built upon community. We will say much more about this in the next chapter, but Boodin hearkened to his "beloved teacher, Josiah Royce" for instilling in him the sense of belonging to "a universal spiritual community and that it is my vocation to participate creatively with the eternal spirit of truth, goodness and beauty, in companionship with all spirit that create in like manner, to spiritualize this temporal world."[77]

Here is Royce's Beloved Community and here rests two important aspects of Boodin's soteriology. First, its real-world directedness; second, its free, voluntary engagement in the spirit of Philippians 2:12. This does not take place in solipsistic self-absorption but interpersonally, building Augustine's City of God. In effect, this is not a state of being but a *process* of *becoming*.

This ultimately gives us an important truth about the interface between deity, humanity, and science. Darwin wondered privately in his notebook why our own thoughts couldn't be just regarded as secretions of the brain; his answer was, "It is our arrogance, it [is] our admiration of ourselves." But he forgot the "immortal vision of Plato," namely that humans (themselves creative) look for meaning in child-like amazement to a cosmos of creative genius. Seen from Boodin's perspective, this must rank as Darwin's supreme metaphysical failure, a failure so profound that despite his ostensible fealty to an ontologically crowned science, he no longer knew what its motivating spirit was. Far from arrogance, science should beckon us to "the poetry of life." We need not be ashamed of our rational capacities, for we have been made so by a creator who seeks to "enlarge the frame of our imagination and admiration, as well as of our material control. And it keeps before the scientist the main theme, that science is for man—not the dethronement of man, but the increased vision of man, power and delight of man."[78] Science is nothing if not *humanitarian*, and power is a sham unless exercised in the humility and responsibility of God-given stewardship. This is the Boodian spirit.

Ever the poet, Boodin ends by asking us to leave behind the old "hobgoblins and scarecrows" of our churches and instead join in a "community of divine spirit" and experience the transformative "laughter" of divinity. This is far from Calvin's TULIP or Jonathan Edwards's "Sinners in the Hands of an Angry God" Sunday sermon. Boodin has no interest in sixteenth-century Reformation theology nor in eighteenth-century sermonizing any more than he thinks the theology of an unrevised thirteenth-century Catholicism that has failed to take time seriously can be taken off Rome's life support to any fruitful spiritual end. Boodin seeks a religion in which people are united as a community seeking only goodness; one in which we are locked in a collaborative embrace of prevenient grace in which our destiny is yet to be created by us and by God. This is something very new, and it forms the final part of Boodin's trilogy.

Religion of Tomorrow

Here Boodin speaks to the community of believers and as such uses the more engaging popular language of religion. If *God and Creation* was a companion to *Three Interpretations of the Universe* for a more general readership, then this volume can be considered a continuation or elaboration upon it. Here, however, it takes on a more distinctly sociological tone. And so it could easily be argued that *Religion of Tomorrow* is the religious companion to *The Social Mind*, which came out only four years before. Either way, the point is that this title is tightly integrated with Boodin's broader concerns of science, religion, and community.

Calling religion "the poetry of life," our poet-philosopher sees it as integral to the human condition and to daily life. Royce seems to haunt most every page, and he invokes a common Roycean concept—*loyalty*—from the outset. Boodin is interested in providing what he regards as a "constructive faith" commensurate with a healthy and vibrant body of engaged and connected persons: Royce's *Beloved Community*.[79] While this is indeed to be a religion "of tomorrow," it cannot charge blindly into the future. Boodin is no iconoclast. Rejecting change for change's sake, he looks to an "ideal social order" cultivating a genuine communal spirit secured in the past but directed toward the future. Just as God and the universe is creative, this community must be animated—even impelled—by a spirit of constructive creativity. In a world that has replaced the cleric's collar for the scientist's lab coat, the new priesthood of the modern world, Boodin takes his position in a hostile world:

> I prefer to believe with the noble idealists of all time that our ideal yearnings, crude though they are in content, are homing instincts, orienting us to our Father's house; that in them somehow the universe which brought us forth gives us its intimations of our place and goal. . . . The world of sense, of solid mechanism, is but an island floating in the larger world of spiritual forces and deriving its direction and significance from it. We are not duped when we believe that the dice of the universe are loaded for right and reason.[80]

The worldview cast here is openly and unapologetically Christian and invokes Jesus Christ not as an abstract philosophy but as a "concrete intuition" for humanity. It is not Christ but the church that has gone astray. By separating the spiritual realm of love and grace from nature it "left the world adrift and opened the way for the triumph of materialism in a godless world."[81] And the Reformation only exacerbated things by further fragmenting the church, destroying what harmony might have been redeemed by opting for a cacophony of creeds resting upon an undiscerning biblical literalism that was dogmatic and thoroughly unfruitful. In words that bear repeating if only because they are as accurate today as when he wrote them nearly eighty years ago, Boodin pronounced a pox on both their houses: "While Catholicism is frankly medievalistic [sic], Protestantism halts between two opinions. This condition cannot help producing both intellectual and moral insincerity. When Protestantism has become frankly modernistic, it has lost the sacramental consciousness of religion and has drifted into a superficial intellectualism and secularism. It has thrown out the baby with the bath."[82] In so doing it paved the way for jingoistic nationalism, capitalism as merely economic materialism, and clannish rivalries and conceits. In such a world chance and necessity seem preferable over such direction gone awry; the problem is that none of this was God's intention but the folly of humankind. If the universe seems bereft of benevolent guidance and control, it is only because we have made it so. At heart the problem of evil is born of the misconception of God itself, a deity that functions with the absolutism of an emperor. As a process thinker, Boodin rejects such a concept. Rather, echoing Augustine, "God made us for Himself," and if so our cooperation is essential. Indeed, in true panenthic/process fashion, "There is but one ultimate reality and that is the mind of God. All our thinking and search moves in God—is God's becoming conscious of Himself in us."[83] In general, we have failed to understand this because our concept of God as an absolute ruler is false. Thinking only in terms of God's omnipotence, we have forgotten that part of being all-powerful is its careful and selective application. The Chinese philosopher Lao Tzu knew God best when he said, "Silence is a source of great strength." God's quiet beckoning and mutual interaction with the creatures he loves can be the greatest expression of his essential attributes. The church has often missed this.

What, then, is the corrective? Instead of constructing Christianity on some outworn feudalist pattern—the king and his fiefdom—it would be more accurate to see God as a creative energetic force "present everywhere by His essence and power," one that "emphasizes the infinite potentiality of personality" embodied in the example of Jesus who "enters into creative communion with us and participates with us in our striving for the best."[84] The problem rests not in his reaching out to us but in our halfhearted attempts at striving to reach him. It is not made any better by proclaiming libelous doctrines of total depravity or speaking of ourselves as passive elements being molded by a master craftsman. The old theology of Isaiah, Jeremiah, and Romans as put to music by Adelaide Addison Pollard in the 1907 hymn

"Have Thine Own Way Lord" is not the right way to relationship with God. Boodin admonishes, "Do not speak of yourself or others as common clay. There is nothing common in the universe when its full-powers are realized. Charcoal and diamonds are made of the same material. The difference is in the *process* [emphasis added]."[85] Rather, our relationship is most fully expressed in "sacramental communion," a "sacrament of citizenship" in the City of God. Instead of attempting to fashion a vision of God based upon an abstract self-existence, it should have been more than sufficient to cast God as love, for that is God's most fundamental essence. Referencing Plato, Boodin reminds his readers that love is not just loving what is good but loving the eternal possession of the good in an infinitely developing world. The key is love ignited and sustained in *personal* relations; a love that is genuine—not power-seeking and manipulative but a striving for goodness and what is best for others. Boodin would be appalled by the attempts all too common today of gaining approval by bandwagon statements of support for popular political, social, and cultural positions. This kind of vacuous virtue signaling is anticipated and decried by Boodin because when "the love of virtue becomes merely abstract and institutional, it loses its real significance and fails to affect and illuminate actual human life."[86] What is sought after is not egoistic individualism or equally egoistic tribal chants eliciting knee-jerk approbation but a nobler society composed of nobler individuals. "Not the individual alone nor social organization alone," he pleads, "but individuals finding their opportunity and incentive in a common life and a common life enriched by creativeness of free individuals—that seems to be the only workable ideal of humanity."[87] This is Royce's Beloved Community.

This is the essence of Boodin's religion of tomorrow. We must learn by living, a dictum that seems full of common sense but too little tried. It took a poet's spirit to tell us this, and eighty years later we are still working to attain it, which is all the more reason to put Boodin's titles—all of them—back on our reading list. But if this is not enough, we should look to the idealistic sociology that undergirds it, and this was provided just a few years earlier in *The Social Mind*, the subject of the next chapter.

Assessment and Conclusion

Rather predictably the trilogy (like all of Boodin's work) was greeted with mixed reviews. As companion volumes, *Three Interpretations* and *God* were typically reviewed together. Some of the objections seem strange. For example,

A Theological Trilogy   133

Clifford Barrett chided Boodin for his poetic appeal to "esthetic appreciation" that he felt detracted from its efforts at "detailed analysis" or critique, but neither book abandoned reason—especially *Three Interpretations*—and it is hard to know exactly what Barrett has in mind here.[88] Barrett also wants to know how the individual can maintain its distinctiveness and creativity while being at the same time (in his words) "never failing to abide within the pattern of the whole." Here the question seems misplaced, for Boodin's wholism is not a subsuming of individuality and creativity like some Hindu reach for Nirvana but rather quite the opposite, namely the only means by which it can be attained—by and through the realization of our fulfilled personality. That our "fullest" attainment can fall short Boodin never denied. Like Barrett, Charles Sawhill Virtue regards Boodin's careful historical examination of three evolutionary cosmic views as "intuitions" that are more "appreciative" than "analytic, poetic rather than logical." It seems amazing how just a little numinous phrasing, idealistic language, or emotive prose here and there can blind certain readers to all the detailed empirical information coming before and after.

But Virtue does raise an interesting question as to why Boodin demonstrates such antipathy to Whitehead given their many similarities. And as mentioned before, that is probably precisely the point. Boodin must have felt the need to distinguish himself from the more famous and followed Whitehead. Some of their similarities have already been mentioned, but clearly Boodin's interest in science and philosophy accords very well with Whitehead's "flights of speculative imagination," to extrapolate from science into metaphysics, that are confirmed by concrete experiences that form two points of empirical perspective. These two broad coinciding projects leave Boodin in similar—though not identical—directions toward a teleological logos. One example is worth noting and is significant for pointing out two things: first is that these differences for Boodin were real; second is that they mattered. In one of his later essays, interestingly titled "Fictions in Science and Philosophy," Boodin takes issue with Whitehead's view of experience as "occasions" that he uses to construct categories of "connectedness" in all of nature, a move Boodin believes Whitehead makes to avoid dualism in nature. Instead Boodin is convinced Whitehead has made commonsense experience an extensive abstraction. "There is no stuff of experience out of which individuals are constituted. Experiencing is a relation," he argues, "not a substantive. It is the substantival conception of experience which makes the relation of experience to its world so mysterious for Whitehead. . . . The paradox is of Whitehead's own making. It disappears if we view experience as relational or prepositional (to use J. Loewenberg's expression) instead of as substantival."[89] Noting that epistemological dualism is unavoidable, he insists that we must

> assimilate the evidence of the larger world into the context of our interpretation. And that effort will always persist so long as we think about our world. The only way to transcend thinking is mysticism, into which F. H. Bradley finally found an escape; and, as I understand Whitehead, he agrees with Bradley's final solution. If I have used Whitehead as a type in discussing the concept of experience, it is because he is the most influential representative at present of the view which I am combatting.[90]

Giving other examples, Boodin concludes, "Whatever value Whitehead's method of extensive abstraction may have in mathematics, it has no relevance to metaphysics."[91]

There is a danger in ending on this somewhat sour note because in so doing it risks overemphasizing the differences between the two philosophers. A few years ago the Library of Congress commemorated the acquisition of an interesting six-page letter from Whitehead to his student Henry Leonard written in 1936. The ensuing symposium brought together Whitehead scholars from many disciplines. One of those, Roland Faber, summarized the letter thus:

> When, in his letter to Leonard, he seems distressed, even annoyed by the tenets and even the very inception of positivism, it is so because of the missing imagination and the muted sense of depth and unprecedented future. If reason becomes reduced to that which is, and the "is" becomes reduced to that which only a method of exploitation through narrow concepts imitates as a purely material reality, one perpetuates not only a reductive primacy of being over becoming, but moreover a reality of merely external particles of matter and a lifeless logic of their mapping—virtually everything Whitehead endeavors to overcome.[92]

In the end, concludes Faber, Whitehead saw the suppression of "aesthetic creativeness" as the principal evil laying ahead for humankind.[93] All of Faber's comments regarding Whitehead apply with equal strength to Boodin. The differences between these two prescient thinkers were more in the paths taken rather than the destination sought.

Focusing away from Boodin and Whitehead, others could be more positive. Boodin's long sympathetic colleague, Hoernlé, called his work "eloquent utterances of a philosophic vision fired with cosmic emotion."[94] Hoernlé thought he was at his best when acknowledging and explaining evil in the world, quoting Boodin's ability to reduce the problem of evil, pain, and suffering in one thought-provoking sentence: "The human spirit wants not only security, but opportunity to attain, and the risk that goes with it." And God was "not simply a theory, but an expression of living faith; not merely a system of thought reflectively elaborated, but an attitude of mind—a conviction to live by."[95] L. J. Russell praised *Three Interpretations* for its clarity of expression and the quality of its prose.[96] Perhaps no one, however, understood Boodin's two books better than his colleague at UCLA James Hayden Tufts, whose review is the most accurate and complete assessment of Boodin's religious cosmology that he considered "a substantial contribution" to the field.[97] When Tufts explains Boodin's cosmic construction he states more than he knows when he says, "In the organic world the field is dynamic and has reference to time. The whole-form must include the life history. The organism is a formative pattern." This idea will take on special significance in chapter 6 and the epilogue. But Boodin knew Tufts understood him. Just a few years earlier he dedicated *The Social Mind* to Tufts, calling him "a sane and sympathetic interpreter of human relations."

Finally is *Religion of Tomorrow*. Harold Bosley takes exception to Boodin's "glib" handling of Protestantism and dismisses it as a generalization too easily made.[98] Then, rather astonishingly, he accuses Boodin of failing to concern himself with the social aspects of instilling religion into human life. One wonders if Bosley even opened much less read the book! Oliver L. Reiser of the University of Pittsburgh felt the book's greatest problem was in harmonizing reason and faith.[99] But nowhere does Boodin ask us to abandon either one, only that faith must indeed precede reason, which the reviewer duly notes. Indeed, Boodin acknowledges that the universe is at heart a *reasonable* universe with God carrying on intelligent experimentation by and through our own experience. In the end, Reiser admits that such harmonization is certainly possible. His real problem is with what he considers Boodin's Christian exclusivity, something he regards as "cultural provincialism." But Boodin is writing about what he knows best; Reiser seems to confuse emphasis with exclusion. Besides, Boodin quite clearly states,

"The presence of God is not a monopoly of any caste or class. It knows no other apostolic succession than the communion of the faithful. It respects no artificial conditions of men."[100] This may fall short of perennialism, but it is much closer to that than to "cultural provincialism." Elsewhere Boodin was more emphatically perennialist when he wrote, "The great souls, the great leaders of man, have often been careless of personal survival. . . . Thus worked Socrates, Confucius, the Buddha, the Christ. They are greater for their complete abandon to the eternal human cause, for losing themselves."[101] In their recorded immortality they are real, not mere myths to us.

Perhaps the most interesting review of all, however, comes from Charles Hartshorne, if only because he is so intimately associated with process theology.[102] At this writing he had already distinguished himself with *Beyond Humanism* (1937) and *Man's Vision of God and the Logic of Theism* (1941). Like Hartshorne, he agrees that God is changing and transforming even as he changes and transforms us, but Hartshorne does not see souls in their immortality as separate identities from God and so he balks at the notion that we are "something entirely outside his own being." Hartshorne recognizes this as Boodin's attempt to avoid pantheism and suggests this could be avoided by admitting that God can be passive as well as active. But Boodin is well aware of God's passive humility in love, what he objects to is Hartshorne's confusion of soul and identity; we seek salvation not subsummation. Hartshorne's view would amount to an effective pantheism. Hartshorne demonstrates his complete misunderstanding of the book when he concludes the review by saying that Boodin offers only "monistic, 'pantheistic,' decidedly nonsocial idealism." One can only reply that Boodin in many places and in many ways has established himself as a pluralist panentheist whose pragmatic idealism is eminently social. One gets the feeling that Hartshorne didn't really read *Religion of Tomorrow* and preferred the skimming method that invariably produces impressionistic and inaccurate results.

In fact, it is precisely the social aspect of Boodin's philosophy that needs to be examined. In many ways it is the final capstone to all his thought, thought that started in concepts of time and space, wove its way into a complete metaphysic, developed into an evolutionary cosmology of creativity, and expressed itself into religious ideals directed toward society. Boodin's sociology of spirit is the focus of the next chapter.

Chapter 5

The Social Mind
Boodin's Sociology of Spirit

Boodin's principal aim in developing his own sociological perspective was to find a framework for constructively dealing with the problems of the world, a world that needed more than ever to think globally and yet personally about the problems it faced. In effect, one could consider Boodin's social philosophy a kind of blueprint for peace. It is, therefore, an unfortunate irony that Boodin's major statement in that regard, *The Social Mind*, was published the very year that Nazi boots marched into Poland in 1939, ushering in World War II. Boodin took note that "a momentous war has broken out in Europe" and that "no lover of democracy can be indifferent" to its outcome. As mentioned earlier in this book, the specter of war haunted this eminently peaceable man and cast an obscuring shadow over his major metaphysical work in 1916, *A Realistic Universe*. But these dark coincidences were certainly brightened by the illuminating mark of his mentor, Josiah Royce. Royce appears in Boodin's writing—especially *The Social Mind*—as a friendly ghost, a constructive force in the social turn of his student's thought. The nature of that influence will begin this chapter. But the essence of this chapter will examine Boodin's sociology through five organizing themes—mind, participation, values, cognition, and will. It ends with perhaps Boodin's most pressing sociological question: Is progress real?

Royce, the Friendly Ghost

There can be little doubt that social thinking was Boodin's Roycean inheritance, and so it would behoove us to examine a bit more carefully what that means. First of all, this emphasis upon society and social relations is in marked distinction to William James, who emphasized the individual. James was more interested in "the mind" and its function in the world than in minds and their social interactions. Just think of James's definition of religion. For him, religion meant "*the feelings, acts, and experiences of individual men in their solitude, so far as they apprehend themselves to stand in relation to whatever they may consider divine* [emphasis in the original]."[1] For James, churches, temples, mosques, and other religious organizations represent secondary emanations of this primary individualistic motive. The turn away from the corporate caused James to regard himself as a "pluralist mystic" who was skeptical of wholeness.[2]

Where this came from is hard to say, but life with his mercurial father Henry was not exactly suited to forming strong communal bonds. In a six-year period, for example, Henry sent young William to at least nine different schools, some at home and some abroad.[3] Such a nomadic education was not likely to foster deep community roots. Royce's background couldn't have been more different. Far from the cosmopolitan centers of America's eastern seaboard and Europe, Royce grew up in Grass Valley, California. In the year of his birth (1855) it had 3,500 inhabitants and was fast growing into a town occupied in quartz gold extraction performed in deep mining operations. This gave Grass Valley a distinctly industrial look and feel. Indeed, Royce's biographer, John Clendening, says that this burgeoning town sitting roughly midway between Sacramento to the southwest and Carson City to the southeast "presented to a boy interested in the growth of new communities a lifetime of future thought."[4]

This sense of community was fostered largely by his mother, Sarah, who kept the family secure and cohesive amidst a rough-and-tumble mining town and a traveling salesman husband. In fact, the little, tidy white home she maintained in Grass Valley (named Avon Farm after her mother's Shakespearean birthplace) provided a serene sense of the English countryside, a cozy contrast to the surroundings of the American West. Within this cocoon young "Josie" could feel isolated at times, although mostly he felt the familial comforts of hearth and home. Within such nourishing soil Royce's Beloved Community could sprout and grow. It might be added that the tumultuous upheaval of the Civil War, the assassination of

Lincoln, and the conflagration of nations in World War I forced Royce to constantly witness the building up of communities only to see them torn down. Such dichotomies doubtless led Royce to see the immediate need of instilling the concept of community and of promoting loyalty as the virtue that could become its lifeblood. Should we wonder, then, that the Beloved Community became, in the words of one of his closest readers, "Royce's most important living ideal"?[5]

What exactly is this living ideal? Some think that the idea of community in Royce emerged only with his *Philosophy of Loyalty* (1908), but it can be readily found as early as 1885 in *The Religious Aspect of Philosophy*.[6] Its animating spirit is loyalty, a concept that Boodin considered "beautiful and convincing." Loyalty, as Boodin explains, is not merely an emotion but a type of conduct that (quoting Royce directly) "consists in the fact that it conceives and values its cause as a reality."[7] Put another way, loyalty becomes a *living* ideal in its unity. "The unity of the world is *not* an ocean in which we are lost," writes Royce, "but a life which is and which needs all our lives in one. Our loyalty defines that unity for us as a living, active unity. We have come to the unity through the understanding of our loyalty."[8] When clearly defined this becomes the Beloved Community of shared memories, shared goals and aspirations, and shared commitment to interpret and refine those commonalities in meaningful and collaborative interactions. Moreover, this Beloved Community is often best expressed in what Royce calls provincialism. Royce explains:

> For me, then, a province shall mean any one part of a national domain, which is, geographically and socially, sufficiently unified to have a true consciousness of its own unity, to feel a pride in its own ideals and customs, and to possess a sense of its distinction from other parts of the country. And by the term "provincialism" I shall mean, first, the tendency of such a province to possess its own customs and ideals; secondly, the totality of these customs and ideals themselves; and thirdly, the love and pride which leads the inhabitants of a province to cherish as their own these traditions, beliefs, and aspirations.[9]

Boodin certainly understood this: "We are made by instinct and capacity primarily for small unities. Happiness must literally begin at home—in the family and neighborhood. But we must be careful to not to cultivate family selfishness or local selfishness or even national selfishness. Such egoism has proved tragic."[10] Provincialism is not to be understood as exclusionary clannishness and cliquishness nor as ideological sectarianism. And Royce clearly distinguishes the provincialism of his Beloved Community from "old sectionalism."[11] Neither Boodin nor Royce were calling for the kind of reactionary nostalgia witnessed in the Southern Agrarians, for example.[12] In fact, quite the opposite. It was Royce's idea of the Beloved Community that formed the central feature of Martin Luther King Jr.'s philosophy of nonviolent social change.[13] It was through E. S. Brightman that a finer point was put on Royce's Beloved Community for King. Brightman limited the idealizing activity of God by taking "the world as it is" as part of God's transformative plan. If change is to be had, it is to come by and through real world conditions, and no one acted on this more effectively or momentously than did King. As Boodin put it, referencing Brightman approvingly, "Cosmic genius works within a world which is not just a logical abstraction of structure but where structure is conditioned by empirical variables."[14]

While Royce was unquestionably a positive influence on Boodin—the wellspring of the poet-philosopher's sociological inclinations—and gave him the communal lens through which cosmic genius could be pursued even if not ever fully achieved, it would be wrong to suggest that Royce's Beloved Community was appropriated wholesale and uncritically. For example, Royce's triadic formula of knowing between Individual A + Individual B mediated via an Interpreter = Community of Interpretation (a Hegelian trait that Royce frequently adopted) formed the epistemological basis for his Beloved Community that Boodin found unimpressive. For Boodin this triadic understanding was too artificial, "the product of the particularization and substantiation of language."[15] Furthermore, Royce's triadic view of knowledge as perception, conception, and interpretation Boodin found too abstract as applied merely to the individual; these must be revised into the human experience of social knowledge by adjusting ourselves to a larger whole of three cognitive judgments that are perceptual, conceptual, and mystical.[16] More broadly, he clearly had Royce in mind when he wrote, "The difficulty with idealistic theories in general [Hegel and Royce included], in spite of the fruitfulness of their empirical intuitions, is that they have been so anxious to arrive at the Absolute that they have slighted the concrete problems of continuity. The Absolute becomes an immense solipsist, with no alter."[17] Boodin also felt in rather the same vein that Royce was too prone to reifying time, as in thinking that the future is something that can be addressed as an external presence. The future is not something "out there" for Boodin, it is ours and based upon our own forward-looking experience.[18]

More will be said of Boodin's differences with Royce, but it is possible to carry those differences too far. Surely Royce would have agreed when Boodin wrote:

> Human nature has not fundamentally changed. It is constituted for a life of personal relations and hence is not satisfied to be a cog in an impersonal machine. By instinct, imagination, and sympathy, human nature is made for small groups, for face-to-face relations and hence is not at home in the artificial, de-humanized leviathan which, like Moloch of old, has only an instrumental interest in the human individual. It may fatten the sheep only in order to devour them. The old inhumanity of man to man has come to seem only more brutal when robbed of the old personal sanctions. And man, when delocalized and depersonalized, reverts easily to the anonymous mob.[19]

Here can be seen in Boodin those characteristics of Royce that are most engaging: the temporalist personal idealist with no little touch of pragmatic experience based upon humanity's natural inclinations toward meaningful, intimate community. While it is common to view Royce as the idealist and Boodin as the pragmatic realist, the fact is that the protégé-like mentor defies thoroughly singular categorization; their philosophies were too thorough and deep to be reducible to a single label. Such descriptive shortcuts are liable to mislead more than inform.

Whatever else may be said of Boodin's debt to Royce's concept of the Beloved Community and his strong commitment to seeing humanity through *social* eyes, he was surely a friendly ghost that haunted nearly every page of *The Social Mind*. And this would spill over into Boodin's religious writing. Just as the church formed an integral part of actualizing Royce's Beloved Community (indeed as it would for Martin Luther King Jr.), so too did Boodin see religion as a distinctly animating force for achieving the teleological goals of the larger cosmic whole—attaining God's will. His *Religion of Tomorrow* makes no sense unless seen as a manifesto for social action and social change in which Royce served as spirit guide and Boodin's silent partner. As we move forward to consider the five organizing themes of Boodin's social philosophy, it would be well to remember that the poet-philosopher's sociology was initiated by the hand of Royce.

A Sociology of Mind:
Participation, Values, Cognition, and Will

The Social Mind can be divided into five organizing themes: chapters 1 through 4 cover the biological basis of society, the nature of groups established as a distinct concept and a sociological principle, and what Boodin calls "social minds"; chapters 5 and 6 cover social participation and its relation to social systems; chapters 7 through 9 discuss values, cognition, and will, respectively. He then applies these to specific social conditions as they appeared at his writing. Of the fifteen chapters, only four ("The Group," "Organization of Will," "The Crisis," and "The Idea of Progress—A Survey") are entirely new. Boodin's earliest foray into sociology came with his essay "The Existence of Social Minds" in the *American Journal of Sociology* in its July 1913 issue and may be considered the centerpiece of *The Social Mind* published thirty-six years later.

It might be well to pause here and ask, What personal influences were brought to bear upon "young Elof" (as Boodin called himself years later)? Far from the European centers of cosmopolitan life that James experienced and equally so from the industrial boom of Royce's Grass Valley, the boy would learn the pragmatic realities of farm life imbued with the mystical idealism of Swedish customs and folklore. It is interesting to see how rural Sweden (ancient and stable) and the mining-centered American West (new and restless) produced such similar social-minded philosophers, philosophers brought together in the heart of New England's staid, stratified, and formal world of scholarship. Fortunately, Boodin left a fairly long and detailed account of his early life, revealing the Roycean provincialism of his own personal Beloved Community. Like Royce, Boodin could appreciate the Beloved Community because he had lived it, and the fact that he could demonstrates its plasticity and its resistance to any kind of constricting formula. But it also reveals its fragility, for the self-sufficient farm of Boodin's day vanished before his very eyes, an experience recounted in rural America today and memorialized in Iris Dement's 1992 nostalgically plaintive song "Our Town." Boodin's highly personal account is descriptive of a bygone era epitomizing Roycean ideals and is reprinted in its entirety in the appendix as " 'Remembrance of a Common Past' (A Glimpse of Provincialism and the Beloved Community of Sweden as Boodin Knew It)."[20]

With this as a backdrop we can delve directly into *The Social Mind*. As already stated the first four chapters give Boodin's exposition of social minds. By *social mind* Boodin means "the synthesis of individual minds into wholes, with new properties."[21] Ever the process philosopher, he places this within an evolutionary context that creates new patterns of new wholeness, a synthesis manifesting the genius of nature. Groups are the foundational reality for humankind because all of our higher attributes are actualized in group relations; in short, sociology precedes psychology and this occurs (as does most everything else) through energy fields. As such, all human conduct must be considered through its whole ensemble of internal and external influences and in so doing its teleological character is revealed in "its forward-looking life." Whatever relationship we may bear biologically to the primates including even the lower social animals (e.g., bees, ants)—all interesting parallels—pale against the fact that humanity's real history begins with its creative intelligence. Here social relations have pride of place in the larger scheme since the group lends its own creativity to the whole by adding new variables to cosmic evolution. If biological evolution has influenced *Homo sapiens*, our social lives have even more profoundly impacted cosmic evolution. Those who would "guide" the biological side of human beings through eugenic proposals—disturbingly popular in the first three decades of the twentieth century—are simply wrong since social conditions are far more important than heredity. The social realities of our existence add a powerful dynamic to the cosmic whole, making positive and negative eugenics—indeed all such reductive manipulations of human beings—not only misguided but dangerous.

In its most positive aspects, our social lives are held together through group loyalty and this should lead us toward a better world. This is not a group loyalty of coercive power over others but of meaningful collaboration and constructive interaction. Sadly, this has often not occurred, as witnessed in the plight of the African American experience. Boodin suggests in this regard that perhaps society itself is in need of psychoanalysis: "When people are willing to examine their class prejudices and their group hatreds in a critical and thoughtful way, a long step has been taken toward a new order. A psychoanalysis which shall expose our unconscious prejudices and introduce good sense into our social relations would indeed be valuable."[22] It's a dangerous proposal, however; Martin Luther King Jr. tried and was assassinated for it. Clearly Boodin wants to not only discern the central features of the social mind but diagnose its pathologies in an effort to make it well or at least make it better than it is.

It is important to realize that Boodin is not abandoning individual experience for some sociological or socialistic idealism. Rather he is building in a new and thoroughly creative dimension through the concept of the social mind. He offers a virtual thesis statement for the entire book when he states: "What I wish to show is that there is a genuine social unity, distinguishable from what we call the unity of individual experience, and if not more real, at least more inclusive than this. The latter may be considered from this point of view as a group of constant traits which we identify in a variety of social situations. What we have in reality is dynamic situations."[23] Instead of starting with individual minds as in psychology, Boodin wants to emphasize intersubjective continuity as an elementary fact. Mind as much as matter must be conceived as a phenomenon of fields existing with their own continuities and their own influences. In this sense mind is not a state of consciousness but an energy. Common purpose and common impulse along with reciprocity and mutual sympathy—for better or for worse—are the stuff of which the social mind is made. When those common purposes and impulses are exclusionary and tribal in nature (it must be realized that tribalism can take many forms, whether in the remotest jungles, the asphalt jungles, or corporate board rooms) we have the makings of social pathology.

Boodin ends his section on the mind by calling the reader to something more than loyalty; he calls for critical and constructive creativity because that attribute perhaps more than any other reflects the animating spirit of the cosmic whole. We must not in our predilection for statistics and mathematics reduce ourselves to mathematical points but see ourselves as dynamic points—centers of initiative—that take loyalty that can make us "share in the great, warm living stream of humanity" and add critical creativity that "can make us a part of the eternal direction of history—prophetic of the kingdom of heaven."[24]

In the next two chapters Boodin deals with participation and social systems. Here he relentlessly attacks behaviorism, which was becoming ever more popular. Here again reductionism rears its ugly head of which operant conditioning, that behaviorism is so fond of, is its most glaring manifestation. Boodin calls this a "jumping jack" theory of human behavior. His assessment is clear: it is not individual behavior that matters but *social* relations that take on social patterns that are of prime importance. Behaviorism treats humans precisely as they should not be treated, as physiological automata. Of course, the fact of social participation can hardly be doubted, but how we participate socially becomes the key. Boodin cautions against thinking we have any advantage over our ancestors in this regard. Can we say, for example, that our acceptance of Darwinian theory is anything other than the fashion of the day? Is the much-touted social virtue of self-expression better than forbearance and self-restraint? Does the suggestion of the crowd offer superior guidance over institutional sanction? Are we self-deluded in thinking we are following our own determined course when, in fact, we are little more than sheep following the herd? These questions are even more relevant today than when Boodin wrote them; they require careful contemplation in an age that seems to value information over being informed, communication over communion. Boodin was right in asking them as a coherent gauge to our own social participation. If this is the measure by which it is gauged, how might it be taken in principle? The answer lies in understanding human belief and conduct in their social matrix, which is the temper, attitudes, and culminated experience of the group and age in which it exists.

Here is where the principle (Boodin likes to call it somewhat grandiosely a "law") of social participation gets interesting. This explanation of social participation seems rather close to the concept of *mentalité* so intimately connected with the Annales school of French historians. The source of this insight might well have come from Boodin's invitation to attend the Ninth International Congress of Philosophy held in the summer of 1937 in Paris. There Boodin shared a session with Marc Bloch, one of the chief architects of the *mentalité*.[25] Like Boodin, the Annales historians understood that we always participate in a number of groups—economic, social, religious, cultural, political—and we find our identity and realization through those groups mediated by their temporal and situational prominence. Thus we find ourselves amidst social constructions. One is reminded of Philipe Ariès (although his most important work came after Boodin) and his study of childhood and later on death, where these natural phenomena are socially constructed in their respective environments through time.[26] For these historians, as for Boodin, time took on significance and meaning not simply as some serial sequence quietly ticking in the background of events but as real living forces based upon social experience. It wasn't until the early 1960s that the medievalists Jacques Le Goff and Georges Duby joined in and refined Lucian Febvre's discussion of "floating" or "imprecise" time during the *longue durée* (to borrow Braudel's phrase) when the hour was estimated by the position of the sun and most people couldn't tell you their exact age.[27] Until then, few historians took time seriously as an operative force in social life that was itself variable; before the Annales school it was largely a serial template upon which events were arranged. Bloch argued that "human time will never conform to the implacable uniformity of clock time. Reality demands that its measurements be suited to the variability of its rhythm, and that its boundaries have wide marginal zones. It is only by this plasticity that history can hope to adapt its classifications."[28] Boodin's social philosophy and concept of social mind might be fruitfully associated with the Annales historians' *mentalité*.[29] It is interesting and suggestive that perhaps Boodin had the larger theoretical and philosophical foundation the Annales historians were always looking for but never quite found. The Annales school was always by its very nature interdisciplinary and had some influence on the continent, but the British were mostly hostile and Americans generally ignored it. But its eventual fragmentation was due in no little part to its lack of a firm theoretical foundation. Whether Boodin's social philosophy might provide it is a matter of further study.

Social participation consists of relationships, and they are first and foremost energy relations rather than things. We have not understood the nature of mind—much less *social* mind—because the Renaissance became fascinated with the physical instead of the social field. Therefore, empirical features of the natural world that operated with regularity (or at least seemed to) were cast as "laws" impressed like edicts upon the cosmos. But Boodin is dubious of the term, admitting that "they are rarely capable of that exact statement which we associate with the term 'law' in the physical sciences."[30] Boodin more modestly suggests that social tendencies can be more generally formulated into fairly definite propositions. Those propositions are the conservation of energy; the principle of least action; action and reaction; inertia; social integration, through-and-through interactions as an interpenetration of common purposes, common bonds past and present; the principle of rhythm, socially and cosmically oriented rebounding between states of equilibrium; equilibrium; degradation; relativity, or social states conditioned by the age in which we live and groups within which we interact; and the principle of parsimony or economy based upon trial and error. These comprise the principal features of interacting social systems. However these may come into play, they are not like cogs and wheels. They are not amenable to any mechanical analogy; social systems are malleable, process-oriented influences upon our groups and ultimately upon us as individuals.

All of this comprises the dynamic categories of social evolution. Variation, selection, and adaptation are its dynamic features. Humans add an additional aspect of moral freedom that includes not just spatial freedom but also time aimed at the future. The creative fact of the whole-form is a space/time interaction in which moral freedom is realized when complexity is harmonized with an ideal commensurate with "the good." The further from this ideal, the closer to pathology we get. Because of their complexity and dynamically changing nature, we should be duly suspicious of applying exacting formula to any of the social sciences.

Our moral freedom is naturally bound up with our values. Boodin defines value as "the congruity of an object of activity with the organized tendencies of the will which seeks realization in terms of the special situation."[31] Society standardizes values for us by its systematic purposes. In this John Dewey and James Hayden Tufts's *Ethics* (1932) is credited with its emphasis on congruity and makes many contributions to value theory. Values are real qualities, giving us genuine insight into the character of the world in conjunction with our wants. Types of values must also be considered. Some values are naturalistic, others are formal. The former comprises biological urges, physical exercise, human interactions, sexual desire, desire for food and the joy of eating, and a host of largely unorganized impulses; still other naturalistic values are utilitarian, such as monetary exchange, supply and demand, and other economic factors of life. Formal values are cognitive activities that search for and seek achievement of truth; certain social, religious, or patriotic beliefs; the yearning for beauty; our ethical and moral precepts. An absolute hierarchy of values is impossible to construct since they vary with peoples, places, and eras. Some are ends in themselves, others are means to ends.

Boodin makes clear that values are derived from and in pursuit of the unity of human nature. In this sense, values are found in human reason, human character, and human sentiment; they cannot be compartmentalized but are involved in the conscious realization of the whole self. Our poet-philosopher concludes, "The organized realization of value implies the temporal solidarity of humanity. In its ideal creativeness and self-criticism there is implied further the unity of humanity, somehow, with an ideal order, which draws us onward in our groping endeavor to realize ourselves as parts of an unseen future, and which makes us 'more than we are and wiser than we know.' "[32]

The next organizing theme is cognition, sociologically most importantly bound up with "the conscious use of experience in adjusting ourselves to a larger whole" of which there are three main types: perceptual (our practical acknowledgment of people, things, and their concrete space/time patterns with their causal implications), conceptual (experience sorted into kinds and types thus establishing expectancies on that basis), and mystical (the sense of larger meaning—even the numinous—in the blending of fluent fields with our full attention to the amassed fields).[33] Boodin further points out that the cognitive process, which is based in instinct, is fused with both presentational and intellectual human capacities. There is not an actual cognitive faculty per se but rather a complete unity of the self as part of our *real living* social process. The social community is never outside the interpretive process but is intrinsic to it. Past, present, or future, we are always part of the interpretive process.

But how are our socialized meanings to be judged in their cognitive contexts? They are validated in the degree to which they can translate the selected fact into its intention and meet the requisite need or situation. Here society itself aids in the correction, revision, and supplementation in its sympathetic cooperation with those meanings and this requires, of course, sympathetic participation. This extends beyond but is inclusive of scientific propositions, which involve a small part of our social meanings. Never taken with his colleagues' increasing tendency to parse words, he adds, "Language is important, but it is not everything. We should have something to say, strange as that may sound to the linguistic fanatics in philosophy. The real drive in our thinking is the feeling for consistency and harmony, not the mere verbal syntax."[34]

The poet-philosopher keeps his feet firmly planted on the ground by rejecting epistemological idealism. It is true that all facts are in experience and find unity in incorporation with our existing body of ideas (Leibnitz's *apperception*), but all existence is not necessarily experiential. Things are selected by us for the interest they generate, but things exist apart and a priori to our interest and they do not by all appearances have their own interests. Things do not require our direct experience to be real, and it is not at all clear that objects have significance for us only as chosen by our interest. These three propositions of epistemological idealism lead to the fourth, namely the unities of interpretation between us suggests a larger unity of interpretation by which and through which all experience exists. This may be expressed as a continuous self-transcending temporal community or as Royce's timeless Absolute. Boodin takes particular aim at the notion that things require our interest to be real. "Even [George] Berkeley, in spite of his paradoxical approach to the problem," writes Boodin, "did not believe that the stellar world ceases to exist on a cloudy night." And what of that transcendent unity? The subject-object relation of our reflective thought is by no means a cosmic requirement; there is no empirical proof of such a proposition. In short, nature is independent of our interpretation of it. The only reasonable alternative would be to succumb to some panpsychism, an alternative Boodin rejects as lacking any proof. The certainties of an idealist Absolute or something akin to it, held together only by the gossamer threads of wishful thinking, are not enough. A cognitive view of the cosmos finds certainties elusive. Plato knew this. We must in our mortal lives claim only probabilities and act on those probabilities *as if* they were real lest we be frozen into complete inaction. "We must be critical of our limitations or we become victims of the will-to-believe," ends Boodin on cognition, "and we must be tolerant of divergence of interpretations or we may fight against the progress of truth."[35] We must be modest interpreters of our cognitive "certainties."

In dealing with will as an organizing theme, Boodin quite expectedly starts with Schopenhauer, whose *The World as Will and Idea* (1819) stood upon the shambles of Napoleon's will to power. Schopenhauer had a stark and frightening reality for the milieu in which Boodin was writing. In fact, the Nazi's goosestepped across Poland as *The Social Mind* ran off the press with Schopenhauer's "will" with them, for Hitler held him dear along with Nietzsche, Wagner, Houston Stewart Chamberlin, and Julius Friedrich Lehmann. Hitler rejected Schopenhauer's pessimism but greatly appreciated his dismantling of Hegel's idealism. Insofar as Hitler had a religion, Schopenhauer was no small part of it. Hitler was no intellectual and even less a philosopher, but he was a reader who treasured his five-volume set of Schopenhauer's works he kept since World War I, and he was surely influenced by the odd percolating stew of racist nationalism that Schopenhauer's ideas contributed to in the brewing discontent of the Vienna and Munich with which he was familiar.[36] But ideas not carefully considered can have devastating consequences. For Boodin, Schopenhauer's complete ignorance of space and time rendered his philosophy a useless mysticism. He calls Schopenhauer's philosophy, "a paralytic carried on the shoulders of a blind giant,"[37] which is frankly an apt description of Hitler's Germany.

Boodin has better things to say about the will. He points out that the evolution of civil society through the establishment of moral control—the creation of practices and beliefs for the good of social welfare—was humankind's greatest and most momentous step forward in history. Not surprisingly biological reductionism has tried its hand at explaining this feature of human sociology. Most of this postdates Boodin, coming from William D. Hamilton and George Price in the 1960s and 1970s with no little popularizing by Richard Dawkins's *The Selfish Gene* (1976). Kin selection (the favoring of reproductive success with one's relatives, sometimes even at the cost to an individual's own survival often taking on the appearance of altruistic behavior) along with Price's equation spelled out the precise conditions under which the interests of the group could trump the interests of the individual.[38] This could be used to explain the origin of the human family and other group dynamics. The problem is, Price based his work upon R. A. Fisher's theorem of the mutation/selection process (basic to neo-Darwinian theory) and that theorem has been shown to be problematic.[39]

All of this is to say that Boodin is right in viewing the will to live and to live socially as a philosophical problem. This is not to say that biology cannot enter in, but perspective must be maintained. Even Oren Harmen in his sympathetic biography of George Price notes,

> Most human behaviors have nothing to do with natural selection in the narrow, biological sense. . . . Selfishness and altruism in human affairs have nothing to do with tadpoles spitting, mole rats digging, and cuckoos sneaking eggs into stranger's nests. Kindness is kindness only if it is meant to be so. . . . There can be no doubt that social life in Nature put into place the substrates that would one day allow the birth in humans of something we call our moral sense: the policing against cheaters, the mechanism for conflict resolution, and the capacities for empathy, shame, jealousy, sympathy, and rage are all stages on this evolutionary journey."[40]

Agreed, but there is no warrant for capitalizing "Nature" into ontological significance; the real significance rests in the teleology implanted therein and thoroughly examined in *Cosmic Evolution*.

Having dealt with some of the biologizing that has accumulated since Boodin wrote, it is certainly accurate to say that however society may have emerged, it did so by means of human intelligence. Boodin highlights the importance of common memories and creative planning of which moral controls were willingly applied. But the treatment of the organization of will is not solely an anthropological question. These ancient social mores and morals have been too incompletely and inequitably applied. Responsibility for the freedom and self-actualizing needs and desires of society must extend from all to all, or as Boodin puts it, "Freedom of the parts must be subordinated to a sense of responsibility for the whole. This is a lesson which democracy must learn or perish."[41] We must become reciprocally socially minded. The ideal social will would organize society so as to meet its greatest creative capacities, for we are made to be creative. It shouldn't be difficult to see that to be creative we must be free and the individual must be recognized but not turned into an *ism* since "individualism" is most often code for exploitation and selfishness. Here ends the five organizing themes of *The Social Mind*.

Here Boodin takes up specific social conditions. The first he calls "the crisis," which in a word is capitalism. Writing in between the disaster of the Great Depression on one side and the explosion of World War II on the other, he is not hopeful for democratic capitalism. But the real issue is not capitalism but humanity. Here Boodin calls for democratic collectivism. This needs to be applied not piecemeal but coherently and comprehensively, even if it must drag business interests along with it kicking and screaming. He remains hopeful for the American nation as a whole, although the poet-philosopher is cognizant of its many problems:

> There is a destiny of the United States of America. It is a land of great variety of climate and resources of nature. Yet it is one people [or should be], forged through stress and strain into one nation. It is a people which has sinned much—sinned against the Indian, sinned against the Negro, sinned against itself—in its ruthless march and yet with a strain of idealism running through its history. It had danced around the golden calf, yet it has never quite forgotten its God. It is a people that has suffered for its sins and must suffer that it may find its soul. It is for us, the living, to enter into the genius of our nation and to create out of its tradition, its resources and problems, a new future, expressive of our noblest humanity.[42]

More than eighty years later we are still searching for our soul. Still, the general outlines of this realistic idealism seem true but perhaps more challenging, not because the problems are necessarily any bigger but because our best tools to meet those challenges as Martin Luther King Jr. understood so well—our ideals and our faith—have since been dulled by an overarching cynicism and skepticism that threaten us as a cancerous pathology. If Boodin was uncomfortable in his world, he would be even more so today. And for good reason. This raises perhaps the most pertinent question of the entire book.

Is Progress Real?

Boodin manages to give a coherent and informed history of the idea of progress in Western civilization in just twenty-seven pages—no small feat for a writer not always given to brevity. While registering some skepticism over John Bagnell Bury's cyclical interpretation of history, he uses his interest in the idea of progress as his guide. While progress is a recurrent theme in Greek thought, it had little influence on Greek historians like Polybius, who emphasized what he called "pragmatic history." Machiavelli reintroduced the concept and Giambattista Vico first systematized cyclical progress in the first half of the eighteenth century. Oswald Spengler investigated the question of progress in his massive two-volume *Decline of the West*. Boodin offers no simplistic answer to progress, saying much depends upon the interest or perspective of the investigator. Two things are clear, however: (1) human history is part of a cosmic whole and (2) it moves through the rhythmic impress of our fallible passions and aspirations upon the canvas of the past. The image we paint thereon is ours to make. Destiny must be considered too, but we are not its puppets; they are made for creativity. In any case, continuous advance may be discerned in biological, technological, and spiritual evolution. But it is not always clear-cut. Modern industrialism, for example, looks like technological progress although it comes at the expense of *social* progress. It is easy to see industrialism as the "new feudalism" based upon the direct or indirect control of the means of production. We can blame technology, but the problem is not our mechanical but our social machine.

Generally speaking, Boodin answers the question of progress in the affirmative. Interestingly, he sees this as a logical outgrowth of evolution itself. For humanity, evolution will be along the lines of greater social and moral freedom, a freedom of cooperation and comradeship because evolution thrives on harmony and cooperation, these attributes being the strongest forces in ensuring survival. The goal of progress is not merely advance but preparation for and the facilitation of creativeness. If we are to achieve it, it must be through our efforts to discern the path forward, and it will come only by the hard "experience of trial and suffering," that is a "time-form and not merely a space-form" that must be a living discovery. What Boodin means by this is that we have been content to theorize about life and go no further than analysis. Our philosophizing "has erected its abstractions into timeless absolutes and forgotten their relativity. To try to substitute abstract categories for first-hand living is like substituting a menu for a real meal, to use an expression by William James. The categories and the menu are only significant as instruments in concrete realization."[43] Here Boodin implicitly recognized James's philosopher's fallacy (i.e., Whitehead's fallacy of misplaced concreteness) and departs from Royce's "universal world time" as an expression of "the finite will."[44]

Is there anything like an applied sociology of progress? In other words, is there a practical way forward to achieve social progress? Boodin is doubtful of any biological solution. He is skeptical of eugenic programs, which in 1939 still had some academic and political popularity in America. He argues that social and environmental factors are of far more importance than attempts to arrive at "good breeding." We must have a more religious inspiration that strives toward the unification of humanity to the common welfare; in so doing the "cosmic spirit of creativeness" will work with us and through us for an attainment of a higher humanity because the evolutionary power of moral unity is stronger than "nature red in tooth and claw." Ultimately this can attain to immortality in the collective unity of a social mind that is commensurate with universal and eternal purposes. In the West, even today the "Greek mind," the "Roman mind," the "Medieval mind," and the "Renaissance mind" are with us in our institutions, our schools, our churches, our cultural souls. They are with us in a thousand ways without our even thinking of it, just as the social minds of the past are at work in the East. Like time itself, social minds work that way. In this sense, Faulkner knowingly or unknowingly echoed Boodin when he wrote, "The past is never dead. It's not even past."

Assessment and Conclusion

Boodin didn't expect much notice from his colleagues of *The Social Mind*. He told Arthur O. Lovejoy that he thought his publisher would have a hard time soliciting reviews because "America is too busy celebrating 'its greatest philosopher' to pay much attention to my work." He was referring to John Dewey. He complained to George Albee, "I have not had students who have propagandized for me as Dewey has had. . . . My work has no doubt made more demands on the reader than that of Dewey's, Bergson's, or Santayana's. There is a good deal of historical accident about fame."[45] Of course "fame" and students can cut both ways. Dewey, who is in many ways best known for his educational reforms, did not receive the endorsement of at least one extremely bright and perceptive student. Dewey's progressive educational techniques enthusiastically employed at the Peabody Model School where Flannery O'Connor attended in Milledgeville, Georgia, elicited only her ire for its fixations on "planning" and for their neglect—even their dismissal—of classical studies. Provoking what Robert Fitzgerald called "her penetration and her scornful humor," O'Connor complained, "They [the teachers at

PMS] would as soon have given us arsenic in the drinking fountains as let us study Greek. I know no history. We studied that hindside foremost, beginning with the daily paper and tracing problems backward." Fortunately, she added, "I'm blessed with Total Non-Retention, which means I have not been harmed by a sorry education."[46] Although Boodin respected his colleague, such comments in an age when Dewey's faults were few would have sweetened Boodin's sour grapes for the pragmatist considerably. In any case, Boodin's concern that his social philosophy was as neglected as his epistemology, cosmology, and metaphysic seems excessive. By comparison, this publication was widely reviewed.

For example, Clifford Barrett gave *The Social Mind* favorable notice in the *New York Times Book Review*, while Ohio State University philosopher Joseph A. Leighton had some reservations but concluded that Boodin had given "a full-blooded expression of a socially creative idealism of values of persons in community." Moses O. Aronson, writing as editor of the *Journal of Social Philosophy*, approved of Boodin's "spiritualistic naturalism" that showed "there truly exists within the framework of nature a cultural level of reality." Privately, Aronson wrote to Boodin praising him for refusing "to pander to contemporary schools [of philosophy]" that he thought was "a mark of true greatness."[47] A. D. Ritchie, writing for *Philosophy*, presented several objections and concerns for Boodin's theoretical formulations but in the end came down favorably, saying that the author "deals with the discontents of the present age and its peculiar political problems in their historical setting." Ritchie also considered the treatment of progress "valuable" and regarded the book as an encouragement during the troubled time of war that "combines great learning, an objective civilized mind, and in addition good sense and sound judgment."[48]

Others, however, were more critical. H. M. Kallen, for example, was unimpressed with Boodin's "quasi-mystical intuition," and Newell L. Sims dismissed the book entirely by claiming that much of its sociology was outdated and he didn't think the beauty of style in the presentation made up for "the general antiquity of much of the subject matter to justify the publication of the book."[49] But one wonders if they were reading the same book as J. O. Hertzler, who admitted that the "group-mind concept has been pretty generally abandoned by social scientists" but nonetheless found Boodin's "manner of presenting the idea . . . a most stimulating one for sociologists." Hertzler was particularly taken with the fact that "group-mind" can often be described as "culture."[50]

But the assertions of Sims and Hertzler (negative and positive) regarding the viability of Boodin's group-mind in sociology needs more careful analysis. Here some broad comparisons and contrasts with Boodin's social theory can be made with that of George Herbert Mead. Mead was a later pragmatist, part of the so-called Chicago School that comprised James Tufts (to whom Boodin dedicated *The Social Mind*), Albion Small, James R. Angell, Edward Ames, Addison Moore, and most notably John Dewey. Interestingly, Boodin was completely silent on Mead. This is probably because of the leading pragmatists Mead was least known in Boodin's day. As Joseph L. Blau has pointed out, "Mead himself made no full-dress statement of his position."[51] Mead's obscurity would not last long however; Werkmeister devoted an entire chapter to "Mead's Philosophy of the Act" in *A History of Philosophical Ideas in America* (1949) and many detailed analyses of his works would follow.[52]

Unlike Boodin, Mead's ideas were deeply influenced by Darwin's evolutionary biology, as were Dewey's, and his philosophy may be distinguished from Boodin's in its denial of transcendence of any kind. His theory of the human mind sought to place it squarely *within* rather than *beyond* nature. Here the story might end, but Mead went further by rejecting traditional individualism in favor of a thoroughly social construction of human activity and of human mind. For Mead, minds and selves arise with the internalization of conversation of gestures. As such, it is the social act of these "conversations" that takes center stage. This kind of social behaviorism should not be confused with the behaviorism of John B. Watson and B. F. Skinner discussed earlier. According to Mead, language is the supreme social act and an important observable interaction of social groups. Individuals, for Mead, must be understood within the context of the social groups in which they live, all of which are many and varied. Humans come to grips with their environment socially, and as such Mead combines biological and psychological approaches as minded individuals organize themselves as social creatures and become aware of themselves in group interaction and communion.

Here may be seen some parallel with Boodin's emphasis on sociology. Like Boodin, Mead attempted to fashion a metaphysic at least in part based upon his theory of time, the relationship of relativity theory in physics to sociology, and how this synthesized into an emergent evolution. He did this in the posthumously published *Philosophy of the Present* (1932) but never fully worked it out, and Blau admits that his "interpretation of relativity . . . has been sharply criticized." But it is worth more than a passing note that, like Boodin, Mead sees novelty as an intrinsic aspect of emergent evolution. Thus both look for a world of creativity, only Boodin's is transcendent and teleological while Mead's is thoroughly natural. Whether seen within the framework of relativity theory or not, Mead sees social interaction and humans as social beings as the lens through which we must peer.

Mead's social theory might have passed into quiet obscurity but for the advocacy of sociologist Herbert Blumer and particularly for his commitment to what he called *symbolic interactionism*.[53] This is, according to sociologists Michael J. Carter and Celene Fuller, "a micro-level theoretical framework and perspective in sociology that addresses how society is created and maintained through repeated interactions among individuals."[54] Far from antiquated, they contend that "despite fragmentation and expansion throughout the years, [symbolic interactionism] is a perspective with historical as well as contemporary significance for the field of sociology. Rather than pointing to the variety of theories and methodologies that have emerged since Mead's work as evidence of the demise of symbolic interactionism, we posit that the diversity of sociological work in the symbolic interactionist tradition is evidence of its utility and well-deserved endurance within the discipline." So the question naturally becomes how close Mead's social theory is to Boodin's. Blumer provides a key. He emphasizes that Mead regarded the self as a *process* not a structure because the self must be reflexive. Thus, the self cannot be merely identified with a psychological or personality structure. Instead humans must work through a process of self-interaction with their fellow beings; the self is thus essentially an acting organism. Blumer points out (writing in 1966) that sociology is dominated by a simplistic action/reaction model that sees given factors under certain conditions producing certain types of behavior. In this view "the human being becomes a mere medium or forum for the factors that produce the behavior." Mead's view is radically different: for him the human being is an "active organism" that perceives, acts, and reacts to dynamic social situations. Action is constructed by the actor rather than simply reacted to by external influences. Blumer embraced Mead's symbolic interaction of interpretation and definition as richer and more fruitful than merely seeing social interaction as a kind of neutral blank slate at the mercy of outside factors. It is in this way that participants are active and ongoing rather than reactive pawns to various factors of the sociologists' making. Blum points out that for Mead human constructions are not separate entities with their own existence. He goes so far as to say that "objects consist of whatever people indicate or refer to." Society consists primarily not in structured relations but in an ongoing process of action.

Blumer is critical of sociology that sees humans as "merely responding organisms" instead of "organisms having selves." In other words, Mead wants to show how human beings fashion their world as active and proactive participants in it. Methodologically it avoids the danger of the "objective approach" in which the observer is a detached part of the players being analyzed. It is all too easy in this situation for observers to literally replace the observed by providing a monolithic and indeed reductionist perspective of their own making.

While we can appreciate Blumer's interest in Mead's philosophy, Boodin can also be found in some of this for four reasons. First, although Boodin certainly did not share Mead's biology or his rejection of the transcendent, he *did* see the central importance of humankind as social beings. Citing the social anthropologist Clark Wissler, he notes that "whatever social phenomena we are investigating in the abstract form exist only in the living group."[55] Second, Boodin agrees with Mead on the fundamental importance of the biological nature of society. But for him the urge for human association and life is "manifested in a new type of wholeness, a spiritual gestalt of common plan or purpose." This kind of transcendence is absent in Mead, but the central importance of social relations and interaction is there for them both. For Mead creativity arises from symbolic interaction; for Boodin creativity arises from a wholistic gestalt that exists on human and cosmic levels. Third, as we have seen, Boodin certainly shares Mead's objections to the kinds of reductions exemplified in Watson and Skinner's behaviorism. Nothing could be more anathema to Mead or to Boodin than psychological behaviorism and its manipulative objectifications or similar sociological schemes that reduce humans to reactive entities dependent on constellations of external factors and forces such as social norms, sentiments, status seeking, and so forth. These do not cause or create our sociological makeup, they derive from it. Finally it must be asked: Can symbolic interactionism be found at all in Boodin? This is a harder question, but ultimately the answer is yes, to a limited extent. Although Boodin was quick to reverse Mead's preeminence of language (for him "first-hand living precedes verbalizing"[56]), Boodin did extensively address what he referred to as holophrastic speech as a precursor to analytic language.[57] Since holophrastic speech substitutes a complex of ideas in one word or phrase, it is by its nature close to if not synonymous with symbolic representation. Boodin admits that we engage in analysis and synthesis when we arrive at symbolic wholes, and arriving at symbolic wholes is an essential part of social cognition. It does not have Mead's emphasis, but it is there.

It would be ill-advised to carry this any further. For one thing, as already pointed out, Boodin never referenced Mead in any of his works. Most of Mead's important works were posthumously published just a few years before *The Social Mind*. But referring to Boodin's sociology as outdated seems presumptuous given the broad collateral work being done by Mead and carried forward by Blumer. Mead and Boodin may not have been growing the same intellectual crops but they were tilling the same field.

However, the most thoroughly critical assessment of *The Social Mind* came rather surprisingly from Alfred Hoernlé, who was usually quite interested in and supportive of Boodin's work. Here, however, Hoernlé takes aim at so many core concepts in the author's presentation that one wonders if he ever understood the poet-philosopher at all. These criticisms are worth investigating in some detail and can serve as a general defense of its leading themes and concepts. Hoernlé makes three main objections to Boodin's argument. First, he claims that the appeal to "human experience" is "ambiguous."[58] He wonders how any one person's experience serves as a collective term for *everyone's* experiences. For example, "the experiences of a nomadic Mongolian shepherd on the Siberian plains are widely different from those of a town-bred Cockney in the East End of London. Hence, a statement which the experience of one man verifies abundantly, may be utterly without verification in the experience of another." Specifically, he questions that Boodin's boyhood experiences on a remote Swedish farm and his emigration to America—"two special sorts of experiences"—can do the heavy lifting of human experience generally. Here, it seems, Hoernlé misunderstands experience in the sense used in *The Social Mind*. Are the "experiences" of making a living, providing for one's self and family, combating common challenges of nature (e.g., weather, disease, and beasts) and society (e.g., hostile neighbors and adverse economic conditions), sexual desire, coordinating socially for mutual benefit, and a host of other universal demands really any different in Siberia or London's East End? Surely they are expressed differently, but the goals of survival and general enhancement of life are the same whether one is battling snow leopards in the Eastern Steppe or lice-infested rodents in London's East End. Is the migration of a nomadic Mongolian tribesman all that different in terms of acclimation to new surroundings and making necessary social and personal living adjustments than a Swedish immigrant to America? Again, they are definitely expressed differently, but the requisite needs toward goals are the same. Besides, nothing in the formulation of human experience requires complete universality, only that they are commonly goal oriented when they do occur.

Hoernlé's second objection is the book's emphasis on creative synthesis, one of Boodin's key concepts throughout his entire career, making this objection all the more stunning. Claiming that creative synthesis is not a historical concept, Hoernlé argues that "there is no presumption here that these wholes are permanent" and that "the 'tendency to form wholes' in the Universe is balanced by a 'tendency to destroy wholes.'" It is clear from this that he obviously has failed to understand the social mind as Boodin presents it, a misunderstanding that might have been corrected by a closer reading of Boodin's final chapter, "Social Immortality." There Hoernlé would have learned how intrinsically the social mind and creativeness is bound up with history. "Not only has Plato given us his own mind," writes Boodin, "but also the genius of the Greek mind which carried him on and to which he in turn helped to give immortal direction. This direction continues as a fundamental theme in human history."[59] Can we say that Moses, Christ, and Muhammed haven't acquired a certain immortality (or at least extended mortality) in the Abrahamic religions of the world or that Buddha or the four Vedas are not exerting some imbedded influence over an increasingly global social mind? Other examples could be given of this kind of historicity. If wholism means anything, it means synergy—the whole is greater than the sum of its parts. It seems atomistic to think of "making" and "unmaking" wholes. Instead, evolution is a whole-making process that by creative activity certain syntheses occur through the canalization of values (canalization being "the form of certain typical tendencies, which are of vital importance in the orientation of life to its environment" and even extendable to those promoting constructive development in the cosmos).[60] Rather than talking of "making and breaking" wholes, it seems better to think of the development and subsuming of wholes.

Not only does Hoernlé fail to understand creativity or wholistic thinking, but he seems accusatory in suggesting largely by implication that "thinkers of this school" (by whom I presume he means *process* thinkers) "read a *direction* into the world process, *viz.*, a direction towards making *new* kinds of wholes, *higher* kinds of wholes, *better* kinds of wholes." Stuck on the "making and unmaking" mechanistic thinking of wholes, Hoernlé seems to see only a series of progressions and regressions at work. But this is only a micro interpretation of history; on a wider scale, this may not hold, and if Boodin's process view is correct, some teleological direction would seem at minimum a warrantable consideration. Some of it may depend on what we mean by progress, but if the idea is leading toward some directional goal, then there seems to be a very strong affirmation of progress in history, not only in religion but in biological evolution. In Boodin's day, for example, orthogenesis—directional, progressive evolution in the life sciences in which groups of organisms develop along parallel lines—was considered an accepted counter to Darwinian theory.[61] But the teleological implications were too much for the neo-Darwinists who campaigned against it and beat it into submission.[62] That notwithstanding, it remains an idea that will not die.[63] In any case, Hoernlé's dismissive attitude seems odd for someone who found Boodin's other works so informative and productive, especially for such a lifelong idealist.

It gets even worse when Hoernlé chides Boodin for his commentary on the current socio-political scene, urging that "it is very unsafe for the philosopher to descend from the plane of general principles to approval or disapproval of particular political events and actions." Philosophy and its followers are not called to safety any more than in Socrates's day. Hoernlé's position would deny philosophy any relevance to our world as actually lived. According to this view, Hoernlé argues that Boodin's impassioned verdict against Nazism (with which he assures the reader he is in full sympathy) cannot hold within the context of his creative synthesis idea because "the belief [in Nazism] is genuine for those who hold it." Sincerity of belief has absolutely nothing to do with the truth of any given belief. Neither is truth related to the success of any given belief. "If you appeal to history," claims Hoernlé, "you must abide by the verdict of history." This, of course, is absurd and Whiggish history at its worst (see prologue). Had Hitler been victorious, none of his twisted views on race would be in any sense validated; victory would only have heightened the tragedy and would not have called anyone to "abide" by the verdict. Since there would have been little real creativity in such a monstrous event, a genuinely creative synthesis would await a different concatenation of circumstances for its realization.

Finally, Hoernlé dismisses Boodin's efforts at formulating constructive applications of his social theory. Not only does he seem skeptical of them but he appears to reject even the attempt to offer them. Specifically, he argues that Boodin's call for a return to subsistence farming and democratic collectivism and planning are contradictory. Furthermore, he insists that Boodin's call for more one-on-one personal interactions and smaller scale community life is mere nostalgia for his boyhood days on the Swedish farm. But there is more than nostalgia in this; there is real recognition that we are failing to live as we were made to live. What Boodin saw so clearly grew even more true with the generation following his death. No one understands Boodin's point better than Wendell Berry. Berry bemoans how modern society has been beset by dehumanizing corporations, how the family farm has been destroyed by conglomerates, and how the many are being led—and largely led astray—by so-called specialists. He writes:

> The disease of the modern character is specialization. Looked at from the standpoint of the social *system*, the aim of specialization may be desirable enough. The aim to see that the responsibilities of government, law, medicine, engineering, agriculture, education, etc., are given into the hands of the most skilled, best prepared people. The difficulties do not appear until we look at specialization from the opposite standpoint—that of individual persons. We then begin to see the grotesquery—indeed, the impossibility—of an idea of community wholeness that divorces itself from any idea of personal wholeness.[64]

Berry points out that some of the worst advice given to farmers has come from "specialists" because that advice did not have individual farmers or their communities in mind at all, only the specialists' special interest groups working on behalf of agribusiness. He says that subsistence farming (whether in his own example of Andean peasants or Boodin's rural Sweden) has as its guiding principle the production of "*enough*, a long-term sufficiency, whereas the governing concept of ours is *profit* or *affluence*, without regards for long-term needs."[65] The ecological consequences are devastatingly predictable. As for "democratic collectivism" and planning being contradictory to such a personalist approach to attaining something akin to the Beloved Community, it is quite probable that these will be utterly required. Berry gives twelve suggestions to remedy the current situation an America, some of which include careful government action and planning. In fact, Berry has actively promoted a farm bill to protect and help build rural communities.[66] All things considered Hoernlé's belief that government action and planning are antithetical to subsistence farming is backwards; if something like a more vibrant rural community peopled with those reconnected to the land and their environment is to be achieved, both will be required. It might be accurate to say that if a poet-philosopher exists today representative of Boodin's spirit, it is Wendell Berry, an idealist and a personalist who *lives* the Beloved Community in Henry County, Kentucky, as close as reason and faith will allow.[67]

The review ends by asking (contra Boodin), "Is it conceivable that the next step in 'creative synthesis' should consist in such extreme splitting-up of wholes which are still growing? Is the formation of large cities and States not an effect of creative synthesis?" The answer is, not necessarily. It seems more likely that the megapolis and the massive, impersonal state upon whose back it rides is a pathology needing correction. Boodin and Berry both think so. Hoernlé can't seem to extricate himself from the idea that what *has* or *is* happening is what *can* or *should* happen. This is the classic is-ought fallacy. Also, the questions he poses are conflations of political, jurisdictional, and purely demographic phenomena with social and relational aspects of human existence that are not exact correlates of one another.

The foundation of Hoernlé's inability to appreciate Boodin's historical sociology rests in his apparent lack of appreciation for Boodin's understanding of time and its wholistic nature. This is worth unpacking because another of his reviewers asks the same question as Hoernlé, namely, if evolution makes wholes, "is it not also a whole-breaking process?"[68] Both reviewers would have benefitted from attending Boodin's faculty research lecture delivered on May 12, 1937, where he put a finer point on this aspect of time. It is important in this precise context and presents Boodin's clearest and most succinct explanation of the space-time continuum:

> If time, change, transformation is real and if there is a space-time structure, that is, not merely a simultaneous order of parts in space but an order of emergent parts in succession, then this structure is normative: it exercises guidance in some degree with reference to the future. The future is not merely relevant to the past and present, but the past and present can only be understood in the light of the future. The future helps to constitute the present and past.
>
> Like the great Greek pioneers, we shall do well to approach the cosmic problem through physiology. In embryology we have the most perfect example of time-space form as a normative structure. Classical genetics has tried to understand life history as a spatial ensemble of factors. It has neglected time or regarded it as accidental. But embryological history is unintelligible unless we recognize the future as giving direction to the past and present. What gives meaning to the one-cell or two-cell or

four-cell stage in embryological history is the future of which it is part. The form of a human life, from its beginning with the fertilized cell, to its maturity, must be conceived not merely as a space-whole of factors, but also as a time-whole, and it is the latter which overlaps, that is, which includes the space-whole at any one moment and makes it significant. We may regard a space-whole as a cross section of a time-whole.

The time-whole looks forward to the future as well as backward to the past. It includes the future in the sense that the form of the whole is a guiding or normative factor in the process. But the time-whole cannot be conceived as an abstract timeless structure like Aristotle's entelechy. It includes the conditions of nature—the external conditions, as well as the internal. Since there can be no life except as part of its milieu and the internal conditions are dependent upon what we call the external conditions, such distinctions are artificial. The outcome is the creative result of all these factors within a forward-looking time-form. Because what we call the external condition of nature—the womb of the mother, the outside environment—are largely uniform in the embryonic history, we are apt to neglect them, but without them there could be no embryonic history. . . .

It is an illusion of our space-bound mind that the direction backward [the past] is fixed while the direction forward is indeterminant. The direction backward is dependent upon the direction forward as truly as the latter is dependent upon the former. The direction backward seems indeed fixed to our momentary point of view. But this is an arbitrary and vanishing demarcation. Past, present, and future are distinctions made by our interest, but they are in fact inseparable, emergent parts of the creative concrete space-time form of nature.[69]

Here is Boodin at his mature and expository best. This is the "real dynamism" he spoke of as he launched his career in 1904 with *Time and Reality* (see chapter 1). Talk of "making" and "breaking" wholes reveals a more serialized view of time. It not surprising that this was not generally understood by Boodin's readers of *The Social Mind*, but Hoernlé's critical and unsympathetic attitude toward the book is certainly more so. Apparently Hoernlé was pleased and approving so long as Boodin spoke broadly and abstractly in *A Realistic Universe, Cosmic Evolution, Three Interpretations of the Uni-

verse, and *God and Creation*. But when Boodin's theoretical propositions are given full application in his sociological philosophy, the depth and extent of Hoernlé's misapprehensions are revealed. As a lifelong friend, is it possible that he understood Boodin the man better than Boodin the philosopher?

There remains one more assessment of Boodin's social philosophy to consider: that of Eugene C. Holmes. In a relatively brief seventy-eight-page essay, Holmes proposes to explain how James Mark Baldwin, George Herbert Mead, and John Elof Boodin used genetic (social/behavioral) psychology "to throw new light on the nature of mind and self," with his purpose "to clarify the implications for social theory which emerged from these analyses of thought and individuality."[70] Remarks will largely be confined to those relating to Boodin, although a few intersections with George Herbert Mead are instructive. Holmes argues that Mead views mind as part of nature while Boodin "conceives of social mind as real only when there is a sufficient 'social fusion' resulting from 'social interaction' that interact and communicate with other individual minds."[71] While that's true enough as a definition of *social* mind, it's important to point out that for Boodin our minds are more than just things stuck in our heads. Minds are better conceived as part of an energy field. In that sense, mind is every bit as much a part of nature as electromagnetism, waves, particles, or anything else; as we will see in the next chapter, Boodin's understanding of mind is more akin to Rupert Sheldrake's understanding of mind, as Sheldrake also sees it as an integral part of nature. Nevertheless, Holmes is right to point out Boodin's connection with Mead in assigning paramount importance to the self in defining itself—even actualizing itself—in a social context with other *social* selves. He also notices the much greater emphasis Mead puts on language, although as we have seen Boodin did not neglect the importance of language. On the whole, Holmes offers interesting observations and linkages between these two philosophers who gave careful consideration to the significance of society in the development of historical and cultural life.

However, it is important to understand the lens through which Holmes examines his subjects and subject matter. Although Holmes was initially influenced by pragmatists like John Dewey, John H. McClendon observes that "his philosophical outlook was forged in the context of the struggle between dialectical materialism and pragmatism" for which the former would decidedly win out, becoming an avowed Marxist while at Howard University, known for his "militant defense of philosophical materialism" and distinguished as "the only professionally trained black philosopher of his time who subscribed to dialectical materialism."[72] His frequent criticism of pragmatism in his dissertation shows that by 1942 he was well on his way toward this worldview. An active Communist Party USA member, he brought forth his radical demand for change in the service of black liberation from his faculty position at Howard.[73] This is mentioned only to point out that dialectical materialism, a term coined by Joseph Dietzgen in 1887 to capture most fully Marx and Engels's ideas, was a metaphysic through which Holmes understood all reality—the primacy of matter governed by discernable laws in opposition to all forms of idealism. Rather than viewing matter as operating mechanistically, the dialectical materialist views those laws as the transformation of quantity into quality, the interpenetration of opposites that develop through contradiction with reconciliations producing new contradictions (sometimes called the negation of negation). In practical terms this translates into class conflict resolving itself in wholly new transformations toward a truly communitarian society. This was the communist view, but of course there are others. Thus can be seen Holmes's interest in the social theories of Baldwin, Mead, and Boodin.

As a developing Marxist, Holmes has little patience for Boodin's teleology and does not think that Boodin's liberal democracy is the answer, because while equitable distribution of wealth, union and agrarian self-determination, economic cooperatives, and conservation of resources are all laudable goals, Boodin offers no real plan for their attainment. For Holmes, "will," "hard thinking," and mere "sentiment" are not enough. And yet Holmes recognizes that Boodin was one of the few philosophers of his day to pay serious attention to the state.[74] He also agrees with Boodin's conclusion that wars of aggression have a fundamentally economic basis and his recognition of the importance and influence of class relations and their ideological power. He nonetheless chides Boodin and Dewey for their timidity regarding political, social, and economic change—for their "refusal to admit complete reconstruction and planning on a large scale."[75] All of these comments and criticisms are logical outgrowths of Holmes's ideological perspective, but it is beyond the scope of this study to explore the respective merits of this position.

However, at least three points need to be made. First, one wonders if Holmes really understands Boodin's cosmic evolution. It is, for example, surprising to see Holmes associate Boodin's cosmic evolution with C. Lloyd Morgan's emergent evolution since Morgan rejected any and all forms of teleology and invoked God or "Spirit" as an assumption on purely instrumentalist grounds (see note 6, chap. 4).[76] That is why Boodin classed him as an emergentist along with Darwin. It is also strange to see Holmes's linkages of Boodin with Comtean positivism.[77] Holmes speaks of Comte's desire to preserve Christian ethics and the positivists' religious impulse, but the trappings of religion and numinous rhetoric must not be conflated with affirmations of transcendent belief. That's why, as we have seen, Huxley was right in calling Comte's positivism "Catholicism *minus* Christianity." Positivists generally had little regard for any spiritual religion, considering it an outmoded stage of social development. If they had a religion, it was, as Renan expressed it, "*la religion de la science*."[78] Expressed somewhat differently as a "religion of humanity," positivism's awkward positioning of alleged "plain" facts with faith created a real dilemma. W. M. Simon argues that positivism "could conceivably only have become a mass movement *qua* religion; but as soon as the disciples attempted to augment their numbers by emphasizing religion, even larger numbers were correspondingly repelled."[79] Thus it was that positivism could never gain a consistently large constituency not because it strictly opposed religion but because it could not enlist an empty shell of religion effectively on its behalf. During Boodin's career its heir, logical positivism, never claimed the pretentions of a popular movement and it even failed to enjoy a large following among American academics.[80] Nothing in the worldly reductionism of positivism resembles this poet-philosopher.

Beyond Holmes's peculiar understanding of positivism, a second and more pressing point is that his treatment of Boodin's social views is at least premature and at most incomplete. Holmes is so focused on Boodin's social theory that he seems to miss the genuineness and importance of his Christian faith expressed outside of *The Social Mind*. Holmes cites Boodin's *God and Creation* in his bibliography but makes no reference to it in the text. In fact, this raises a lurking methodological concern: Holmes cites eight different Boodin sources, but in the text itself only *The Social Mind* is referenced. More consideration of other sources might have yielded a more thorough and better contextualized analysis. Had he waited one more year Holmes would have had Boodin's *Religion of Tomorrow* to consider. If Holmes faults Boodin for having no plan of action to realize the goals of his liberal democracy, it is rectified here. Not that it would have persuaded Holmes, but looking at the *total* corpus of Boodin's writings it certainly was premature and presumptuous of Holmes to suggest none was forthcoming.

Holmes is conspicuously silent on the historical role religion and teleology has played in the sociology of civic activism, but some of the greatest civil rights advocates for social change used faith-based spiritual convictions to galvanize support on behalf of significant social and political change: first in the moral suasion of Protestants against slavery dating from the Second

Great Awakening (ca. 1795–1835) and later in the much broader activism of the Social Gospel typified in the unionization and worker rights efforts of Washington Gladden and the urban reforms of Walter Rauschenbusch, then in Holmes's own generation given a deeper global voice in Mahatma Ghandi's ecumenical Hinduism, Archbishop Desmond Tu Tu's expansive Anglo-Catholicism, and the Rev. Martin Luther King Jr.'s inspirational Protestantism. Most pertinent to America, Diana L. Hayes has written persuasively of King's commitment to the Beloved Community in the context of deep and abiding religious and spiritual roots that go back to Africa, with Beverly J. Lanzetta even calling King a "social mystic."[81] When Boodin wrote, "We are not duped when we believe that the dice of the universe are loaded for right and reason,"[82] he echoed King's famous statement, "The arc of the moral universe is long, but it bends towards justice." These two men were very different in historical significance and personality—Boodin a quiet and even withdrawn academic, King an eloquent and unyielding firebrand—but both had very similar faith-based sensibilities. Like Boodin, King saw no conflict between science and religion, only "superstition disguised as religion and materialism disguised as science." Furthermore, Boodin surely agreed with King's assertion that "religion has to be interpreted for each age; stated in terms that that age can understand. But the essential purpose of religion remains the same. It is not to perpetuate dogma or theology; but to produce living witnesses and testimonies to the power of God in human experience."[83] That is nearly a thesis statement for Boodin's *Religion of Tomorrow*. Boodin would accomplish this quietly from the safe confines of his study in salubrious West Coast solitude; King would do it in the sweltering streets of the American South amidst snarling police dogs, state troopers, and taunting reactionaries fueled on ignorance, hatred, and misplaced nostalgia. Different as they were, it is nonetheless remarkable that in history sometimes the lion does indeed lay down with the lamb. Again, though, Holmes is premature; the Boodin/King nexus (a nexus of ideas and inclinations, not personal acquaintance) lay ahead. King would not even deliver his first speech at the Ebenezer Baptist Church for another five years.

The final point is Holmes's treatment of teleology as a philosophical concept. Although his objections to it are fairly obvious, the specifics hardly are. He hints that the many teleological views presented in the early twentieth century were satisfactorily met by William Pepperell Montague, Roy Wood Sellars, and George Santayana (all materialists). He states this as an established fact without any explanation. Before taking Holmes's word for it, see the discussion of Sellars in chapter 3, which may serve as a measure of its effectiveness. In any case, the dogged persistence of a wide range of teleological views suggest otherwise, views throughout the century (especially in the life sciences) and in our own time, from theologians like Teilhard de Chardin to neurobiologists like John C. Eccles and most recently biologist Rupert Sheldrake.[84] Then, associating Boodin with Hans Driesch (whom Boodin criticizes extensively in *A Realistic Universe* and *Three Interpretations of the Universe*), Holmes claims he falls prey to the "vitalistic fallacy."[85] Holmes is apparently associating this "fallacy" with orthogenesis, an idea in biology of progressive evolution that, as we have already seen, has had a persistent life of its own. Holmes also faults Boodin and Benjamin Kidd for believing that the "mere fact that life had begun" in the distant past suggests "that there was some volitional harmony or coordination between life and cosmic conditions."[86] There are few philosophers, theologians, or cosmologists who would preface the beginning of life with the adjective "mere"—the presence of life does indeed demand an answer, and for all but Holmes it is *not* a "mere" fact that speaks for itself. This notwithstanding, "some volitional harmony" is certainly suggested in the cosmic fine-tuning of the universe and the anthropic principle. Although most thoroughly presented by cosmologists John D. Barrow and Frank Tipler in 1986, it is an idea given in coherent form at least as long ago as evolutionary biologist Alfred Russel Wallace's *Man's Place in the Universe* (1903).[87] Whatever its theistic and teleological implications, it undoubtedly reversed the anti-anthropic Copernican principle and generally infused idealism with a new lease on life. "The Copernican turn," remarks Friedel Weinert, "led to a loss of physical centrality. But human existence is still precious in a dual sense. We are the intelligent species within our cosmic neighborhood. As such we do not depend on any physical centrality. Through the force of abstract reasoning human minds crisscross the universe. We know more from thinking than from seeing. This is a worthier kind of centrality. It is rational centrality."[88] This should not slight Boodin's commitment to empiricism, for rationality itself involves empirical inquiry; besides, his was never a monistic view.

More could be said of Holmes's handling of Boodin, but as important as Holmes's essay is in the relatively meager Boodin historiography, the foregoing is sufficient to raise questions concerning using his examination of the poet-philosopher as anything resembling a complete or definitive assessment of his ideas. In fairness to Holmes, his essay is not aimed at comprehensiveness; instead it is an extremely limited study of Boodin's use of the genetic method in developing social theory and philosophy. Such a micro analysis is valuable as far as it goes, but it fails to capture the details and richness that a more panoramic view might offer. His effort is significant not so much in his precision of analysis as in being one of the first and few philosophers of the generation following Boodin to pay him any heed at all. That in itself makes this contribution interesting and worthwhile.

Whatever the reactions elicited by this book, *The Social Mind*, in its application of Boodin's unique space-time framework for a wholistic process philosophy applicable to human and cosmic history actualized through creative synergy, really is the culmination of his life's work in philosophy. Almost every element of the poet-philosopher's intellectual repertoire is there—his pragmatic realism, his tempered idealism, his special concept of time, his science-based empiricism, his commitment to creativity, his concept of the Beloved Community. In some ways *The Social Mind* is the most thorough presentation of his thought. It is certainly its most mature expression. Nevertheless, it would be presumptuous to assume that it alone conveys the richness of Boodin's philosophy. To know *The Social Mind* is not to know the whole of Boodin.

The direct examination of Boodin's thought may fittingly end here. In the next and final chapter some attempt will be made to place our poet-philosopher in a wider context. That he swam against a professional tide of increasing reductionism, materialism, and linguistified analytic philosophy uncongenial to a metaphysician of pragmatic and idealist inclinations will become readily apparent. But Boodin had—and *has*—allies too. Whatever tide he swam against, we shall see that his greatest ally has been science itself.

Chapter 6

Boodin in a Hostile World

John Elof Boodin's career spanned the first half of the twentieth century. This was both an advantage and a disadvantage to an academic of no little skill and sagacity. It is our purpose here to examine exactly what those advantages and disadvantages were and indeed *are* because they in large measure explain the otherwise mysterious obscurity of a philosopher who should be far better known than he is today. While some of this is bound up with the philosophical profession itself, more generally it may be ascribed to a general flow of intellectual currents in the West that are inimical to Boodin's brand of idealistic, panenthic, pragmatic empirical realism presented in a process metaphysical system of thought that emphatically affirms the evolutionary cosmos in which we live. Boodin looked to science to provide guidance toward the real nature things while the world he increasingly found himself in had fixed its hypnotized gaze upon science as the ultimate arbiter of truth. Boodin was interested in science as its faithful student, but the world was infatuated with scientism, and it still is if Richard Dawkins and similar media celebrities are any indication. Boodin not only took time seriously, but he took God seriously in an age that took only itself seriously, and if the postmodernists are to be truly believed, not even that. The twentieth century found its intellectual foundations in Victorian-era reductionisms heralded in by Darwin and his X Club apostles and gave what the Annales historians called the *mentalité* of the era that remains with us today. It is an era of tremendous technological achievement sullied by a fractured and schizophrenic humanism torn between nihilism and narcissism. Boodin fought—and fought hard—against these trends. And fortunately, then and now Boodin has never been entirely alone; he has allies. I want to critically examine all of this and then see where Boodin stands today. In this chapter we shall see if there is still room for poetry in philosophy.

The Fortunes and Misfortunes of History

John S. Haller Jr. has recently given us a valuable picture of intellectual life at the busy and often contentious intersection of science and religion just prior to and during Boodin's formative years as a philosopher/scholar. His *Fictions of Certitude* provides the descriptive and analytic cartography that maps the hopes, fears, triumphs, and challenges of a generation that felt that they had landed in a terra incognita between the old religious order of their fathers and the new science-based order of a world that seemingly couldn't accommodate both, a world they were helping to bring about. As Haller explains:

> The inclusion of science in the search for certitude had the effect of freeing religion from biblical literalism as the sole manifestation of the Bible's transcendent truths. The combined science/religion approach neutralized the staking of certitude on scripture alone, the fear of punishment, or the hope of reward in the afterlife. It opened the door to a combination of utilitarianism, the model of Jesus's life, and the presumption of a teleology implanted in the natural world that justified faith from Design. Depending on its relative weight in the equation, science either subordinated faith to the certainty of science or was itself subordinated to the certainty of religion.[1]

This was the world a young immigrant Swede found when he arrived in America; it was all the more keenly felt when he arrived at Harvard to study under Josiah Royce and William James. Boodin sought to craft a complete philosophical system by walking confidently through that door opened by the new science/religion approach. He sought to avoid the subordination of either science or religion by taking the best he could find from both Royce and James as well as the leaders of a new scientific era—the mathematics of Henri Poincaré, the theoretical physics of Hermann Weyl, the relativity theory of Albert Einstein, the quantum physics of Werner Heisenberg and Erwin Schrödinger, and many others. By doing so he went about constructing a new intellectual model yielding certainty, a certainty derived from a new physics and, as we shall see, a new paradigm for the life sciences. It would be a certainty in which science and religion became synergistic aspects of a process philosophy based upon a renewed faith in the value and importance of humanity rooted in a thoroughly teleological universe. Such an approach was ill-suited to the old *sola scriptura* approach of previous generations and produced a genuine and liberal religious conviction that sprang more from Plato than from Paul.

But this raises an interesting question. It is relatively easy to find Royce in Boodin—the Absolute, the idealism, the Beloved Community are all there even if they are expressed differently. But what of James? Haller's accurate description of James is, on the face of it, hard to find in Boodin:

> James rejected the God of orthodoxy because he found Him devoid of any practical purpose or meaning. Mere existence commanded no reverence. Instead, God needed to be a co-laborer with people in building a moral universe. . . . Such a finite God, explained philosopher David Paulsen, was "pragmatically richer than belief in any absolutely unlimited God in that it provides greater virility and impetus to our moral endeavors." James admitted to having "no living sense of commerce with a God. I envy those who have for I know that addition of such a sense would help me immensely." Busying himself instead with the prospect of improving the human condition individually and communally, he chose pluralism to either a death-of-God approach or placing in God too much control. The relevance of the traditional all-knowing and all-powerful God was irrelevant to human existence, but belief in God resulted in a moral and ethically more meaningful existence. If God was the missing factor in bringing unification to human ideals and choices, then he urged humans to use it.[2]

And Boodin *did* use it. The uncertainty and tentativeness of the Jamesian God was resolved in pluralistic panentheism expressed as a compelling process theology. The whole point of *The Social Mind* and *Religion of Tomorrow* was to do exactly what Hoernlé denied: making teleology and God coherent by investing religion with what James would have called its "cash-value." This was an important aspect of Boodin's social philosophy. Therefore, Boodin's arrival on the scene amidst these uncertainties was his good fortune since it provided him with the raw materials of resolving the age-old science/religion "conflict" into a thoroughgoing metaphysic that reconciled both.

But, at the same time, it was also Boodin's misfortune to live and work amidst the rise of a powerful secularizing and reducing force in the West, namely modernity. Its roots run deep in the Renaissance discovery of a human-centered universe, the Enlightenment's enthroning of reason (quite literally during the French Revolution) as an exclusive avenue only leading toward that human-centered world, in Kant's declaration *"Supre aude,"* in positivism masquerading as science and its subsequent elevation from a method to an ontological category as science, in the *Humanist Manifesto*, and so on. As Thaddeus J. Kozinski writes, "We late-moderns can add private self-creation, moralistic therapeutic deism, and a globalistic market and multi-cultural state eschewing objective moral and spiritual truth to ensure perpetual peace and stability."[3] Of course, "objective" moral values *are* permitted so long as they affirm self-professed humanistic standards and go no further. In the sense described here Boodin was no modern but he was contending with the *mentalité* of modernity.

How might Boodin be situated within this thorny thicket of modernity? The key is his relationship with pragmatism. Was Boodin a pragmatist? The answer is yes and no. Of course, our poet-philosopher was in many respects a pragmatist, and while he adored James and wrote with great sympathy for pragmatic views, it should be remembered that he always resisted the efforts of his mentor/friend to join openly in the campaign. Boodin refused the label "pragmatist" and felt that his writing about pragmatism—however supportive—was resented precisely because it was written as an outsider.[4] Why did Boodin refuse James's invitations? Although he never really gives an answer, the most likely explanation somewhat surprisingly comes from G. K. Chesterton. Although Chesterton affirmed and defended "the pragmatist method as a preliminary guide to truth," he admitted that "there is an extreme application of it which involves the absence of all truth whatever." He refers here to James's rejection of the Absolute. "The pragmatist [i.e., James] tells a man to think what he must think and never mind the Absolute. But precisely one of the things that he must think is the Absolute. This philosophy, indeed, is a kind of verbal paradox. Pragmatism is a matter of human needs; and one of the first human needs is to be something more than a pragmatist."[5] Whatever else one may think of this assessment, I believe it was Boodin's assessment even if he did not subscribe to Royce's view of the Absolute. For Boodin, a transcendent Absolute was always more than it was for James, a mere instrumentality. Once this essential difference is understood it is easy to see how modernism and the poet-philosopher would have serious issues. Boodin expressed this most eloquently at the end of his *Religion of Tomorrow*, saying:

> It is an infinite privilege to be born into the world of spirit even for an instant, to wake even in a small way to the beauty and meaning of God. I thank God that I can share in the creation of a new world, and help, as my strength goes, to bring order into chaos within me and about me. If it is only for a brief moment, I would not give it for all the aeons of stars and galaxies, for these are after all only the preparation and the frame for such a moment of living with God. I shall not fear the vast cosmic expanses, for the spirit of God pervades all space. And I shall not mourn my brief moment in the ages of time, for God is working, constructing and redeeming throughout all time. Wherever I am and whenever I am, I am with God and His love sufficeth. He has not brought me here and now for naught. My destiny is in Him and that is enough for me.[6]

This was the God that eluded James. This was the God Boodin first glimpsed as a boy in Sweden and grew in intimacy as his process theology deepened.

Having set Boodin within the context of modernity, it might useful to look at the kind of philosophy with which he had to contend. Here we must again look at Roy Wood Sellars, for not only was he in many ways modernism's most significant representative on the American philosophy scene, he was also in some respects Boodin's shadow figure, a negative philosophical doppelganger embodied by the assumptions of the era. We needn't repeat the ground already covered in chapter 3. The commentary there should set the stage for the difficulties encountered here. Sellars's critical realism (sometimes *emergent* realism) became a large part of the American philosophical scene and was further elucidated later on by Ian Barbour and Roy Bhasker. Unlike Sellars, Barbour joined Arthur Peacocke and John Polkinghorne in applying critical realism to science in non-reductionist and theistic ways. As such, it is difficult to give a thoroughgoing assessment of critical realism because there are so many different varieties. But Sellars's materialism posits a distinctly secular version, and it is primarily known in this context. With regard to Boodin, this must be our main focus. Sellars frequently writes as though critical realism is a settled issue. He insists that critical realism regards human knowing as a direct knowledge of objects mediated by "logical ideas." But, observes Richard McDonough, "it is hard to see how knowledge can be both mediated and direct. The claim that one perceives independent objects via one's sensations but does not perceive those sensations themselves is a fair negative point, but seems to require a more robust positive account of the precise role of sensations in the perception of external objects." Sellars's version of critical realism is intriguing, but many feel it requires further clarification.[7]

Other problems arise in Sellars's sometimes contradictory statements. For example, he attacks the positivists and Hume because they reduce "perceiving to sensing [and therefore] can have neither things nor selves."[8] But later on he writes, "the capacities of the brain-mind are known through the use of sense-data in external observations."[9] How is this any different from the positivists and Hume? If his addition of participative consciousness is supposed to add anything new, it seems unclear how or why in any significant way since sense-data assumes participative consciousness. Sense-data and consciousness in and of themselves as explanations for the mind-body problem seem like either a useless tautology or another reduction of the "nothing but" variety of physicalism. Then there is Sellars's tendency to make unsubstantiated claims, as when he states, "The developments in physics seem to me to have largely helped to make a philosophical materialism possible, while physicists are assuring the public that it liquidates materialism. . . . What they should mean is the tautological statement that the new physics liquidates the old physics."[10] Heisenberg certainly had more than the "old physics" in mind when he made reference to quantum theory vindicating Plato over the atomists (see his *Physics and Philosophy* [1958] and *Natural Law and the Structure of Matter* [1970] referenced in the introduction).

Our critique of Sellars might end there, but he has continued to receive posthumous recognition and acclaim. Pouwel Slurink, admitting that "naturalism is fashionable as never before," suggests a serious reexamination of Sellars's evolutionary naturalism.[11] Slurink finds Sellars of special interest because he claims that "most of his writings come well before the era of the revival of Darwinism," an uncritical acceptance of the so-called eclipse of Darwinism theory discussed in chapter 3. To reiterate, there is little to recommend any serious eclipse of Darwinian evolution at least in Anglo-American science. When paleontologist Robert Broom attended the zoological section of the British Association in London in 1931, he noted that the "majority supported Darwinism as the prime factor."[12] If the "eclipse" theory were correct, the long shadow should have been amply cast at that meeting because no major works comprising the modern evolutionary synthesis had yet reached the profession. In 1931 the ink was barely dry on Ronald Fisher's *Genetical Theory of Natural Selection* and this was six years before Dobzhansky's *Genetics and the Origin of Species*, eleven years before Julian Huxley's *Evolution: The Modern Synthesis* as well as Ernst Mayr's *Systematics and the Origin of Species*, and thirteen years before George Gaylord Simpson's *Tempo and Mode in Evolution*. Slurink applauds Sellars's *Evolutionary Naturalism* by emphasizing that (quoting Sellars), "'organization is objectively significant and causally effective' and that 'function and structure go together at every level.' Evolution is defined as 'the active rise of new wholes with new properties.' 'Evolution means that there are levels in nature, that the higher is an outgrowth of the lower, that A and B integrated are more than A and B separate.'" No wonder Boodin felt that his bones had been picked clean by Sellars, or so he told Arthur O. Lovejoy. Slurink thinks that Sellars's biological emphasis as "the clue to an adequate naturalism" is what makes his philosophy worth revisiting, but how is this new or innovative? Darwinism *is* naturalism and has *always* been a form of naturalistic reductionism, a blind materialism. Sellars merely put new clothes on an old swine to promote what Boodin called "pig-trough philosophy."

Of course, to be a good doppelganger for Boodin, Sellars would have needed a philosophy of religion. And indeed, he had one. His books *Next Step in Religion* (1918) and *Religion Coming of Age* (1928) are contrary predecessors of Boodin's *God and Creation* and *Religion of Tomorrow*. The former effort George A. Coe thought is not philosophy or science, but preaching and proselytizing. Weak on sources but high on claims, Sellars's impersonal view of nature, of which he is sure humanity is a part, gets caught up in its own "nothing but" reductionism: "If nature includes man," noted Coe, "and man creates and values, then the universe is not a closed system that springs from the nature of its parts, nor is evolution utterly indifferent to good and evil."[13] Boodin surely agreed.

In *Religion Coming of Age* Sellars describes spiritual development as having

> passed through stages which are more or less distinct. It began in a cry of supplication to the mysterious powers of life and death; it was caught up into the magnetic influence of the belief in another world and thereby became a religion of salvation, yearning for Heaven and fearful of Hell; finally, it is in our own day clearly shifting to a religion of this life with its problems

and possibilities. Religion and humanity have been organic to one another. The progress of one has been bound up with the growth of the other. In a rough way, we may speak of these stages of religion as corresponding to the childhood of the race, its romantic adolescence, and its maturity. Religion is coming of Age.[14]

Philosopher William de Burgh explains that, according to Sellars, "The sign of maturity is the shedding of otherworldliness and the replacement of an outworn theology by a new faith, grounded on the inductive sciences" based upon self-realization.[15] Here again, it seems, we have an old swine dressed in new clothes, this time wearing the suit of Comtian positivism. Sellars may deny the connection all he wants, but this seems like just another way of stating Comte's law of three stages only slightly revised: the world conceived superstitiously; the world conceived theologically; and the world conceived scientifically. This is easily commensurate with Comte's rejection of metaphysics and thus the different parsing of stages, although the overall thrust is the same—an overarching historical scientism. De Burgh thinks this is a formulation that Whitehead could hardly approve of and concludes that Sellars's humanist perspective of a wholly nontranscendent self-actualization "is scarcely an inspiring vision."

It should be said, however, that this was not the majority opinion of Sellars's philosophy or theology. Leading *New York Sun* critic James G. Huneker considered Sellars's *Next Step in Religion* along with Joseph Conrad's *Arrow of Gold* the two best books of the year, and it was the general success of both his books on religion that prompted his authorship of the first *Humanist Manifesto* (1933). And the assessment of his philosophical work though a little threadbare has drawn recent praise for making physicalism "a meaningful metaphysical thesis with substantive cognitive content."[16] Clearly, Sellars is riding the crest of a wave swelled by all that modernity holds most dear—science as an ontological guide in all things, affirmation of the fact/value dichotomy, and a humanistic faith.

Although I have attempted to show that Sellars's views are often problematic, frequently indefensible, and at times unwarranted, the refusal of the majority to address or even to recognize them go a long way in explaining why Sellars remains an entry in most general ready-reference sources in philosophy and why Boodin is absent. It is probably less the power or merits of the arguments themselves than their respective compatibilities or incompatibilities within the milieu of firmly held intellectual prejudices in which they reside that makes the difference. Sometimes certain perspectives and voices for those perspectives are simply not heard against the chorus of those who hold the majority, in this case the assumption that science can and should answer all questions with the rest being left to a promissory note. Those who think I overstate the case should read Austin L. Hughes's insightful article on the subject.[17] Of course this alone cannot explain why Boodin's pragmatic, panenthic idealism has failed to be more appreciated; in terms of the Boodin/Sellars nexus, however, I believe it is paramount because the foundation of not only all Sellars's scientific certainties but even his most passionate humanism rests on Darwinism as an all-explanatory metaphysic. It's not that Boodin was uninformed of the latest developments in science or philosophy—he was a close student of both—it's that he read the emerging new physics and its implications differently just as he read biology in ways more aligned with less popular alternatives of Darwinism; as we shall see in the final section of this chapter, he read it in such a way that is only now beginning to be fully appreciated.

The Friendlier Side of a Hostile World

As mentioned, Boodin was not alone; he knew a kindred spirit when he saw one too. He found one in Shailer Mathews. Only six years Boodin's senior, Mathews was invited by William Rainey Harper to join the University of Chicago in 1894, where he could pursue his research interests and more importantly evangelize for a new era of democratic theology and religious faith. He would find this opportunity particularly congenial to his passion for a liberal Christianity infused by a commitment to social goals and objectives. There he would become dean of the prestigious Chicago Divinity School from 1908 until 1933, carrying on the work of Walter Rauschenbusch, who promoted the Social Gospel movement. As dean of the school and as a prominent Christian scholar in his own right, Mathews could serve as an important proponent of the Social Gospel, believing that history was inherently moved by what Lester Ward called a *telic* principle of progress; for him, work toward social betterment was indeed God's work.[18] Although Boodin was never a part of the Social Gospel movement, both he and Mathews faced challenges arising from different quarters that made distinctions without a difference. Just as our poet-philosopher found himself dismissed and otherwise ignored by an academy that was turning away from metaphysics and religious interests, Mathews found himself attacked by the rise of a new wave of Christian fundamentalists led by men like Harry Rimmer and Willliam Bell Riley. Beleaguered from their respective positions on the American scene, Boodin found much to commend in Mathews's *Spiritual Interpretation of History* (1916). His enthusiastic essay review found the dean's treatment "lucid and attractive throughout" with conclusions "based on an unusually rich and varied mastery of the field of human experience."[19] He agreed with Mathews that all "monistic interpretations of history," such as those of Karl Marx, are too simplistic and highlighted the book's theme of historical process as *social process*. Mathews's idea that human personality is not passive but an active creative factor in history echoed his own thoughts. Both exuded a liberal, hopeful attitude derived from a common faith in a spiritually guided history designed for human improvement.

Boodin also had close colleagues who took his own work seriously. His departmental associate at UCLA, James Hayden Tufts (who had joined with John Dewey in writing *Ethics*), showed an unusually keen understanding and appreciation of Boodin's philosophy. Two especially close friends and colleagues were Hugh Miller and Donald Ayers Piatt. It was Miller who introduced the novelist John Steinbeck to Boodin's writing, and following Boodin's death it was Piatt who took on the responsibility of gathering his unpublished papers and seeing them to publication. Boodin always felt unfairly ignored, at times even misrepresented, and that his ideas sometimes were co-opted, but he was neither an outcast nor a pariah. His presidency of the Western Philosophical Association and many invitations to lecture demonstrate otherwise.

Boodin was not completely forgotten in death either. Kenneth Ray Sutton's unpublished dissertation resurrected the poet-philosopher's ideas in a study that examined his central concept of creation in detail. It fortunately came in 1969 when those who knew and remembered Boodin were still alive, and Sutton made extensive use of the recollections of Bernard Mollenhauer with whom the author corresponded extensively. Correctly noting that Boodin cast creation within fields of cosmic control, Sutton shows how the best science of the day was used in the construction of a complete philosophical system. The specifics of Boodin's philosophy have been covered in previous chapters, but Sutton does make the interesting observation that the philosopher/historian Frederick Copleston made room for Boodin in his *History of Philosophy* as "the philosopher who carried evolutionary idealism into the first half of the twentieth century."[20] The dissertation gives a good overview of Boodin's leading ideas but does not always fully contextualize them. For example, scant attention is paid to how

Cosmic Evolution relates to other evolutionary theories and Darwinism is barely mentioned. Also, little attention is paid to other philosophers and their systems of thought—Kant is mentioned only in a brief biographical context, and Hegel, Schopenhauer, Comte, Spencer, and other philosophers are absent. Nevertheless, Sutton performed an important service in keeping Boodin's ideas alive, noting his thorough consistency and coherence as a philosopher. As the poet-philosopher receded into the memory of American thought, his final verdict pleaded his case: "John Elof Boodin's creationism was a highly consistent philosophy. It certainly deserves the scholarly and critical attention of philosophers, teachers of philosophy, and students of philosophy."[21] And, one might add, historians of American social and religious thought and culture.

It cannot be said that Sutton's call was widely heard. But there are two notable exceptions. James Wayne Dye of Northern Illinois University set his sights on a careful review of Boodin's theory of consciousness.[22] Pointing out that his idealism was counterbalanced by the British empiricists, French mathematician Henri Poincaré, and the American pragmatists Peirce and James, Dye proceeds to give an overall sympathetic account of Boodin's philosophical formulations as they relate to consciousness. He correctly highlights Boodin's replacement of being with dynamic energy and proceeds with a thorough examination of *A Realistic Universe*. The materialist view of consciousness has the problem of explaining "an unintelligible *saltus*" between two entirely different categories. Only *process* can do this. A parallelism of one-to-one correspondence between mental and physical phenomena has the problem of treating both as extendable, one to the other. Boodin resolves this by treating consciousness as a distinct type of reality. Dye's close reading of Boodin on consciousness reveals the poet-philosopher's own evolved concept. In fact, Dye observes that by the time of his 1931 second edition to *A Realistic Universe*, "awareness" now served where "consciousness" once did because it "has come to seem hopelessly contaminated by previous philosophers identifying it [consciousness] with mind."[23]

Dye argues that Boodin originally emphasized consciousness as immediate experience, abstracting consciousness from mental and physical continuities that comprise the individual, conflicting with our common sense of possessing our own consciousness; in this way only the verbal common sense of the term is rejected while all that comes from direct experience is retained. If this seems odd, it is just as likely that in the future our reluctance to universalize consciousness as a blind and dogmatic assumption equivalent to the Cartesian denial of consciousness to animals will obtain.

"In his defense," writes Dye, "we might observe that on the level of simple introspection we seem to own our own time and space at least as clearly as we do consciousness, yet we have accepted intellectual accounts of time and space which ignore this fact. Consciousness here is treated as an analogue of absolute time and space." But Boodin's further study of the rapidly unfolding theory of relativity in physics prompts him to relativize consciousness by linking it with the individual contexts in which it occurs, thus shifting the emphasis from universal constancy to variability. These contexts are known only behaviorally, causing Boodin's pivot from immediate experience to observed behavior whereby awareness is abstracted. Identity of conscious awareness is still affirmed, but different levels are now introduced in accordance with the complexity of different behavior responses.

Of course, Dye notes that a constant, universalized consciousness doesn't seem to have empirical evidence to support it. But he also notes that the evidence could be turned in Boodin's favor by identifying consciousness with "sensitiveness or selectivity," which establishes a genuine philosophical universal by being related to a trait that is arguably present even in the most rudimentary of events. With Boodin "awareness" becomes, according to Dye, a basic feature of elementary particles as well as human behavior. If so, with "universalized awareness," Dye then asks provocatively, "do we not open the door for infinite levels of awareness corresponding to the structural complexity of the underlying events?" The answer has to be yes, and perhaps an excursion into panpsychism. But Boodin was always opposed to this idea and would have been quick to point out that he was *not* reducing the universe to metaphysical mind-stuff as the panpsychists do. Dye further points out that Boodin resisted Whitehead's "radical psychologization of nature," thinking it an animistic expression. Dye thinks this move would have made for a more advantageous ontological simplification, while Boodin nevertheless retained mind as a separate kind of energy. While this may seem a valid point, it will be shown in the final section that Boodin's handling of mind in this fashion will accord well with biologist Rupert Sheldrake.

Whatever its problems, Dye still suggests that Boodin's theory of consciousness has much to recommend it. It retains four distinct advantages: (1) it harmonizes with his other previously established theories; (2) it provides a tidy solution to winding up with endless forms of consciousness; (3) it meaningfully answers the potential complications of mixing consciousness with mind-stuff, one of them being lapsing into panpsychism; and (4) it retains viable teleological and religious implications instead of succumbing to epiphenomenalism or other form of reductionism. Dye believes that

Boodin's use of consciousness in his cosmology "complicates the theory" and does not apparently add anything to its explanatory power. But that was a verdict rendered nearly fifty years ago. With the rise of more wholistic approaches to science rendered more plausible from the 1990s to the present, it seems that Boodin is on the right side of history. In any case, Dye ends up praising Boodin's "creative innovation" in tackling the difficult problem of addressing the operational significance of consciousness that remains of perennial interest.

After Dye's essay, Boodin disappeared from the philosophical landscape until Randall Auxier revived the poet-philosopher in 2001 with his introduction to the reprint of *Time and Reality* mentioned earlier. Auxier's entry in *The Dictionary of Modern American Philosophers* in 2005 again made Boodin accessible in an important ready-reference source. This has served as Boodin's lifeline to American thought, and without Auxier he almost certainly would have sunk irretrievably into the sea of obscurity.

Most recently, however, attention has been brought to bear on Boodin in connection with his old materialist doppelganger, Roy Wood Sellars. It shouldn't be surprising that, in an age more comfortable in Wood's world than in Boodin's, an example be made, albeit in a gentle but determined manner, of how an idealist pragmatist/process philosopher got it "wrong." The next section assesses this claim in detail.

The Gentle but Dull Scalpel of Matthias Neuber

Matthias Neuber insists that Boodin was wrong to focus on George Santayana's *Skepticism and Animal Faith* (1922) when he should have directed his attentions to Roy Wood Sellars's concept of critical realism.[24] Had he done so, Neuber claims, Boodin's critique would not have worked. It is easy to say a critique *not* aimed at a third party–imposed target would "not have worked"; of course not, it was directed at Santayana *not* Sellars. More germane, however, is Neuber's claim that Boodin's characterization of *critical* realism as making "a bifurcation of thing and environment as though they were only externally related" is, as "a programmatic characterization . . . false." Neuber's curious addition of "programmatic" seems to be an effort to insulate his assessment against objections by some special pleading of an indeterminate nature; what exactly he means by "programmatic" is unclear. "To be sure," Neuber argues "the bifurcation of thing and environment figured prominently within the framework of *new* realism. However, from the standpoint of critical realism, there is no such bifurcation. It was particularly Sellars who, in his *Evolutionary Naturalism*, insisted upon the interrelatedness of physical objects, perceiving organisms and their environment." But this is too simplistic for two reasons. First of all, Sellars himself talks of the "parallelism" between "a behavior attitude" and "the *content* of perception." For Sellars there is an "absence of any cognitive relation between the physical existent known and the prepositional knowledge about it."[25] This sounds like the kind of bifurcation—the physical existent known and the knowledge about it—Boodin is talking about. But in *Cosmic Evolution* Boodin puts this bifurcation another way that more directly addresses Sellars's materialism:

> This is what materialism does; and modern science is wedded to the ideal of mechanics. It tries to atomize nature and reduce everything to external relations. It treats nature as merely objective and ignores the subjective factor, thus making a false bifurcation at the outset. It confines itself to certain monotonous aspects of nature; and, in its onesided emphasis, it fails to understand even these, because these can be truly understood only in the wholeness of nature. Thus aligning itself with materialism, science has landed in logical bankruptcy. In its groveling emphasis of the mechanical aspects of nature, it has missed the soul.[26]

Claims of material "interrelatedness" hardly addresses the issue; it *is*, after all, possible to speak of "interrelatedness" specifically (item-by-item) and still bifurcate generally (by epistemological categories).

But there is a second reason to object to Neuber's characterization. It is based upon Sellars's following claim: "If the physical existent is extramental [outside the mind], it is nonsense to speak of a cognitive relation between it and the act of referred knowledge. Such a relation could only be transcendental and non-natural. And like all transcendental relations we soon find that it is absolutely unnecessary."[27] This coincides with Boodin's definition of critical realism, which is *not* quoted by Neuber, namely "that the qualities and relations which we sense are mostly due to the percipient organism and do not exist in objective nature as we perceived them." Whereas in contradistinction *functional* realism, by Boodin's definition, "holds that the qualities and relations which we perceive are functions of objective nature and the percipient organism in perspective relation to one another."[28] The cosmos is an interrelated whole, a mutual interaction of energy fields in a plural universe. Although Neuber wants to argue that Boodin got "lost in the isms," it seems more likely that Sellars got caught up in his own materialism, which is that the physical processes of the brain cause, upon certain development, mental traits and *qualia* itself to emerge (this is not strictly speaking epiphenomenal but rather Sellars's own brand of emergent brain-state materialism). In any case, it is still a form of reductionism.

Now here we have another kind of bifurcation lurking in Sellars's epistemology. In order to make his emergent brain-state materialism work he proposed conscious (those of sensory/emotional responses, *qualia*, etc.) and non-conscious brain states (neural and physiological functions). As John Kuiper points out,

> It is impossible to differentiate (which this theory requires us to do) between brain as conscious and the brain as non-conscious except by reference to mind (which this theory refuses to do). The distinction between conscious and non-conscious brain processes presupposes mind; it cannot explain it. I see no way of avoiding or resolving this impossibility: the concept of a percipient mind is needed to explain what it means for something to be "directly presented," that is, to explain consciousness, but the only way to explain consciousness, according to this theory, is to refer it to the brain. But to explain how the brain can have consciousness within it brings us right back to the fact of immediate awareness, which is unintelligible except on the assumption of a subject or self which has the power of being aware, that is, of being conscious.[29]

So Neuber's criticism of Boodin does not obtain. In fact, Boodin's focus on Santayana was likely premised on his belief that this was the more formidable argument. Furthermore, Boodin's neglect of direct engagement with emergent brain-state materialism was based upon the conviction that Sellars had been effectively defeated by his functional realism. And he was right. After all, Boodin had seen this sort of thing before with William Pepperell Montague's monistic realism. He answered Montague and the *new realists* in 1907 by proposing a pluralistic model of energetic centers of various grades and kinds based upon the differences they make to our reflective purposes (see glossary).

But there is another aspect of Neuber's analysis that bears comment. In his review of Boodin's 1934 article "Functional Realism," he suggests a deliberate distancing from pragmatism. Indeed, according to Neuber, "the major purpose of the paper is obviously to establish functional realism as an autonomous position." There is a good explanation for this. Even by the opening decades of the twentieth century (certainly by 1925) it was clear to Boodin that, although Peirce and James established pragmatism on a firm foundation of empirically based experiential inquiry, Dewey had since taken it in a very different direction. For example, in *Cosmic Evolution* Boodin praises James's "teleological character of mind" and states in comparison that "when Dewey says that mind is the applying of future results to present situations, he ignores the temporal character of the process, the creative advance of nature, for, aside from physical nature which we can take as practically uniform, the future results cannot be thus stated in advance."[30] And further on Boodin criticizes Dewey's procedural approach to thought as "artificial" and "barren." Boodin adds,

> Dewey's scheme [of a complete act of thought] is as far from an account of the process of discovery in a creative mind as is Aristotle's syllogism, though in justice to Aristotle and the logicians of our own day, it ought to be added that they do not pretend to give an account of the psychological process of thought but merely of the formal implications of propositions. Such schemes are *a posteriori* abstractions at best; and while they may have value in clarifying the mind of the spectator and in arranging the products of thought, they do not furnish the rationale of the creative process.[31]

Boodin rejects Dewey's theory of mind as a "veiled Hegelianism" and in renouncing the absolute is "trying to make the sticks [of his naturalistic dialectical theory] actually generate the rose-bush by their own friction."

Yet Boodin's approach toward Dewey was nuanced. He quoted A. W. Moore ("a member of the Dewey school") on consciousness as "a function of a social process, in which my body or brain or mind is only one factor" with some approval. But he adds:

> The theory thus stated does not try to define the nature of the social situation, neither does it discriminate between situations where the motive is individual, and where the social aspects, such as language, science, etc., are strictly instrumental and the situations where the motive is consciously social. In so far as we use the concept social to characterize all our experience, we

have obviously failed to give the differentia between what we may term the individual consciousness on the one hand and the group mind on the other. Moreover, the word "function" is ambiguous. Are my thinking and the physician's thinking in regard to my headache, identical states of consciousness? Or do they merely figure with reference to a common problem? Evidently the latter is all that can be meant in this case. It still remains, therefore, to explain the nature of that social context in which both our minds figure. Does this amount to a common social unity, including both minds and having an existence of its own, or are we simply two numerically distinct minds thinking of the same object?[32]

None of this proved Boodin's crucial point, namely that a social mind exists composed of certain cognitive processes and "distinguishable from what we call the unity of individual experience, and if not more real, at least more inclusive than this."[33]

Nevertheless, Boodin appreciated Dewey's theory of value and wrote approvingly of his *Ethics* (1913) coauthored with James Tufts.[34] But privately Boodin complained to Arthur O. Lovejoy, "I have been picked clean to the skeleton so far as my material suggestions are concerned. (New realism, Dewey, Sellars, etc.) The skeleton or system alone remains."[35] It must be remembered that although Boodin pulled his comments regarding Dewey together in *The Social Mind*, all of them had been expressed much earlier in "The Existence of Social Minds," *American Journal of Sociology* (1913) and "Value and Social Interpretation" (1915). Boodin took note of his own system of metaphysics as "the only individual system which has originated on American soil!" but except for Hoernlé and E. V. Huntington of Harvard, it had been largely ignored. Boodin also complained to Lovejoy that much of Dewey's work had been derivative of his own. Though he stopped short of charging plagiarism, Boodin told Lovejoy that when he read Donald A. Piatt's chapter on "Dewey's Logical Theory" in *The Philosophy of John Dewey* (1939), "it was so like my functional realism for the most part that I scarcely realized that I was reading about Dewey." But then softening his tone he added, "But that had always been the catholic character of pragmatism."[36] For all of Boodin's ill feelings toward Dewey, they shared important philosophical ideas—the emphasis upon the social side of pragmatism, the identification of truth with the test for truth, and the conviction that certain ethical truths are real and can be found in genuine experience. In this sense, it is not surprising that Boodin would speak well of Dewey and Tufts's *Ethics*. None of this withstanding, however, Boodin had deeper reservations about the direction of pragmatism. When he discussed it in *Cosmic Evolution*, he insisted, "I am using pragmatism in its earlier form, before it started to clear away 'misunderstandings.' In its later phase it is difficult to distinguish it from classical empiricism. It has become merely an exaltation of scientific method, of which it has no particular monopoly."[37] There can be little doubt that Boodin is speaking of John Dewey here.

As it turns out, Boodin had taken notice of something largely missed by his colleagues. This has been hinted at but not been clearly elaborated upon until the work of British philosopher Howard O. Mounce over seventy year later. Mounce's thesis is as follows:

> In the twentieth century, Pragmatism has figured, sometimes in alliance with Logical Positivism, as a species of Scientific Positivism or Naturalism, the very doctrine against which Pragmatism, in the hands of Peirce and James, was very frequently directed. It was John Dewey who was largely responsible for this change. This does not mean that he reverted to a reductionist Empiricism. His aim, rather, was to transform Empiricism, using the insights of Peirce and James. But what he produced, in effect, was a defence of the Scientific Positivism which they had attacked. In short, we are confronted by two Pragmatisms.[38]

So we can see how Boodin could mention Dewey and Sellars together in the same breath as having picked him "clean to the skeleton." The commitment of Dewey and Sellars to realism gave them much in common and, as we shall see, their mutual commitments to Darwinism and positivism. Since Sellars's Darwinism in particular and positivism in general have already been covered in chapter 3, attention can be focused solely on Dewey.

Dewey's turn toward positivism should not be surprising. He signaled this direction in his fawning treatment of Darwin, who, as we have already seen, established positivism as the reigning dogma of Anglo-American science.[39] Despite some claims that Darwinian evolution can be found in Peirce and James (a conclusion arrived at only by conflating Darwinian evolution with evolution as a general concept), and it can be demonstrated that when confronted with a distinctly un-Darwinian form of evolution by the cofounder of natural selection, Alfred Russel Wallace, these leading pragmatists clearly endorsed it. The occasion was Wallace's much-touted American tour commencing in New York City on October 23, 1886, and continuing across the North American continent through August 8, 1887. The visit was prompted by an invitation by Boston's Lowell Institute to give a series of eight Lowell Lectures where Wallace presented his uniquely teleological brand of evolution. The opportunity was undoubtedly prompted by Wallace's fame at having developed quite independently of Darwin a theory of natural selection. But Wallace was also known as an intrepid explorer who lived among native peoples of South America from 1848 to 1852 and also the peoples of the Malay Archipelago (today known as the East Asian Maritimes) from 1854 to 1862, compiling his adventures into a tremendously popular scientific travel narrative titled *The Malay Archipelago* (1869). As if to further solidify his scientific reputation in 1876 he published *The Geographical Distribution of Animals*, making him widely regarded as the father of modern biogeography. With Darwin's death in 1882, historian Martin Fichman aptly proclaimed Wallace as "England's elder statesman of evolutionary biology."[40]

When Wallace delivered the first Lowell Lecture on November 1, 1886, it was to a packed house; all eyes and ears were fixed upon the one person who could explain the modern theory of common descent by means of natural selection. But what they would get would be far from the Darwinian version of evolution proceeding largely by chance modifications sorted out by a blind natural selection assisted by two subsidiary theories—sexual selection and a neo-Lamarckian version of heredity Darwin called pangenesis. Wallace, on the other hand, dismissed sexual selection and pangenesis as unnecessary. At least from a purely empirical point of view Wallace argued that natural selection could do most of the heavy lifting, but *not all*. Since the April 1869 issue of the *Quarterly Review* Wallace believed that an "Overruling Intelligence" was necessary to explain the unique higher mental capacities of human beings. Wallace expanded his teleological evolution in three important books: *Darwinism* (1889), in which he extended his transcendent teleology to the origin of life and the beginning of sentient beings; *Man's Place in the Universe* (1903), an argument for the special place of humans based upon cosmological fine-tuning; and *The World of Life: A Manifestation of Creative Power, Directive Mind and Ultimate Purpose* (1910), Wallace's grand statement of a new, evolutionary natural theology. While Wallace by now was a committed spiritualist, it goes too far to ascribe his theistic evolutionary ideas simply to spiritualism since his interests in phreno-mesmerism and the plebian socialism of Robert Owen, all important influences prior to Wallace's first spiritualist experiences on July 22, 1865, moved him away from methodological naturalism and toward a rational idealism ill-suited to Darwin's scientistic reductionisms.[41]

In any case, what that audience heard was not a rehash of Darwin's *Origin of Species* or a Thomas Henry Huxley style "secular sermon" on scientism but a natural theology based upon the latest findings of biological science. In this sense the reactions of Charles Sanders Peirce and William James to Wallace's lectures are important. What is clear is that both Peirce and James agreed with Wallace that the militant scientific naturalism being promoted by Darwin's X Club proselytizers like Huxley, George Busk, and John Tyndall was "misguided and dangerous."[42] Peirce, in particular, accepted Wallace's reconstitution of natural selection within a broader teleological framework as "superior to Darwin's." To Peirce, natural selection is undoubtedly true but incapable to accounting for all of life's diversity, a view he ridiculed as "pseudo-evolutionism."

So what does Peirce and James's encounter with Wallace say about pragmatism and the rising tide of scientific naturalism? It seems clear that when faced with a choice, Peirce and James sided with Wallace. Peirce was particularly vocal in denigrating Darwin's *Origin* as little more than giving the "politico-economical views of progress" a scientific veneer. He was obviously uncomfortable with the ruthlessness of the theory and suggested that Darwin could have rephrased his title from "the struggle for existence" to "the Devil take the hindmost!" In the end, Peirce simply called it "the Gospel of Greed."[43] Peirce would surely have agreed with Oswald Spengler's assessment that Darwin's theory simply drew its air from the stifling "atmosphere of the English factory." James is harder to gauge for two reasons. First of all (within limits) James had a rather annoying tendency to extol the virtues of whatever idea happened in front of him at the time. Second, James thought that pragmatism should be freely used and open to all, he could see only in Dewey's praise of Darwin yet another application of the "cash-value" of the pragmatic view. But Dewey and James had real differences. Voluntarism and fideism that James imbued from the French philosopher Charles-Bernard Renouvier gave James a less positivist tone.[44] But just ten years after approving Wallace's Lowell lecture emphasizing the teleological nature of evolution, James would give his own Lowell Lecture by declaring that "Darwin opens our minds to the power of chance happening to bring forth 'fit' results if only they have time to add themselves together."[45] At best it can be said that James appreciated Wallace's more idealist approach to evolution, and, like Peirce, never accepted that natural selection could be a sufficient explanation for the evolution of mind.[46] In this sense Peirce and James were one with Wallace.

But all this leads back to the beginning, back to Neuber, who faults Boodin for getting lost in the "isms." Again, taken together it seems that if any are caught up in isms it is Dewey and Sellars. Like Boodin, Peirce was sure of Darwinism and it wasn't good; James at times waffled but overall fell in with Peirce. Dewey, however, took pragmatism into the desolate realm of Sellars's uncongenial land of hidebound empiricism and materialism, and *that* was a place the older pragmatists never intended to go. (It is perhaps worth noting that Hilary Putnam recalled his teacher, Hans Reichenbach, somewhat surprisingly complained that Dewey was "too positivist."[47]) Of the three pragmatists discussed here, only Boodin (and much later Mounce) saw clearly what had happened. Dewey and Sellars got caught up in the "isms"—Darwinism and its two companions, scientism and positivism.

I have already explained Sellars's many confusions, but Dewey's case is equally informing if not surprising. Darwinism, taken on its face, is at heart a self-regulating and blind sorting process. If Darwinism has no goal and no intent, then reform within such a scenario becomes oxymoronic. But Darwinian evolution has been no stranger to a wide array of social reformers despite the fact that the theory itself discounts the notion of evolutionary "improvements" by means of purposeful tampering. The eugenicists of the early twentieth century are obvious exemplars of this. It is worth noting that Dewey questioned both the science and the goals of early twentieth-century eugenicists.[48] But even Dewey—too wise and thoughtful to be caught in the eugenics web—suffered his own contradictions by wedding so much of his social agenda to Darwinian ideas, as historian Gertrude Himmelfarb explains:

> Even John Dewey, who prided himself on applying the evolutionary principle to ethics and politics as much as to metaphysics and epistemology, was guilty of this basic inversion of Darwinism. Having earlier objected to "evolutionary ethics" on the curious grounds that by banishing God from "the heart of things," it had also banished the ethical ideal from the life of man—"whatever exiles theology makes ethics and expatriate"—he later accepted it for the equally curious reason that, in showing the origin and development of man in his environment, it showed intelligence

how to fashion that environment for man's great satisfaction. Not the unintended effects of unconscious struggle, which was the point of Darwinism, but the deliberate effect of intelligent control was Dewey's prescription.[49]

Dewey was not the first or last progressive reformer to get caught in logical claws of Darwin's naturalism.

In the end, what can be said of Neuber's contemporary analysis of Boodin? Neuber's effort to seriously examine Boodin's ideas is commendable. He even admits that "Boodin's contributions to the complex discussion of realism, pragmatism, and their mutual relationship are worth reconsidering."[50] This well-intentioned and lightly critical approach is laudable, but his corrective surgery is performed with too dull a scalpel and is commenced without a preliminary investigation as to whether surgery is needed at all. Neuber should have heeded the advice of Hippocrates who always cautioned, "*Primum non nocere* (First, do no harm)." His analysis only ended up harming Sellars and Dewey. You always hurt the ones you love.

Boodin Today

The question naturally arises: Where might we place Boodin today within the larger landscape of science and philosophy? His biographer cites Teilhard de Chardin and Whitehead as possessing marked similarities to the poet-philosopher. Because both have followers today, it is worthwhile to look at each to compare and contrast with Boodin. Here the reader will have to excuse the frequent use of personal pronouns since what is about to be said is more impressionistic and speculative than definitive. According to Charles H. Nelson, Teilhard and Whitehead "had much in common with Boodin in their desire to follow through on the idealistic-theistic implications of evolutionary theory."[51]

At first glance, Teilhard's approach to evolution would seem broadly akin to Boodin's; both developed non-mechanistic teleological theories of cosmic proportions focused ultimately on their meaning and relationships to humanity. There, however, the similarity ends. Teilhard was no academic but rather a Jesuit paleontologist whose principal work, *The Phenomenon of Man*, was written in the 1930s but not published until 1955, the year of his death. The book was not available to Anglo-American readers until an English translation was prepared by Bernard Wall in 1959, so Teilhard's fame was not achieved until well after Boodin's death. The cleric's conception of evolution was cast in explicitly Christian terms unlike Boodin's more strictly philosophical categories. Also, Boodin had nothing like Teilhard's triune conception of evolution as prelife (material), life (the biosphere), and thought (the noosphere). Evolution proceeds teleologically through each stage culminating in its ultimate destiny, the "Omega Point." In some ways, Teilhard's evolution is more reminiscent of Bergson's *élan vital* and certainly more pantheistic than Boodin's. Everything in evolution is the product of God's action and no matter is inert but rather imbued with the presence of God and to be considered sacred. We share in evolutionary development by drawing together in Christ's agape love.[52] While Boodin never had an opportunity to comment on Teilhard's theory, it seems likely he would have regarded it as a form of Christian mysticism lacking any firm empirical moorings. Although Teilhard has retained a consistent following (especially among New Agers), others utterly reject his ideas not on scientific but religious grounds. Catholic mathematician Wolfgang Smith, for example, scoffs at his conflating pantheism: "Notwithstanding all the considerations brought forth by Teilhard de Chardin, there is then, after all, a difference between the Way of the Cross and the metamorphosis of insects! Nor is the cosmos at large destined to complexity itself into the Mystical Body. And it is above all a false and dangerous idea that the voice which presently beckons humanity to pursue the twin goals of technological progress and 'socialization' is in truth the Voice of Christ."[53] Taken in its entirety, the Teilhard/Boodin connection has little to recommend it; mere theism is not similarity.

With Whitehead and Boodin, however, it is quite different. Here we have philosophers of parallel strength and vision. Because a great deal has already been said about Whitehead in relation to Boodin's thought, I shall not dwell on it here. But given the broad similarities between the two, one might wonder why it is that Whitehead's process philosophy prevails today over Boodin's. Both, of course, provide complete and coherent philosophies that include thorough metaphysical systems. But it seems to me that the prevalence of Whitehead is largely circumstantial. To begin with, Boodin never had anything like Whitehead and Russell's monumental *Principia Mathematica*. But that alone cannot explain the endurance of Whitehead. In fact, Ronny Desmet and Andrew Irvine argue that Whitehead's "genuine modesty and aversion to public controversy made him invisible at the philosophical firmament dominated by the brilliance of Ludwig Wittgenstein."[54] Nevertheless, Whitehead is regarded as the leading founder/proponent of process philosophy I think largely for two reasons, both of which lie outside of the field itself. First is the influence of Charles Hartshorne where he dominated the Divinity School of the University of Chicago. Hartshorne took Whitehead's ideas and molded them into a process theology that by the 1960s was picked up by John B. Cobb Jr. and utilized in his influential Christian natural theology. It cannot be said that Boodin launched anything comparable. Some of the reservations expressed in Hartshorne's review of Boodin's *Religion of Tomorrow* suggest some philosophical differences between the two (see chapter 4). However, I think Boodin's circumstances were far different as well. At UCLA, he was very far from Whitehead's Cambridge or later Harvard. The 1930s were formative years for UCLA and the philosophy department didn't award its first PhDs until 1942, three years after Boodin's retirement. Thus, Boodin didn't have an army of protégés to campaign on his behalf. By comparison it might be said that Whitehead had friends in high places.

But Whitehead had a second sphere of influence, and as we shall see this one has a direct bearing on Boodin. I'm referring here to organicism, a term coined by organicism's biologist founder, William E. Ritter, and defined in his *Unity of Organism* (1918): "*The whole organism is as essential to an explanation of its elements as its elements are to an explanation of the organism* [italics in the original]."[55] The word was coined by Ritter in his seminal work and today is closest to systems theory, which biologist Ludwig von Bertalanffy first proposed in 1928 and continued to develop through the 1960s. Of historical interest and contemporary with Boodin is the Theoretical Biology Club, a group of scientists who organized under the impetus of Joseph Needham and Joseph Woodger in April of 1932 and continued, at least in its first permutation, until 1938. Besides Needham and Woodger, it consisted of John Bernal (a pioneer in crystallography in molecular biology), mathematician John Wrinch, and C. H. "Hal" Waddington (a polymath embryologist and geneticist). Of these "members" (it was never a formally constituted organization), Waddington became its most important and sustained proponent. What is more important, Woodger converted Needham, who in turn inspired Hal to study the philosophy of Whitehead; it was through them that a "third way" for the life sciences was developed vis-à-vis Whitehead's non-reductionist process thinking.[56]

The "third way" refers to a new ground for biology between mechanistic explanations of evolution (dominated by the neo-Darwinists) on the one hand and the vitalists (led chiefly by biologist Hans Driesch and philosopher Henri Bergson) on the other. For the neo-Darwinists evolution was a gene-centric model of blind change through time operating primarily (though not exclusively) upon individual organisms; for the vitalists matter was inert and needed some animating force—Bergson's *élan vital*, Driesch's *entelechia*, or some *nisus formativus*—to enliven or vivify matter. Needham specifically charged mechanistic biology with Whitehead's fallacy of misplaced concreteness because neo-Darwinists were forever going beyond observations to extrapolating "facts" that failed to take account of the interconnectedness of the given phenomenon.[57] The organicists saw serious problems with both ideas. The mechanistic approach was too given to piecemeal explanations and overly simplistic "just so" stories to be believed and the vitalist argument lacked any empirical proof. The organicists harkened to Whitehead in rejecting the form/matter perspective for an organization/energy model emphasizing interconnected fields. They saw genetics as working on *potentialities* while development as seen particularly in embryology as working on *actualities*, and Waddington especially used canalization, a term directly appropriated from Whitehead, to show how evolution could be "canalized" or restricted along a nonrandom pre-generated path.[58] Indeed, for Waddington—and this formed a significant source of neo-Darwinian opposition—canalization could account for immediate intergenerational fitness and had neo-Lamarckian implications.

Ernst Mayr and other leaders of neo-Darwinism rejected organicism and pushed against its emphasis upon the epigenetic landscape. Mayr saw Waddington's Whiteheadian focus on epigenetics as too speculative and he told Hal privately in 1959 that his ideas were filled with too many "metaphysical concepts that have no place in our modern thinking."[59] This is rather ironic since Mayr himself was fond of engaging in his own metaphysics tinged with its own contradictory positivist assumptions, insisting that Darwin's *Origin* freed Western civilization once and for all from the shackles preventing progress toward its scientific inheritance—the end of creationism and natural theology, the end of anthropocentrism, the demise of essentialist thinking, the enthronement of probabilistic emergence, and the abandonment of teleology.[60] In fact, historian John C. Greene and Mayr wrangled over the disingenuousness of this position, Greene claiming that Mayr and most of his neo-Darwinian colleagues used vitalistic and teleological language totally incompatible with their mechanistic worldview because it served a psychological function "to lend meaning and value to processes and scientific studies that would otherwise, given that formal philosophy, seem destitute of meaning and value."[61] Nevertheless, in 1961 Mayr wrote an influential essay that talked about biology as two distinct domains differentiated by their approaches to causality: evolutionary biologists can either argue from proximate causes or, as functional biologists, they can argue from ultimate causes. This creates a situation where, depending upon their perspective, biologists can easily talk past each other and use the same words to discuss entirely different phenomena. He then went on to completely reject final causation as a legitimate explanation in modern biology. Waddington, who had developed over the past few years a series of illustrations aimed at mapping his conception of the epigenetic landscape, was compelled to respond. He did so in the most prestigious journal of the sciences, *Nature*:

> It is becoming inadequate to point out, as Mayr does, that natural selection is not purposive. In itself it is of course no more purposive than is the process of formation of interatomic chemical bonds. But just as the latter process is the basic mechanism underlying the protein syntheses which are integrated into the quasi-finalistic mechanism of embryonic development, so natural selection is the basic mechanism of another type of quasi-finalistic mechanism, that of evolution. The need at the present time is to use our newly won insights into the nature of quasi-finalistic mechanisms to deepen our understanding of evolutionary processes.[62]

Waddington insisted that Mayr's account was incomplete in that he ignored developmental biology. His use of the term *quasi-final* was used to avoid the taint of overt teleology, but he obviously was proposing a far less reductionist version of evolution than the neo-Darwinists. In the end, Waddington's pleas were largely ignored, so much so that by the 1970s Hal became as disgruntled as Boodin had some thirty years before at all the farm yard philosophizing. He expressed his dissatisfaction by turning Boodin's attack against "pig-trough" materialism into an acronym directed primarily against the neo-Darwinists—COWDUNG (Conventional Wisdom of the Dominant Group).[63]

Although there seems no evidence that Boodin corresponded with any members of the Theoretical Biology Club, it seems clear that these organicists agreed with Boodin that Darwinian evolution was too reductionist, too incomplete, and too mechanical to be of much real value in the life sciences. In fact, Boodin's discussion of time in his lecture titled "Man in His World" with its exemplary emphasis upon embryology could have almost been written by Hal himself.

Of course, the big difference between Boodin and the organicists was the former's reliance upon a transcendent teleology in evolution, though as we have seen it was not of the vitalists' variety, which Boodin always criticized. For this organicist, Whiteheadian—perhaps better stated *Boodian*—perspective, we need to examine the ideas of biologist Rupert Sheldrake. It is here that today's best representative for the poet-philosopher can be found.

Since 1981 Cambridge trained biologist Rupert Sheldrake has proposed two complementary theories of evolution, both of which are based upon Whiteheadian philosophy and the biological model of organicism. The first is formative causation; the second is morphic resonance. Formative causation "proposes that nature is habitual. All animals and plants draw upon and contribute to a collective memory of their species. Crystals and molecules also follow the habits of their kind. Cosmic evolution involves an interplay of habit and creativity."[64] Morphic resonance is the idea that "habits" move and function through canalized (à la Hal Waddington) energy fields. Morphic resonance is the theory that memory operating through morphic fields is inherent in nature. Rather than thinking in terms of fixed and eternal laws of nature, it is more accurate to view nature as comprising evolving habits that are predisposed toward creativity.[65] Sheldrake is a thoroughgoing process thinker who, as Boodin would say, takes time seriously. Evolution at biological and cosmic levels may be thought of as occurring through morphic fields, which as Sheldrake explains, is

> a field within and around a morphic unit [a unit or form of organization] that organizes its characteristic structure and pattern of activity. Morphic fields underlie the form and behavior of holons [a term borrowed from Arthur Koestler] . . . at all levels of complexity. The term *morphic field* includes morphogenetic, behavioral, social, cultural, and mental fields. Morphic fields are shaped and stabilized by morphic resonance from previous similar morphic units, which were under the influence of fields of the same kind. They consequently contain a kind of cumulative memory and tend to become increasingly habitual.[66]

Morphic fields are inherently creative. Preexisting morphic fields demonstrate "creativity in a weak sense of the word. The end points or goals or attractors given by the fields remain the same; what are new are the ways of reaching them. This kind of creativity is commonly expressed by words such as *adaptability, flexibility, ingeniousness,* and *resourcefulness.* The appearance of entirely new fields with their new goals or attractors involves a higher order of creativity and originality."[67] And this creativity through "wholes" in nature need not be restricted to living organisms but can include crystals, molecules, atoms, and subatomic particles. Sheldrake explains that "the morphogenetic fields of crystals and molecules are probability structures in the same sense as the electronic orbitals in the morphogenetic fields of atoms are probability structures."[68] The morphogenetic fields, in effect, restrict the vast number of possible arrangements otherwise permitted by their constituent molecules.

Although influenced by the organicists, Sheldrake approaches teleology unafraid. He is honest and reasonable regarding the epistemic reach of formative causation; he does not claim too much for it. For example, it cannot explain the ultimate source or origination of new forms and new patterns of behavior arising within these morphic fields, and it is equally incapable of explaining subjective experience. Both of those explanations lie beyond the reach of natural science and belong in metaphysics. Nevertheless, formative causation is open to four possible conclusions: modified materialism or physicalism; the conscious self, or the ability to "enter into" morphic fields presumably through some quantum aspects of the brain; an imminence within life similar to Bergson's *élan vital* that could even exist in hierarchies within nature; and finally, transcendent reality that proposes a conscious being as the source of the universe in which the "wholeness" of organisms at all levels are a reflection of the transcendent unity upon which they depend and from which they were derived. Of these four, Sheldrake, an openly perennialist Christian, comes down on the side of transcendent realism because only this last "metaphysical position affirms the causal efficacy of the conscious self, *and* the existence of a hierarchy of creative agencies imminent within nature, *and* the reality of a transcendent source of the universe."[69] Sheldrake's proposals are admittedly controversial, but after forty years he has only been met with dogmatic skepticism. Nevertheless, the jury remains out on Sheldrake but there are certainly reasons to take his ideas seriously. As one recent analyst has concluded, Sheldrake's new kind of biology "remains a source of stimulation for new generations of researchers, both through the questions it asks and the possibilities it proposes."[70]

The parallels with Boodin's philosophy should now be obvious. Writing well before Sheldrake, Boodin almost eerily anticipates his central ideas:

> I can see only one possible way in which the organism [or anything else in nature] can function as a whole.... The organism is more than a collection of factors. It is a whole.

> It remembers, it intuits, it looks forward according to its own genius. The multicellular organism is more than interacting cells. It organizes these into an energetic unity with a pattern of its own. There is, if you like, a psychological element or aspect to the process; but this principle is immanent in the process as a whole, and it is by virtue of it that the collection of factors is a whole. You may retort: Why not say that a crystal or even a molecule or an atom possesses soul? They act as wholes and have unique characteristics. I have no objection to extending the meaning of soul to this extent. There is the emergence of a new whole-forming pattern in all these cases. But we must recognize different levels of soul. The soul which emerges in organic genesis is not personality with imagination and thought. That is a later emergence in the social milieu. The organic soul has its own characteristics—its own sense of wholeness, its own memory and foresight, its own methods of adapting means to ends. It emerges as a result of the synthesis of two streams of heredity in the milieu of its environment—somatic and cosmic. It controls the interaction of parts and follows a logic and a sense of its own.[71]

The essential elements of Sheldrake's ideas are here: the creative wholeness of the universe, from living organisms to crystals, molecules, and other things and the idea of a natural "memory" or "intuition" that occurs at multiple levels and is even extendible to social relations; all are transpiring within a transcendently guided evolutionary process. Boodin even alludes to canalization when he makes reference to "the dynamic field" becoming "grooved through habit."[72] In contrast to the physicalists who believe that our thoughts, actions, and emotions are controlled exclusively by our supercomputer brain, Sheldrake argues that the mind, through formative causation, is extended well beyond the organ inside our head. "We are linked to our environment and to each other," explains Sheldrake. "Likewise, our minds pervade our bodies, and our body images are where we experience them, in our bodies, not just our heads." He adds that "the mind-in-the-brain theory turns out to have very little evidence in its favor. It contradicts immediate experience."[73] Boodin too rejects the mind-in-the-brain idea and says, "We must conceive mind as a field of energy, which in turn owes its characteristics to the intersection of the life stream with the structure of the cosmos, for it is in creative adaptation to the cosmos that the organism evolves for mind and becomes charged with mind."[74] Even in papers Boodin had been working on at the end of his life but never published, wholistic cosmic field theories canalized in teleological communal histories not unlike Sheldrake are clearly in evidence.[75]

If, as I firmly believe, Boodin's closest representative today is Sheldrake, then one possibility that might occur to some readers can be ruled out—the association of Boodin with the modern intelligent design (ID) movement. Sheldrake rejects ID because its proponents are too wedded to the machine analogy of design. William Paley's watch as an example of design has been appropriated by Michael Behe in the bacterial flagellum as an "outboard motor."[76] Nothing like this, ID proponents insist, could come about by random mutation and natural selection. But this kind of argument is inimical to Sheldrake. He points out:

> The problem with the design argument is that the metaphor of a designer presupposes an external mind. Humans design machines, buildings and works of art. In a similar way the God of mechanistic theology, or the Intelligent Designer, is supposed to have designed the details of living organisms. Yet we are not forced to choose between chance and an external intelligence. There is another possibility. Living organisms may have an internal creativity, as we do ourselves. When we have a new idea or find a new way of doing something, we do not design the idea first, and then put it into our own minds. New ideas just happen, and no one knows how or why. Humans have an inherent creativity that is expressed in larger or smaller ways. Machines require external designers; organisms do not.[77]

Boodin would surely agree.

ID is frequently attacked as a pseudoscience, but that is not its real problem. Its real problem is it is so utterly captivated by the desire to be considered a real science that it has fallen prey to its own brand of scientism. Erkki Vesa Rope Kojonen has accurately identified this as ID's central difficulty. He states quite accurately that "the influence of scientism on the ID debate is implicitly visible in the overt focus on arguing over whether ID is part of the natural sciences or not, and in the undervaluing of arguments and forms of rationality that are not part of the natural sciences, such as philosophy and theology."[78] Kojonen's point can be readily confirmed by simply looking at ID's own special pleadings for its claim to science.[79] ID's self-proclaimed definition according to their leading spokesperson, Stephen C. Meyer, is that "the theory of intelligent design holds that there are tell-tale features of living systems and the universe that are best explained by an intelligent cause—that is, the conscious choice of a rational agent—rather than an undirected process."[80] The examples typically given are bacteria and complex and intricate organisms and even ecosystems that function with machine-like precision. Of course, precision is in the eye of the beholder, but here nature is given over to a concatenation of contrivances the inferences of which are really more metaphysical in nature than scientific. Things can be complicated, but the next step to a designer—whether it's called a mind, a telic process, or something similar is all essentially the same idea—that ID takes is really a metaphysical inference. Behe writes approvingly of Paley's watch example, admitting only that the cleric pushed his point too far and added some examples that were downright silly.[81] But it must be remembered that Paley's argument was as the title of his book, *Natural Theology*, not science per se. ID proponents either cannot seem to tell the difference or simply prefer the prestige and status of science while downplaying philosophy, theology, or any of the humanities. As it is, then, the "rational agent"—God, spirit, telic force—is reduced to a diligent craftsman (a kind of demiurge) or worse, an assembly line worker. Reductionisms can come in many forms, even when in search of transcendent guidance in nature. If we are to chide the neo-Darwinists for turning their metaphysical commitments into scientific proclamations, then the same applies to ID proponents. As Greene has observed, "All are entitled to say their say so long as they do not represent their way of making experience as a whole intelligible as the finding of science or a necessary deduction therefrom. By the same token religious persons [ID advocates included] should not represent their religious beliefs [or metaphysical convictions] as 'creation science.' Scientism is equally objectionable at both ends of the spectrum."[82] Boodin consistently rejected scientism and reductionism in any and all forms and there seems little in ID theory that suggests anything of interest for our poet-philosopher. Boodin always argued for transcendent teleology in the biological and cosmic worlds, but (like Sheldrake) it was on metaphysical grounds. Science can surely inform metaphysics but it cannot *become* metaphysics.

Here a relevant postscript is worth adding. Given Boodin's career-long commitment to a universe of progress and teleological significance, did he ever second-guess himself? Did his faith ever waver? The answer seems to be a qualified yes. In one of his last published works, an essay titled "The Philosophy of History," he wrote:

> I confess that, though I believe that the cosmos is a whole and acts as a whole through the multiplicity of histories, I do not understand what sort of whole it is. It is true that a tree must be known through its fruit. But what is the fruit? I like to dwell on the creation of beauty and order. But what about the disorder? Perhaps blindness and chance are part of the evolution of order. But there is so much disorder and suffering that seem to lead nowhere.

This from a man who spent book after book, paper after paper, telling his readers exactly what kind of whole the cosmos is and that in order to have gain there must be suffering and we accept the risk along with the freedom that comes in the bargain. But he collects himself a few lines later, reminding us that we are children of God, "of free creative personalities, working to realize a new order and a new happiness—that is something we must strive to bring to pass. And this striving is part of the cosmos, its urge in us. It is, I believe, its deepest urge."[83] Written late in life, when age and infirmity may have been crowding in (just a few years later he would suffer a debilitating stroke), this waffling is out of character, and we would be hasty in drawing too many—if *any*—real conclusions from it. Like us all, he should be allowed his uncertainties; it shows a humility and fallibilism to which none of us are immune.

Epilogue

What are the implications of Boodin's thought? Having traveled down this long road together, there should be some take-home lessons to be learned and some profitable ideas to be gleaned from a philosopher who is all but forgotten. I am going to suggest three broad areas where Boodin has relevance for today.

Science

The first to be examined is science. As should be clear by now, Boodin was interested in the latest scientific research and findings, not only in the rapidly developing field of physics but in biology as well. Andrew Reck's dismissal of Boodin's philosophy on the basis of science in 1967 shows the riskiness of making sweeping and definitive statements about the value of someone's work. Reck treats science as a fixed and cumulative endeavor rather than a dynamic enterprise. Science changes, and, as it turns out, the physics of today is nothing like the physics of over fifty years ago. Much of it began with theoretical physicist David Bohm's implicit order released to the public thirteen years after Reck's rather damning declaration.[1] Far from a "retreat" from wholistic concepts or from the reductionism implied in that entry, current physics takes the idea far more seriously today. As physicist/philosopher Bernard d'Espagnat has concluded:

> The idea that Being is somehow prior to the mind-matter splitting becomes defensible even in front of a scientific audience. Consequently, the idea that the mind may vaguely "recall" something of Being does not look any longer irremediably absurd. So

that, even though it is not "reachable" the "Being" in question appears to be an "I don't know what" to which it is conceivable that the human mind is not altogether extraneous. A Being, in other words, that may constitute for it a horizon. What I mean is that, perhaps, the archetypes of some of our feelings, great longing, love, etc., are hidden there. Northing, of course, proves this to be the case. Nothing even *suggests* it is. But nothing proves, nay, nothing serious any more suggests it is not. People who, explicitly or not, ground their thinking on the concepts of classical physics are quite naturally incited by the latter to consider that we are foreign to the world. . . . But an analysis grounded on the data of contemporary physics lead, as we see, to a conception that is quite different and highly less pessimistic.

From all this it follows that, today, even setting into operation an (indispensable) acute critical turn of mind no more results in discrediting the spiritual impetus that moves mankind. An impetus Einstein adequately evoked . . . when he concisely wrote that men want to live the "whole of what is" as something endowed with both unity and meaning.[2]

And this surely was Boodin's interest and goal throughout his career as a philosopher—an examination and explication of "the spiritual impetus," indeed the quest to know and to live the "whole of what is." If Boodin's philosophy tells us anything it is that a nonreductive science can lead to rich and rewarding inquiry into areas involving the nature and meaning of existence and our place in it. It can tell us that but no more; it cannot tell us exactly *what* that nature or meaning is. Boodin, of course, was interested in precisely that. He buttressed his philosophical conclusion with the empirical data of the sciences, but for Boodin science was a tool to an end, not an end in itself. Science is not designed to address Boodin's ends. In order to do that we must go beyond and delve deeper into the import and meaning of the poet-philosopher for us today. We must go to other kinds of investigation of deeper epistemological reach.

History/Philosophy

Within the context of this study, history and philosophy are intimately related. Philosophy is about ideas, and insofar as human beings think, organize, and express those ideas they become part of our collective past. As such, they *become* history. Now I am keenly aware that I am not a philosopher, but if I may be permitted the luxury of analyzing and commenting as an observer on the outside looking in, perhaps both disciplines will have something to gain. Understanding Boodin by studying the science that underpinned his ideas will not take us very far; in fact, I believe the overall epistemic reach of science is fairly short. It is certainly not unimportant to ground our ideas in the empirical data of science, but data are more descriptive than explanatory. History and philosophy, however, are quite different; if they do *anything* worthwhile, they must explain.

Apropos here is the long and often torturous debate over evolution. Boodin's process philosophy was inherently evolutionary on multiple levels, but clearly also the most controversial starting point is his relationship to Darwinism. Much of Boodin's writing occurred before or during the period when discoveries in genetics were being applied to Darwinian theory thus yielding what we know today as the evolutionary or neo-Darwinian synthesis. Boodin consistently rejected Darwinian evolution as a mechanistic version of emergence that expected too much from chance and necessity. Its materialistic implications rankled him. What is important to remember here is that Darwinism is not merely a set of propositions about the diversification of species; it is a set of propositions cast within a framework of methodological naturalism leading toward what Maurice Mandelbaum revealingly calls "undogmatic atheism." There are, of course, Darwinian theists, but examination of their beliefs show these generally to be mere declarations that are, in fact, incoherent add-ons to the theory itself. As I have explained earlier, this shouldn't be surprising given the fact that Darwinian evolution is really a form of positivism. While this is controversial, it has considerable historiographical support. The neo-Darwinian synthesis did nothing whatsoever to alter this, and it remains as positivist today as it was in Victorian times, though, of course, it is expressed differently. It may more generally be considered a form of philosophical naturalism. One must agree with historian George Levine when he writes of the Victorian science flag wavers, "The world looks to be one thing; astonishingly, it is really another. Ironically, even naturalism's antimetaphysical epistemology is built on fundamental metaphysical assumptions, and the deep ethical urgency of the naturalists' project is bogged down in the impossibility of deducing an 'ought' from an 'is.' To write like a scientific naturalist is to create a world that is fundamentally self-contradictory."[3] This problem vexed Huxley, but by the time Ernst Mayr was writing as the sole surviving grandmaster of the neo-Darwinian synthesis, the qualms has dissipated into a scientism that was only interested in the *is*.

The point is, Darwinian evolution is a set of scientific proposals that serve as scaffolding for its positivism. Surely this scaffolding has shifted and has been reconfigured through history to shore up those parts of the positivist edifice that appear most in need of support, but the structure itself is *always* there; in fact, that is the reason for the scaffolding in the first place. So, then, it is not out of the question to regard Darwinism and its synthesis as really a metaphysic.[4] In this sense, Boodin's, and for that matter Sheldrake's, evolutionary views are essentially competing metaphysics with Darwinism and its neo-Darwinian twin. The big difference is that Boodin and Sheldrake both recognize when they are talking science and when they are talking metaphysics. Boodin and Sheldrake also recognize the value and importance of sound metaphysical statements and systems. The Darwinian evolutionists do not; they are thoroughly enmeshed in "science talk" even when they are really talking metaphysics. Because of this they invariably tend to be poor metaphysicians in ways that Boodin and Sheldrake are not. The criticisms leveled here against the Darwinists can also be charged against their intelligent design opponents who, as I've already pointed out, have fallen prey to their own brand of scientism.

It is possible to put a finer historical/philosophical point on this, as Hilde Hein had done many years ago. Unfortunately, Hein's insightful papers have—like Boodin—been virtually forgotten.[5] Had it not been for Erik Peterson's outstanding book on the vitalist/organicist debate (*The Life Organic*), I would have missed her valuable addition to our understanding of the history and philosophy of science.[6] Hein proposes that concepts like vitalism and mechanism are not genuine scientific theories and they are not amenable to experimental verification. "Labels like 'vitalism' and 'mechanism,' do not refer to theories about nature." She argues, they "refer rather to theories on the meta-level, to theories about theories about nature. It is only theories about nature that we require that they be scientific; theories on the meta-level fulfill different functions."[7] Meta-theoretical commitments such as these cannot be resolved by an appeal to facts because each respective meta-theoretical commitment has its a priori interpretive filters through which these "facts" are always passed; confirmative "evidence" for the one will always be seen as nonconfirmatory or perhaps even falsifying for the other. Indeed, for Hein, meta-theoretical disputes are not constructions of varying pieces of scientific data but are, in fact, "differences in fundamental philosophical commitments."[8] Viewed in this light I see no reason Darwinism, Darwinian theism, organicism, orthogenesis, structuralism, or some other teleological version of evolution cannot each be seen as competing meta-theories about organic and/or cosmic change.

Hein is not suggesting that meta-theories are useless and unproductive accoutrements that should be abandoned. She would be wrong if she did. At the very least they serve heuristic functions. I'm not sure they could be avoided in any case; science and philosophies of nature do not occur in intellectual vacuums. But we should be clear about what exactly we're doing when we have these conversations. In this sense, endless evolutionary arguments over the "best" science are simply wrongheaded. If Boodin has shown anything it is that we need evolutionary theories founded not upon misattributions of "science" but upon sound philosophy.

Here Boodin may definitely show the way. I have proposed that mere tweaking of the extended evolutionary synthesis as proposed by many evo-devo proponents, welcome as some of their suggestions are, will do little to remedy the intellectual prejudices currently plaguing it. A complete overhaul is suggested by applying principles with biology in a multidisciplinary, non-reductionist philosophical framework. That could be done by building on the concept of organismic-systems biology, a component of general systems theory associated with the organicist biologist Ludwig von Bertalanffy, and cosmic evolution as proposed by Boodin to form the outlines of an entirely new evolutionary synthesis.[9] My suggestion is offered not as a complete theory by any means but as a model proposal for further study. There is certainly no reason that many of Boodin's original formulations could not be updated by including morphic resonance and formative causation around which they might be revised. Wider analysis of Boodin's theory of time and his own panenthic process theology could fill out evolution with a richness it simply couldn't fulfill until now. Whether my historical/philosophical approach could be built into a lasting metatheoretical structure remains to be seen. At the very least the Boodin/Sheldrake connection may have already gotten us to something similar.

These issues are essentially historical in nature, but where has philosophy itself stood in this? As a historian I am not in an ideal position to comment in any detail on how the profession has handled these issues, although the obscurity of Boodin may be suggestive. As Peterson points out, in the summer of 1918 the American Philosophical Association did convene a five-person committee to examine the vitalism/mechanism controversy. When the report was issued, the verdict was straightforward: "Vitalism was bad science at best, probably poor philosophy as well."[10] The facts are a bit more complicated since for philosophers like Boodin the question wasn't really between mechanism and vitalism but between larger issues and categories of significance particularly as he had laid them out in *Three Interpretations*. Others apparently agreed. Professor Pratt complained, for example, that the association convened and addressed the matter under false pretenses. It was supposed to be an open discussion of vitalism and mechanism and, in fact, "it had turned out to be a pæn of unanimity for mechanism."[11] The idealist Hoernlé who had been a part of the proceedings defended vitalism at least in principle by addressing the larger metaphysical question (Hein would have been pleased) and noting that "mechanical categories are inapplicable even in the realm of biology, that biological facts, in so far as they are distinctively biological, can not be subsumed under mechanical categories."[12] He then proceeded to launch into an attack of "nothing but" types of explanations to which the mechanists were prone. Boodin would have agreed.

Clearly, however, the majority opinion sided with mechanism and in the ensuing years those like Boodin who protested were pushed aside as irrelevant to the "real business" of philosophical progress. A few lone voices like philosopher Robert Kreyche, who as we have seen protested against the popular reductionisms of Roy Wood Sellars, chided his colleagues for their "betrayal of wisdom." He bemoaned the field for falling "in the hands of linguistic technicians"; for its too ready acceptance of the behaviorists' denial of self-transcendence; for its rejection of the best that's in pragmatism (the integration of philosophy and life), but joining in pragmatism's inability to "sustain a sense of purpose that goes beyond the limited goal of our immediate experience"; and for becoming so "self-conscious concerning the problem of method" that many in the profession "have abandoned philosophy in order to become handmaidens of science."[13] Kreyche takes particular aim at Dewey for subordinating the methods of philosophy to the methods of science and his attempt at reconstructing philosophy as an "earth-bound naturalism that makes little provision for the human spirit."[14] It is easy to applaud Dewey for his democratic spirit and social engagement, but overall he seems a laudable but in some sense tragically incomplete figure. Boodin would have agreed with this, including Kreyche's comments on pragmatism (it should be recalled that Boodin adopted many pragmatist positions but never became a *pragmatist*). This was the state of affairs in philosophy some twenty years after Boodin's death. Can it be wondered then that Boodin, whose complete metaphysical system might have provided an ameliorating influence over these impositions, was ignored and worse, dismissed? It would appear that philosophers of the period had fallen into a hubris all their own.

How true this is today I cannot say, but others more acquainted with the profession have commented on its current state.[15]

Modern Culture

It remains to suggest Boodin's relevance to our modern culture, popular and otherwise. I can name three areas. First, we must get beyond the creationist/science debate and begin a conversation over the respective merits of competing metaphysical systems. Arguments over "faith versus science" and accusatory tirades against "unscientific" Americans are not helpful to the public discourse.[16] Unfortunately, as we have seen, in an era when we need philosophy more than ever, we have seen the rise of scientistic "philosophical know-nothingism" (I wish I could take credit for the phrase but it belongs to Erik Peterson)[17] and an increasing close-mindedness over what amounts to serious metaphysical questions. Even though I have been critical of intelligent design, I agree with philosopher Thomas Nagel when he complains, "Even if one is not drawn to the alternative of an explanation by the actions of a designer, the problems that these iconoclasts pose for the orthodox scientific consensus should be taken seriously. They do not deserve the scorn with which they are commonly met. It is manifestly unfair."[18] Of course the intelligent design proponents bring much of this on themselves by staking their claim emphatically to be a science, making them, in effect, rivals to the neo-Darwinians who are both suitors to the acknowledged cultural prom queen, namely, an all-consuming Science of ontological proportions. The sorting out of metaphysical versus scientific claims is urgently needed. As Auxier has observed with regard to the endless (indeed *fruitless*) evolution debates, there is a conspicuous "absence of an idealist voice in the present fray."[19] This, of course, is not surprising when both sides are courting Science (capitalization intended) with the fawning infatuations of a lovestruck adolescent. Until they are disenthralled from their pursuit, any expression of genuine idealism would be the equivalent of bringing your father along on a date. Surely a reexamination of Boodin's thought would be a step in the right direction.

To clarify by way of example, consider what we know about evolution. According to the neo-Darwinian synthesis, organic evolution consists of four components: (1) the random process of mutation with DNA; (2) random drift, determining which mutations are retained and which are discarded (most mutations must cross a fractional threshold before they can become fixed in a population, and thus most do not); (3) natural selection, which is not random because it acts as an eliminator in the process of fitness challenges presented to it; and (4) the environment (climate parameters of precipitation, temperature, even wind variables. As a whole none of this can be directional or goal directed). Three of the four points would appear to be random, without direction of any kind, but we don't *know* that. It could be that design and intentionality even under the neo-Darwinian model occur at the quantum level; this is Alvin Plantinga's suggestion.[20] But what we know today suggests that this is an incomplete and superficial model. As James Shapiro has pointed out, epigenetic factors show that "cellular and genomic capacities of eukaryotic cells have proven to be capable of executing rapid macroevolutionary change under a variety of conditions."[21] Furthermore, "the concept of abundant 'selfish' or 'junk' DNA in complex genomes is mistaken," a product of gene-centric tunnel vision typical of the neo-Darwinian perspective. What this "science" of speciation can tell us about the nature of evolution is unclear. These capacities could be strictly internal to the organisms themselves; this is what Shapiro and Denis Noble believe. This is pure conjecture on their part; it has little to do with "science" as such. It could just as easily suggest a return to some form of vitalism. This is the conclusion of physiologist/biologist J. Scott Turner. The question is, is the internal capacity of organisms for the kind of rapid macroevolutionary transformations a "nothing but" purely material capacity or something more? Turner would see Shapiro and Noble's claims as ahistorical. For him, "That older conception [of the organism] embodies in the organism an ideal of autonomy, integration, purposefulness, and intentionality. In this conception, the organism is life's unique expression, and this makes the organism a quintessentially vitalist idea."[22] Perhaps. Or maybe Boodin and Sheldrake are closer to the truth. One thing is sure—science can inform these questions but it cannot answer them. *That* requires careful philosophical examination and discussion; we need more and not less philosophy. We will not get there by confusing scientific descriptions with metaphysical explanations.

A second relevant area is the need for what can only be called a better social ecology. Royce expressed this in his Beloved Community, and Boodin expanded on it in *The Social Mind*. This is not just better stewardship over the planet (needed as it is), but it is bringing together the wholeness of knowledge. Actually, the problem just outlined between contentious ideologies not knowing the difference between scientific proof and metaphysical argument is indicative of a cultural divide that has grown up in modern society. Wendell Berry understands this and takes aim when he writes, "It is clearly bad for the sciences and the arts to be divided in to 'two cultures.' It is bad for scientists to be working without a sense of obligation to cultural tradition. It is bad for artists and scholars [philosophers and historians] to be working without a sense of obligation to the world beyond the artifacts of culture. It is bad for both of these cultures to be operating strictly according to 'professional standards,' without local affection or community responsibility, much less any vision of an eternal order to which we all are subordinate and under obligation."[23] Boodin speaks to this in his social philosophy; it could serve as a precis for his book. Although some of the problems he addresses seems dated, his overall perspective and guiding theory of community wholeness isn't.

Implied in Berry's statement is the third and final area of Boodin's relevance as witnessed in his theological trilogy. Boodin was fond of saying that we need to take time seriously. I would add we need to take religion seriously, at the very least in a way that perhaps doesn't require church-going, mosque-going, or temple-going religiosity (although all are fulfilling actualizations of the Beloved Community) but takes the efforts to come to grips with principles and ideals larger than ourselves seriously. We can be ecumenical in this, taking religious precepts and impulses in their widest possible applications instead of regarding them as quaint superstitions. And yet when it comes to religious ideals and aspirations, we seem to have fallen into either a blissful ignorance or Bunyon's slough of despond. Martin Luther King Jr., arguably the greatest philosopher-activist of the twentieth century, understood this better than most. King, who took religion most seriously, used faith as the unifying agent to bring society into wholistic union. He tragically died before his work was completed, but in the years since, we seem to have cast his credo aside for something different. In any case, revisiting Boodin's theology and its wholistic, process, and social emphases would lead us in a constructive direction. His faith was heady but honest, passionate but practical, heaven bound but earth focused. Boodin's *Religion of Tomorrow* still awaits the dawning of that day.

So what specifically might be done to resurrect the best Boodin has to offer? In answer, I would suggest a reprinting of *The Social Mind* and *Religion of Tomorrow*; both books are out of print and obtainable only by interlibrary loan or through an antiquarian book aggregator. Both books are important parts of Boodin's overall philosophy and worth having a wide readership. Beyond this perhaps a book of readings intelligently informed with an introduction would be excellent additions to any intellectual history course seminar and presumably philosophy as well. Without making too many constricting recommendations, I would say that his essay "The Reinstatement of Teleology," his lecture titled "Man in His World," and his two-part essay "Fictions in Science and Philosophy" should be included. His 1913 teleology paper appears as part of his *A Realistic Universe*, but certain revisions were made that take some of the punch out of the original; it is better taken undiluted from the context of a larger work. Boodin's faculty lecture "Man in His World" is a consummate summary of his mature philosophy and should be better known. His two-part essay "Fictions in Science and Philosophy" should be added if only because so many of the fictions discussed therein are still with us today.

Obviously, I think we have much to learn from John Elof Boodin. I certainly learned a great deal. If nothing else the poet-philosopher taught this old dog a new trick: not to forget that predecessors haunt ideas, even the seemingly "latest" ones. Until Boodin focused his attentions on the philosophical problems attending mechanism and the mechanistic outlook, I failed to see an ill-suited and fractured natural theology lurking behind intelligent design that at the same time wanted desperately to be a "Science" honored and respected. The inherent contradiction was made evident in Boodin's *Three Interpretations* though he didn't mention Paley once. He didn't need to. When I saw the affinities between Boodin and Sheldrake it all came together. The neo-Paleyan special pleadings of intelligent design—whether Behe's sophisticated flagellar "motors" or his "irreducibly complex" biological systems compared to mouse traps—end up looking as threadbare and inadequate today as metaphysical explanations as they did two centuries ago.

I can only doff my cap to this kind, affable, and at times brilliant man. I certainly have no definitive answers to the nature of organic life or how it has diversified nor can I offer solutions to our many social problems. But the reductionist answers do not seem coherent or satisfying. Although Boodin's career transpired out of place (in a small college in the Midwest followed by a fledgling university in the far West) and out of time (an age of scientism and materialism that is still with us), he offers some intriguing possibilities that deserve resurrection. I can only hope others will read this and rethink the reductionisms plaguing our intellectual landscape. We should get out our shovels to unearth these long-entrenched ideas, although some have become so ossified a pick-axe may be required. Boodin offered alternatives. Over the years I found him an affable companion. In the end, Boodin became not a subject of study, he became a friend. May he will be for you too.

Appendix

"Remembrance of a Common Past"

A Glimpse of Provincialism and the Beloved Community of Sweden as Boodin Knew It

The rural neighborhood, in the highland of southern Sweden, where I was born and lived my childhood and early youth, came very near being self-sufficient. My family raised the grain on the farm, took it to the mill on the river nearby, brought back flour, and made it into bread. We raised our own vegetables. Milk was furnished by our own cows and we made our own butter. We sheared our own sheep and spun the wool and wove it, then took it to the fuller on the river. The leather for shoes came from hides of our own cattle, which we slaughtered at home. We then sent the hides to the tanner, a little farther away. When we needed clothes the tailer of the neighborhood came to our house with his apprentices and stayed until the large family was provided. When we needed shoes, the shoemaker and his apprentices likewise came and lived with us till they had finished. The village blacksmith provided us with the implements—plows,[1] harrows, scythes, axes which we needed for the farm. We had no modern machinery. We cut our own wood from our own forest of pine, birch, and oak. We had a very considerable pine forest and sometimes sold a park of mature timber to the outside world. We carted it for the most part ourselves to the station six miles away, and since my older brother wanted me as a child with him on his trips, I saw the mysterious trains come from the unknown and go to the unknown. It was a great thrill to watch them but I did not ride on one until I went to college at the age of sixteen. The groceries which we needed we bought from a neighbor's store. The public primary school was in the neighborhood and my father was one of the directors. But the grammar school was next to the church three miles away and I walked back and forth. My further preparation for college was by the kindhearted Pastor, C. D. Sjöfors, and his wife (blessings on them!), who took me into their family.

We all went to church on Sunday and extended our acquaintance with the parish. The parish school was the center of the communal life as well as of religion. The parish affairs were disposed of at communal meetings, which selected their own chairman. The pastor represented the state as well as the church. He baptized the infants, churched the mothers, confirmed the young (after they had finished grammar school and had received in addition instruction in the catechism from himself), gave the sacrament to those confirmed, visited the sick, and buried the dead. He was the head of the schools. He represented culture and authority in a very kindly way. He and the teachers were elected by the voters of the parish, but with permanency of tenure.

While the parish was the corporate community, it was rather a secondary than a primary community, at least for me as a child. It stretched three miles in each direction from the church. Many of the people were comparatively strange to one another, though of course they were conscious of belonging together as a parish. The neighborhood was the intimate community. Since the same families had lived there for generations and intermarried, it was largely a kinship community. Nobody was very rich and no one was allowed to starve. At worst, the parish saw to that. There was no great distinction between rich and poor. At my home the family and servants ate at the same table and lived in the same house. The servants were really the most independent, as they were the most important members of the family. When the tailor and the shoemaker—each with his apprentices—came, they were members of the family. When the tailor had gone on a spree and had delirium tremens, my brothers were sent to take care of him. When the miller or shoemaker was ill, my family watched over them and sent them things to eat. Everybody who was ill or in need in the neighborhood was the concern of my father and mother. All the unfortunates were, in a large way, members of our family and could come and stay as long as they wished. My father was Father Elias to the whole neighborhood. He loved all, rich or poor, but especially the poor. When they were in distress or sickness or dying, they wanted him and he did not fail them.

As a boy, I took part in everything around the farm, as I was able, but especially I remember bringing the cattle home at night and being a messenger boy to those in need. My family managed to send me to college for a couple of years, then my father died; and since we were a large family, I decided at the age of nearly eighteen to emigrate to America. I was uprooted from my community. However kind the new world has been and whatever my success within it, the loss of my own community has always haunted me. The change meant a change of language, a change of history, a change of tradition. It was a complete cleavage with my world before eighteen. For some years while my mother was living and some of my family remained at the old home, I made a visit occasionally, but I was no longer part of the community. And now the community itself is gone. My kindred have died or moved away. The tailor, shoemaker, and blacksmith have, for the most part, been crowded out by the factories. The new farmers have no past. They know nothing about the legends of the place—the hill within which a giant sits on his chest of gold and silver, the trolls who lived in the lake (now drained and cultivated) and who long ago did the harvesting for my ancestors annually, at night, for a bowl of porridge. Nor do they know about many monuments still remaining of the Old Stone Age. You cannot have community unless people are rooted in the remembrance of a common past.

Glossary of Important Terms

Behaviorism: First proposed by psychologist John B. Watson in 1913, it associated human and animal actions with observable responses to external stimuli, suggesting a scientific basis for manipulation and control of living subjects that was empirical and deterministic. Watson's protégé B. F. Skinner proposed an extreme version that argued that belief, intention, and desire were "mentalists" illusions and therefore unscientific. This has important implications for metaphysics and the "hard problem" of consciousness. Biologist and systems theorist Ludwig von Bertalanffy wrote a scathing analysis of behaviorism, *Robots, Men and Minds* (1967), charging it with reducing human action to mere conditioning so that "the ethical ideal became a scientific dogma." Although it was quite popular in Boodin's day, the poet-philosopher rejected behaviorism in principle and in application.

Cosmic evolution: Boodin's most complete metaphysical statement is embodied in the book thus titled and published in 1925. Combining the latest discoveries arising from the new physics of relativity and quantum theory with a process cosmology, he was able to build a wholly new type of evolutionary theory quite distinct from the emergent evolutionism of Charles Darwin (see *Darwinism*). Boodin's system comprises a cosmic structure of eternal coexistent fields of energy at multiple levels. It is not dissipated by loss of energy nor subject to the stochastic influences of emergence. It is a closed system of dynamic spatiotemporal equilibrium that has an overall teleological character and is, therefore, distinguishable from the Gaia hypothesis of James Lovelock and Lynn Margulis.

Critical realism: An elaboration upon the new realists' reaction to idealism, it is largely associated with an epistemology introduced by Roy Wood Sellars in his book of that title in 1916. For Sellars, knowing is interpretive of objects.

As such, it affirms the reality of independent physical things, but they are not directly and uniformly presented to us in all perceptual situations; they are mediated through operations, symbols, meanings, and categories. Sellars's emphasis upon the findings of the natural sciences is stressed to a degree that is arguably excessive (see *scientism*). For example, his critical realism attempts to hold behaviorism at arm's length but overall agrees with the behaviorist project (see *behaviorism*). Furthermore, Sellars uncritically accepts Darwinism and opposes modifying it with teleological speculations. Boodin opposed Sellars's critical realism because he believed it creates a "bifurcation of thing and environment as though they were only externally related" and is based upon "dogmatic assumptions" that lead in the end to a kind of "universal agnosticism." The distance between Boodin and Sellars can be seen in Peter H. Hare's revealing quotation of Sellars that "Matter is . . . existent in its own right. And I shall think it in terms of the category of substance" rather than process. Although Sellars denies reductionism, his critical realism integrates with his evolutionary naturalism in ways that Robert J. Kreyche argues produce a pervading simplistic and reductionist collection of emergent substances.

Darwinism: This must be understood in two senses. In its most descriptive sense it is a theory of the development and diversification of life primarily but not solely by means of natural selection. However, in its broader sense it presumes to be a positivistic worldview that can explain the origin and nature of life itself. Its most distinctive feature asserts that the components of evolution when taken together are blind and purposeless (see *teleology*). As Darwin confessed in his *Autobiography*, "There seems to be no more design in the variability of organic beings and in the action of natural selection, than in the course which the wind blows" (Nora Barlow, ed., *The Autobiography of Charles Darwin* [1958; repr., New York: W. W. Norton, 1969], 73). Boodin thought that Darwinism in its first sense was at best incomplete, a limited and superficial account of life; in its second sense he regarded it as simply wrong, an outmoded mechanistic emergentism based upon false assumptions from which it drew erroneous conclusions.

Epistemology: How it is that we *know* things. This is the theory of knowledge largely concerned with its nature and derivation, the means by which it can be verified, and its reliability. Like most process philosophers, Boodin's epistemology as presented in *A Realistic Universe* (1916; rev. ed. 1931) is essentially relational. First outlined in *Truth and Reality* (1911), Boodin called this the *postulate of totality*, namely, "facts are part of one world in such a way that every fact can, under certain conditions, make a difference to other facts" ("Pragmatic Realism—The Five Attributes," *Mind* 22, no. 88 [Oct. 1913]: 509).

Idealism: In philosophy this is the view that while the external world does indeed exist, it does not exist entirely independent of the mind. It thus has a tendency to privilege mental things over physical things and presents itself in opposition to materialism.

Materialism: While there are many kinds of materialisms (e.g., central-state, dialectical, eliminative, revisionary, historical), it is essentially the view that all reality is composed of matter. Even where attributes such as mind are seemingly not material themselves, materialism seeks to reduce them to their "nothing but" physical components (e.g., mind is "nothing but" brain, thought is "nothing but" neural activity, behavior is "nothing but" response to environmental stimuli). It owes its beginning to Leucippus and Democritus in the fifth century BC. Although scientific materialism as espoused by Richard Dawkins, Daniel Dennett, Michael Ruse, and others operates within this broadly reductionist framework, alternatives have been proposed by eliminative materialists such as Paul and Patricia Churchland, Wilfrid Sellars (son of Roy Wood Sellars), Paul Feyerabend, and Richard Rorty (among others) and John Bickle's middle-ground revisionary physicalism. Each must be evaluated on a case-by-case basis.

Naturalism: This can be of two types: reductive and non-reductive. The former is the position that everything belongs to the observable, testable world we sense and perceive. The latter is a more liberal form that (like reductive naturalism) agrees that nature is a single realm without incursions of outside forces but recasts nature in much broader and extensive terms. Boodin may be said to adhere to this more liberal form. He discussed this most fully in his essay "Functional Realism" (*Philosophical Review* 43, no. 2 [Mar. 1934]: 178), where he states, "Since the ideal [mental and emotional] aspects of our experience of nature are as real, as much part of nature, as the sensory, the ideal aspects imply a mutuality with nature that is appropriate to their character. If we are interested in what sort of universe makes our sense aspects possible, we should be interested in what sort of universe makes our spiritual experience possible." Thus, like most process philosophers, Boodin argues for an essentially all-inclusive nature that can even allow for paranormal phenomena.

New Realism: A reaction to idealism (especially that of Josiah Royce) led by Ralph Barton Perry, whose metaphysics Arthur O. Lovejoy called "a soft-spoken, if not a 'tender-minded,' materialism," and supported by William Pepperell Montague whom Lovejoy praised as "a first-rate silversmith in philosophy" that held the curious position of "a Platonic as well as an animistic materialist" resulting in a rather "sketchily argued" monism. When Montague proposed that the perceived object should be considered identical in substance with the physical object because it was composed of the same energy, Boodin challenged this monistic realism in 1907 with a pluralistic theory of energy centers of various grades and kinds based upon the differences they make to our reflective purposes (see his "The New Realism"). Boodin's opposition to Sellars's critical realism may be seen as a variation on his response to the new realists and a reaction against the materialists' project in general.

Ontology: A branch of metaphysics concerning what exists. The distinctions between "real" versus "apparent" existence, what and which entities belong to various categories (physical, mathematical, universal, abstract, etc.), and the assumptions underlying existence are all proper considerations of ontology.

Panentheism: This states that all parts of reality [*pan*] are included in [*en*] and creatively synthesized by the one all-inclusive eternally creative whole of reality [*theos*]. Contrary to pantheism, panentheism regards theos as greater than and inclusive of all parts of reality or the universe. In short, God is *in* the world not one *with* the world. Thus can be seen the teleological nature of Boodin's cosmos. Here God is deeply personal and eminently transcendent. Alfred North Whitehead, Charles Hartshorne, and many process philosophers and theologians (Boodin among them) hold to panentheism.

Pantheism: The view that God or spirit permeates all of physical and organic reality. Here God is impersonal and nontranscendent. Henri Bergson and Teilhard de Chardin may be considered pantheists.

Positivism: Historically it has had two general iterations. The first was presented by its founder, Auguste Comte, in his *Cours de philosophie positive* in six volumes (1830–1842), condensed and translated by Harriet Martineau as *The Positive Philosophy of Auguste Comte* in 1853. Comte regarded human history as advancing in three distinct phases he called "The Law of Three Stages": first theologically (dominated by superstition), next metaphysically (dominated by "a fiction of abstractions"), finally scientifically (guided by rational investigation and inquiry), culminating in sociology, an encompassing secular humanism of common welfare. Comte's system rests on a foundation of hard verificationism and rejection of all metaphysics as meaningless. Many regard Darwin's theory of evolution as a form of positivism (see *Darwinism*). The second iteration came as logical positivism in the 1920s and 1930s, particularly with the Vienna Circle led by Moritz Schlick, for whom Ernst Mach had prepared the way. Schlick and his disciples sought to add logical rigor using modern mathematics and physics to ideas first proposed by Hume and Comte, yet as a philosophy their ideas remained largely indistinguishable from their French predecessors. By the mid-1970s Karl Popper could claim in his autobiography *Unended Quest* (see bibliography) to have "killed" logical positivism by pointing out its untenable verificationist stance, an intellectual murder committed with the 1935 publication of his *Logik der Forschung* (translated in 1959 as *The Logic of Scientific Discovery*). Bernard Phillips, James K. Feibleman, and Hilary Putnam have also been severe critics of logical positivism.

Pragmatic realism: A term introduced by Boodin in *Truth and Reality* (1911). Instead of the dogmatic method pursued by the old idealism and materialism alike, this substituted the critical method called pragmatism. Boodin understood this to mean carrying the scientific spirit into metaphysics. It means the willingness to acknowledge reality for what it is, what it always means for us, and what difference it makes to our reflective purposes.

Pragmatism: Introduced by Charles Sanders Peirce in 1878. It was soon adopted by William James as the idea that truth must have a close relationship with successful action based upon experience. Based upon what "works," its subjectivity places priority on empirical verification to ground itself. This caused a division between idealistic pragmatists like F. C. S. Schiller and science-inspired humanistic pragmatists like John Dewey. Although Boodin adopted many pragmatic arguments and perspectives, he never announced himself as a pragmatist as James had always wanted him to do.

Process philosophy: Any brief explanation of process philosophy risks oversimplification, but succinctly put it is a worldview focused on relation and change—being as becoming. While it is goal directed, the goal toward which it is directed is not forever stable. It is inherently affirmative and aspirational rather than fixed and deterministic. This, along with process philosophy's view of nature (see *naturalism*), allows for a variety of theological positions of which Boodin's is one in which God is not a decreeing magistrate but a relational presence seeking the creative potentialities of all humanity. In addition, process philosophy typically rejects the Cartesian notion stated in his *Principles of Philosophy* that substance is "an existent thing which requires nothing other than itself to exist." Although process thinking dates back to Heraclitus of Ephesus (ca. 560 BC), its most significant modern proponent is Alfred North Whitehead. Others include Randall E. Auxier, Joseph A. Bracken, John B. Cobb, David Ray Griffin, Charles Hartshorne, Arthur Peacocke, and Marjorie Suchocki, to name a few.

Scientism: Science is fundamentally a method of inquiry and investigation that can be a valuable tool in examining and elucidating philosophical questions. Whitehead and Boodin used science in this manner. However, when science is elevated to absolute epistemic preeminence (i.e., scientism), it becomes a worldview—a belief system—that has proven difficult to defend. Sellars's brand of critical realism and positivism can be regarded as forms of scientism expressed differently. There are a number of issues surrounding the extension of science in this manner, but the principal one is falling victim to a form of epistemic monism.

Tautology: Generally regarded as a logical fallacy. It is essentially an argument that simply restates its premise in its conclusion. Many consider Darwin's adoption of the Spencerian phrase "survival of the fittest" a classic tautology because fitness is defined as those that survive as well as the test for survival. Who survives? Those that are fittest. Who are the fittest? Those that survive. While this may be true enough, it doesn't explain very much.

Teleology: Teleology is about purposes in nature. These very often can be found in teleological descriptions that involve "in order to" statements relating phenomena in a goal-directed or goal-seeking way. Etymologically the term is rooted in *telos* (τέλος in Greek) meaning end or goal. This can be understood in two senses. In one sense goal-directedness may be extrinsic or intrinsic in nature. For example, a chimpanzee may fashion a stick *in order to* extract ants to gain a meal. The goal-directed nature of the stick is extrinsic in this case. Birds, on the other hand, migrate *in order to* increase food resources and nesting site availability. Here the goal-directed nature is intrinsic to the birds' instincts. But for purposes of understanding Boodin's philosophy it is more important to understand goal-directedness not as extrinsic or intrinsic but as transcendent (originating over and above strictly empirical agencies) or non-transcendent (originating from purely utilitarian aspects of nature as in the chimp's stick or birds' migrations). Because process philosophy is by definition concerned with ever-changing relational goals, it tends to be more amenable to transcendent teleology, often melding with non-transcendent teleology in a seamless web of relational possibilities and capacities. Historically speaking, natural philosophers had always considered nature to involve some transcendent goal or purpose-driven aspect, beginning with the ancient Ionian Anaxagoras (ca. 500 BC) continuing on to Plato and Aristotle and throughout much of Western philosophy and science. Even well into the nineteenth century most naturalists assumed some form of transcendent goal-directness in nature (e.g., Richard Owen, John Herschel, William Whewell). Charles Darwin's evolutionary theory (see *Darwinism*) changed all that by turning design and purpose over to chance.

Time: For Boodin, time is not all of reality but an independent variable that is a preeminent attribute of reality. It is dynamic non-being (fluid and fleeting), inherent in a universe fundamentally understood as one of process (see *process philosophy*). As a process philosopher time is the flux of change and an essential part of Boodin's entire metaphysical system. It is worth noting that Boodin's *Time and Reality* (1904) demonstrates his engagement with process thinking many years before Whitehead's Gifford Lectures in 1927–1928 that would lead to his *Process and Reality* considered the magnum opus of process philosophy.

Notes

Prologue

1. Barbara Tuchman, *Practicing History: Selected Essays* (New York: Alfred A. Knopf, 1981), 81.

2. Most of the biographical information for this study comes from Charles H. Nelson, *John Elof Boodin: Philosopher-Poet* (New York: Philosophical Library, 1987).

3. See John Elof Boodin's autobiographical statement as part of his "Nature and Reason," in *Contemporary American Philosophy*, eds. George P. Adams and William Pepperrell Montague (New York: Russell and Russell, 1962), 136.

4. Boodin, 136.

5. Boodin identified as a "philosopher-poet," and he told writer George Albee (brother of Richard who had been one of Boodin's students) who was preparing an article on his life, to emphasize this theme (Nelson, *John Elof Boodin*, 20). But it seems to me that this order modifies the wrong noun. Boodin was a philosopher first and a poet by predilection and style. I have, therefore, reversed the order to poet-philosopher much as Michael Roberson has done in his "Nietzsche's Poet-Philosopher: Toward a Poetics of Response-ability, Possibility, and the Future," *Mosaic* 45 (Mar. 2012): 187–202.

6. Boodin has left a richly detailed summary of his friendship with James in "William James as I Knew Him," reprinted in Linda Simon, *William James Remembered* (Lincoln: University of Nebraska Press, 1996), 207–32, qtd. at 207.

7. Some have argued for the sole influence of James's *Principles* in shaping Dewey's shift toward a pragmatic theory of knowledge, but this has been questioned because Dewey was already working on this before 1890. Nevertheless, the impact of *Principles* upon Dewey's early thought cannot be denied. See Michael Buxton, "The Influence of William James on John Dewey's Early Work," *Journal of the History of Ideas* 45 (July–Sept. 1984): 451–63.

8. Robert D. Richardson, *William James: In the Maelstrom of American Modernism; A Biography* (Boston: Houghton Mifflin, 2006), 6.

9. Boodin, "William James as I Knew Him," 207–8.
10. Boodin, 214.
11. Boodin, 222–23.
12. Richardson, *William James*, 488.
13. John J. McDermott, "Suffering, Reflection, and Community: The Philosophy of Josiah Royce," in *The Basic Writings of Josiah Royce*, ed. and intro. John J. McDermott, 2 vols. (Chicago: University of Chicago Press, 1969) 1:4.
14. For a complete discussion, see Randall E. Auxier, *Time, Will, and Purpose: Living Ideas from the Philosophy of Josiah Royce* (Chicago: Open Court, 2013), 103–23. See especially his "Royce's Pragmatic Temper" made up of the primacy of *experience*, philosophical reflection from genuine doubt as a means of critical problem-solving, and value derived from practice and practical consequences, 109.
15. Auxier believes that Royce has no equal as a philosopher of religion, including William Ernest Hocking and Charles Hartshorne. See his *Time, Will, and Purpose*, 32. This will be explored further in chapter 4.
16. Boodin, "Nature and Reason," 139.
17. See the complete letter in *The Letters of Josiah Royce*, ed. and intro. John Clendening (Chicago: Chicago University Press, 1970), 552.
18. See John K. Roth's introduction in *The Philosophy of Josiah Royce*, ed. John K. Roth (New York: John Y. Crowell, 1971), 5.
19. Auxier gives an interesting comment on this in comparison with Hocking. See his *Time, Will, and Purpose*, 28, 365.
20. Boodin, "Nature and Reason," 139.
21. Nelson, *John Elof Boodin*, 142.
22. Nelson, 131.
23. For a complete discussion, see Jeffrey Wayne Yeager, "The Social Mind: John Elof Boodin's Influence on John Steinbeck's Phalanx Writings, 1935–1942," *Steinbeck Review* 10, no. 1 (2013): 31–46. It should be mentioned that although Boodin's book *The Social Mind* was written in 1939, he had written on this topic much earlier. See, for example, his article, "The Existence of Social Minds," *American Journal of Sociology* 19 (July 1913): 1–47.
24. Nelson, *John Elof Boodin*, 95.
25. R. F. A. Hoernlé, review of *Cosmic Evolution*, by John Elof Boodin, *Journal of Philosophy* 24 (Mar. 17, 1927): 160.
26. See the chapter "The Cosmic Philosophy of John Elof Boodin," in *Recent American Philosophy: Studies of Ten Representative Thinkers*, by Andrew J. Reck (New York: Pantheon Books, 1964), 123–53. This is actually a reprinting of Andrew J. Reck's "The Philosophy of John Elof Boodin," *Review of Metaphysics* 15 (Sept. 1961): 148–73.
27. Andrew Reck, s.v. "Boodin, John Elof," in *The Encyclopedia of Philosophy*, vol. 1 (New York: Macmillan, 1967).

28. Nelson, *John Elof Boodin*, 169. The complete letter from Child to Nelson (July 5, 1979) is reproduced in Appendix A of Nelson's biography, 167–70.

29. R. F. Alfred Hoernlé, review of *The Social Mind*, *Mind* 50 (Oct. 1941): 393–401.

30. Bruce Kuklick, *A History of Philosophy in America, 1720–2000* (Oxford: Clarendon Press, 2001), 244.

31. Nelson, *John Elof Boodin*, 20.

32. Auxier, *Time, Will, and Purpose*, 1.

33. Auxier, 108.

34. See Auxier's entry on s.v. "Boodin, John Elof," in *The Dictionary of Modern American Philosophers*, vol. 1 (Bristol, UK: Thoemmes Continuum, 2006). See also his introduction to Boodin's *Truth and Reality: An Introduction to the Theory of Knowledge*, Early Defenders of Pragmatism, vol. 2, ed. John R. Shook (1911; repr., Bristol, UK: Thoemmes Press, 2001).

35. Auxier, *Time, Will, and Purpose*, x–xi.

36. John K. Roth, introduction to *The Philosophy of Josiah Royce* (New York: Thomas Y. Crowell, 1971), 9.

37. Auxier has an interesting section on this: "Defective Community Persons," *Time, Will, and Purpose*, 271–73.

38. Boodin, *Truth and Reality*, 251–68, passim.

39. Herbert Butterfield, *The Whig Interpretation of History* (1931; repr., New York: W. W. Norton, 1965), v.

40. Herbert Butterfield, *The Origins of Modern Science*, rev. ed. (New York: Macmillan, 1962), ix.

41. Walter Hooper, *C. S. Lewis: A Companion & Guide* (New York: HarperSanFrancisco, 1996), 553. Lewis attributed the concept of chronological snobbery to his longtime friend Owen Barfield, who "taught me not to patronize the past." Lewis, who was introduced to the idea prior to his conversion to theism in 1929, was influenced by it for the remainder of his life and often referenced it with variations on the theme. For example, not long before his death he bemoaned the declining state of the arts at Cambridge and Oxford, especially due to the influence of F. R. Leavis's demand for social relevance and literary criticism, by writing to J. B. Priestly (Sept. 18, 1962):

> The actual history of Eng. Lit. as a "Subject" has been a great disappointment to me. My hope was that it would be primarily a historical study that wd. lift people out of (so to speak) their chronological provincialism by plunging them into the thought and feeling of ages other than their own: for the arts are the best Time Machine we have. But all that side of it has been destroyed at Cambridge and is now being destroyed at Oxford too. . . . It now has a strangle hold on the

schools as well as the universities (and the High Brow press). It is too open an avowed to be called a plot. It is much more like a political party—or the Inquisition. (*The Collected Letters of C. S. Lewis*, ed. Walter Hooper, 3 vols. [New York: HarperSanFrancisco, 2007] 3:1371.)

42. John E. Boodin, "Time and Reality," *Psychological Review*, monographic suppl., vol. 6, no. 3 (New York: Macmillan, 1904), 102.

Introduction

1. An excellent summary is available in Laura J. Snyder, *The Philosophical Breakfast Club* (New York: Broadway Books, 2011), 1–3.

2. A good discussion is available in Henry Thomas, s.v. "Comte, Auguste," in *Biographical Encyclopedia of Philosophy* (Garden City, NY: Doubleday, 1965).

3. Edward J. Larson, *Evolution: The Remarkable History of a Scientific Theory* (New York: Modern Library, 2004), 69. Natural selection didn't have to imply atheism, pantheism, deism, or anything else. In fact, Wallace would ultimately break with Darwin in 1869 and call upon the necessity of an "Overruling Intelligence" to explain the origin of life, intellect of humans, and certain features of complexity in nature. What Wallace was essentially working toward was a new, refurbished natural theology to incorporate evolution by means of natural selection over the species fixist ideas of William Paley. For a complete discussion, see Michael A. Flannery, *Nature's Prophet: Alfred Russel Wallace and His Evolution from Natural Selection to Natural Theology* (Tuscaloosa: University of Alabama Press, 2018). Although Wallace never retreated from his concept of natural selection, Darwin wanted nothing to do with what he viewed as unwarranted teleology. No wonder that Ross A. Slotten titled his biography of Wallace *A Heretic in Darwin's Court* (New York: Columbia University Press, 2004).

4. Curtis Johnson, *Darwin's Dice: The Idea of Chance in the Thought of Charles Darwin* (Oxford: Oxford University Press, 2015), xii.

5. Johnson, xxiii.

6. Quoted in Flannery, *Nature's Prophet*, 55–56.

7. Johnson, *Darwin's Dice*, 226.

8. Charles Darwin, *Descent of Man and Selection in Relation to Sex* (1871; repr., New York: Barnes & Noble, 2004), 102–6.

9. Neal C. Gillespie, *Charles Darwin and the Problem of Creation* (Chicago: University of Chicago Press, 1979), 8, 41–47, 53–54.

10. John C. Greene, *Science, Ideology, and World View: Essays in the History of Evolutionary Biology* (Berkeley: University of California Press, 1981), 153. See also Silvan S. Schweber, "The Young Darwin," *Journal of the History of Biology* 12 (Spring 1979): 175–92; Frank Burch Brown, "The Evolution of Darwin's Theism,"

Journal of the History of Biology 19 (Spring 1986): 1–45; and George Levine, "By Knowledge Possessed: Darwin, Nature, and Victorian Narrative," *New Literary History* 24 (Spring 1993): 363–91.

11. Greene, *Science, Ideology, and World View*, 140.

12. Sydney Eisen, "Huxley and the Positivists," *Victorian Studies* 7 (June 1964): 337–58.

13. Gillespie, *Charles Darwin and the Problem*, 153.

14. Thomas S. Kuhn, *The Structure of Scientific Revolutions*, 2nd ed. (Chicago: University of Chicago Press, 1970).

15. See Greene's extended critique, *Science, Ideology, and World View*, 30–59.

16. Arthur Koestler, *The Sleepwalkers: A History of Man's Changing Vision of the Universe* (1959; repr., London: Arkana, 1989), 523.

17. Koestler, 550.

18. Alfred North Whitehead, *Adventures in Ideas* (1933; repr., New York: Free Press, 1967), 125.

19. Philip Paul Weiner, "Some Metaphysical Assumptions and Problems of Neo-Positivism," *Journal of Philosophy* 32 (Mar. 28, 1935): 174–81.

20. Brian G. Henning, "Recovering the Adventure of Ideas: In Defense of Metaphysics as Revisable, Systematic, Speculative Philosophy," *Journal of Speculative Philosophy* 29, no. 4 (2015): 437–56.

21. Hilary Putnam, "Pragmatism," *Proceedings of the Aristotelian Society*, New Series 95 (1995): 291–306.

22. Greene, *Science, Ideology, and World View*, 188.

23. Letter from Darwin to William Graham, July 3, 1881, "Letter no. 13230," accessed Nov. 15, 2021, https://www.darwinproject.ac.uk/letter/?docId=letters/DCP-LETT-13230.xml.

24. Werner Heisenberg, *Natural Law and the Structure of Matter* (London: Rebel Press, 1970), 32–33. In fact, Heisenberg was more than willing to discuss the philosophical implications of his work late in life. See his *Physics and Philosophy: The Revolution in Modern Science* (1958; repr., New York: Harper Perennial Modern Thought, 2007).

25. Dyson has written and spoken on the complementarity of science, religion, and philosophy many times and in many places. One excellent example is his essay "Varieties of Human Experience" that begins with William James, includes the poetry of William Blake, and ends with some refreshing epistemic modesty in exploring the special cognitive worlds of the autistic, which suggest "that there may be more things and heaven and earth than we are capable of understanding." See Dyson's *A Many-Colored Glass: Reflections on the Place of Life in the Universe* (Charlottesville: University of Virginia Press, 2007), 131–54.

26. Stanley Aronowitz, *Science as Power: Discourse and Ideology in Modern Society* (Minneapolis: University of Minnesota Press, 1988).

27. Aronowitz, 12.

28. Aronowitz, 17.

29. James Hannam, *The Genesis of Science: How the Christian Middle Ages Launched the Scientific Revolution* (Washington, DC: Regnery Publishing, 2011), 190.

30. See Steven Shapin, *Never Pure: Historical Studies of Science as if It Was Produced by People with Bodies, Situated in Time, Space, Culture, and Society, and Struggling for Credibility and Authority* (Baltimore, MD: Johns Hopkins University Press, 2010). Shapin's sixteen essays call for a more fallibilist approach to science. His insights are definitely worth considering, but they go far beyond the scope of this study.

31. Hilary Putnam, *The Collapse of the Fact/Value Dichotomy and Other Essays* (Cambridge, MA: Harvard University Press, 2002), 145.

32. James Woelfel, "Challengers of Scientism Past and Present: William James and Marilynne Robinson," *American Journal of Theology & Philosophy* 34 (May 2013): 175–87.

33. Hilary Putnam and Ruth Anna Putnam, *Pragmatism as a Way of Life: The Lasting Legacy of William James and John Dewey* (Cambridge, MA: Belknap Press [Harvard University Press], 2017), 231.

34. Gert J. J. Biesta, "How to Use Pragmatism Pragmatically? Suggestions for the Twenty-First Century," *Education and Culture* 25, no. 2 (2009): 34–45.

35. Dewey's *The Quest for Certainly* (1929) quoted in Putnam and Putnam, *Pragmatism as a Way of Life*, 293.

36. John Elof Boodin, *Studies in Philosophy: The Posthumous Papers of John Elof Boodin*, ed. John Ayres Piatt (Los Angeles: University of California Press, 1957), 153–55.

Chapter 1

1. Randall E. Auxier, *Time, Will, and Purpose: Living Ideas from the Philosophy of Josiah Royce* (Chicago: Open Court Press, 2013), 51.

2. Lewis Mumford, *Technics & Civilization* (1934; repr., Chicago: University of Chicago Press, 2010), 14

3. Quoted in Rosalind Williams, "Lewis Mumford's *Technics and Civilization*," *Technology and Culture* 43 (Jan. 2002): 149.

4. Williams, 142–43.

5. Williams, 144. *Mentalité* has been defined by its originator, French historian Marc Bloch, as "wherever fidelity to a belief is to be found, all evidences agree that it is but one aspect of the general life of a group. It is like a knot in which are intertwined a host of divergent characteristics of the structure and mentality of a society." See his *The Historian's Craft*, trans. Peter Putnam (New York: Vintage Books, 1953), 32. (See also chap. 5.)

6. Williams, "Lewis Mumford's *Technics*," 148.

7. Stefan Tanaka, *History Without Chronology* (Ann Arbor, MI: Lever Press, 2019). https://doi.org/10.3998/mpub.11418981.

8. Tanaka, 28.
9. Quoted in Tanaka, 32.
10. Tanaka, 33.
11. Tanaka, 43.
12. Alfred North Whitehead, *Science and the Modern World* (1925; repr., New York: Free Press, 1967), 50.
13. Whitehead, 58.
14. Jim Al-Khalili, *The World According to Physics* (Princeton, NJ: Princeton University Press, 2020), 162–63.
15. On James's opposition to Newtonian time, see Jonathan Bricklin, *The Illusion of Will, Self, and Time: William James's Reluctant Guide to Enlightenment* (Albany: State University of New York Press, 2015), 213.
16. Peter N. Carroll and David W. Noble, *The Free and the Unfree: A New History of the United States*, 2nd ed. (New York: Penguin Books, 1988), 58–59.
17. Thomas J. Schlereth, *Victorian America: Transformations in Everyday Life* (New York: Harper Perennial, 1991), 29–31.
18. Quoted in Carroll and Noble, *The Free and the Unfree*, 153.
19. Stated this way, however, James's putative priority can be misleading. Royce pointed to purposes as future plans in an unpublished essay in 1880, "On Purposes in Thought," and as such establishes him (including Peirce) as "the first full-fledged process philosopher in the U.S., and certainly temporalist from at least 1880 forward." *Time, Will, and Purpose*, 40.
20. See letter from James to Robertson, Nov. 9, 1887, in Ralph Barton Perry, *The Thought and Character of William James*, 2 vols. (Boston: Little, Brown, 1935), 2:84–85.
21. All subsequent references to James's "The Perception of Time" will come from the original article, William James, "The Perception of Time," *Journal of Speculative Philosophy* 20 (Oct. 1886): 374–407.
22. James, 378.
23. Robert D. Richardson's misreading here is rather surprising given his otherwise close and careful analysis of James's life and work. I do agree with Richardson that James had used time and space experience phenomenologically. See his *William James: In the Maelstrom of American Modernism: A Biography* (Boston: Houghton Mifflin, 2006), 273.
24. James, "The Perception of Time," 397.
25. Quoted in Bricklin, *The Illusion of Will*, 213.
26. Josiah Royce, "Immortality," in *The Basic Writings*, 1:390–91.
27. James told Royce that when he was preparing his Gifford Lectures, which would form part of his *Varieties of Religious Experience*, that he did so "with one eye on the page and one eye on you." That "you" was Royce's Absolute—the "Eternal Consciousness"—whereby all temporal processes can be analogized with the finite consciousness of time. See Bricklin, *The Illusion of Will*, 244–45, 330 n.4.
28. Auxier, *Time, Will, and Purpose*, 176.

29. Royce, "Immortality," 1:390.
30. Tanaka, *History Without Chronology*, 61.
31. Tanaka, 92.
32. Alfred North Whitehead, *Process and Reality*, corrected ed., eds. David Ray Griffin and Donald W. Sherburne (New York: Free Press, 1978), 18.
33. Quoted in Richardson, *William James*, 428.
34. Quoted in Richardson, 427.
35. Milič Čapek, "Time and Eternity in Royce and Bergson," *Revue Internationale de Philosophie* 79, nos. 79/80 (1967): 22–45.
36. Čapek, 44.
37. Čapek, 37.
38. Suzanne Guerlac, *Thinking in Time: An Introduction to Henri Bergson* (Ithaca, NY: Cornell University Press, 2006), 79–81. Bergson is here responding to Kant who argued that freedom was independent of time and space. Here Guerlac argues that Kant's error placed "freedom completely beyond our reach," something Bergson corrected by placing it "in our most immediate experience" (102).
39. Auxier summarizes it best when he writes, "Royce is a pragmatist, as well as a personalist and a process philosopher. His position on panexperientialism is as strongly stated as a pragmatist can allow, which is to say that he holds experience to be a universal postulate." See his *Time, Will, and Purpose*, 258.
40. Boodin, *Time and Reality*, iii.
41. Boodin, 7.
42. Boodin, 8.
43. Boodin, 19–20.
44. Boodin, 24.
45. Boodin, 28–31.
46. Historians should appreciate Boodin's explication of time here, especially when considered against the absurdity of those would attempt to write counterfactual history—the "what if" historians who speculate about a Confederate victory in the Civil War or if Hitler would have prevailed in Europe in World War II. The most ridiculous recent example is Peter J. Bowler's *Darwin Deleted: Imagining a World without Darwin* (Chicago: University of Chicago Press, 2013). Who knows or can even guess at what biology without Darwin might have been since the entire concatenation of historical Boolean operators would be disrupted in the process. David Hackett Fischer humorously captures the irrationality of counterfactual history by noting "the counterfactualists . . . might profit from the advice of those learned, if unlovable, logicians, Tweedledum and Tweedledee: 'I know what you're thinking about,' said Tweedledum; 'but it isn't so, no how.' 'Contrariswise,' continued Tweedledee, 'if it was so it might be; and if it were so, it would be; but as it isn't, it ain't. That's logic.'" See his *Historians' Fallacies: Toward a Logic of Historical Thought* (New York: Harper Torchbooks, 1970), 21. The real fallacy here, of course, is the attempt to do history by disregarding time.
47. Boodin, *Time and Reality*, 59.

48. Boodin, 72–73.
49. Boodin, 74.
50. Randall E. Auxier, s.v. "Boodin, John Elof," in *The Bloomsbury Encyclopedia of Philosophers in America from 1600 to the Present*, ed. John R. Shook (London: Bloomsbury Academic, 2016).
51. See Boodin's "The Reality of the Ideal with Special Reference to the Religious Ideal," *The Unit* [Grinnell College] 5 (1900): 97–109.
52. For a thorough discussion see Flannery, "The Process Theology of John Elof Boodin," *Religions* 14, no. 2 (2023), https://www.mdpi.com/2077-1444/14/2/238.
53. Boodin, *Time and Reality*, 96.
54. Boodin, 118.
55. Published as "The Concept of Time," *Journal of Philosophy, Psychology and Scientific Methods* 2 (July 6, 1905): 365–72.
56. See B. H. Bode's review in *Philosophical Review* 14 (Nov. 1905): 730–31; and Percy Hughes's review in *Journal of Philosophy, Psychology and Scientific Methods* 2 (Apr. 13, 1905): 218–20.
57. Walter Rosenblith, "On Cybernetics and the Human Brain," *American Scholar* 35 (Spring 1966): 246.
58. Charles H. Nelson, *John Elof Boodin: Philosopher-Poet* (New York: Philosophical Library, 1978), 66.
59. Boodin, *Truth and Reality*, 11.
60. Boodin, 29, 37.
61. Boodin, 44–64.
62. Boodin, 83.
63. Boodin, 85.
64. Boodin, 116–17.
65. Boodin, 118–19.
66. Boodin, 126–45. These are discussed at length, but their designations are fairly self-evident and need not be explicated in detail.
67. Boodin, 155.
68. Boodin, 161.
69. Boodin, 167.
70. Boodin, 183.
71. This is perhaps more clearly stated in Boodin, "The Ought and Reality," *International Journal of Ethics* 17 (July 1907): 454–74.
72. Boodin, *Truth and Reality*, 253–54.
73. Boodin, 260.
74. Boodin, 288.
75. Boodin, 291–92.
76. Boodin, 300.
77. Will Durant, *The Lessons of History* (New York: Simon & Schuster, 1968), 49. Interestingly, it was Durant's engaging style and wit that induced Hilary Putnam to turn to philosophy as a profession. For me, the influence of Durant directed me

to history. In either case, it was Durant's remarkable facility for bringing ideas to life that caught our attention. He remains recommended reading for that reason.

78. Durant, 51. Written over fifty years ago, this statement's validity may seem questionable but it certainly has not been falsified based upon comparative crime rates per 100,000. See United States Crime Rates 1960–2019, Disaster Center, accessed Dec. 11, 2021, https://www.disastercenter.com/crime/uscrime.htm. The Disaster Center is a civic organization that is part of the Library of Congress.

79. Boodin, *Time and Reality*, 322.

80. Boodin, 324.

81. The cash-value metaphor has always been controversial. But George Cotkin has ably defended it as consonant with James's style and pragmatic philosophy and deeply resonant with him personally. See his "William James and the Cash-Value Metaphor," *ETC: A Review of General Semantics* 42 (Spring 1985): 37–46. As Cotkins states, "While certainly typical of the colloquial language that James loved to playfully employ, the cash-value metaphor figured centrally because it worked, the *sine qua non* of the pragmatic temper. Secondly, the metaphor . . . had a deep, personal resonance to James. It served to link his personal financial problems with the cash realities inherent in the format of lecturing that he used to test and present his ideas as well as to help support himself" (39). In this context it is easy to see that James's metaphor had an especially deep resonance with ideas and concepts given the equivalent of monetary value (what could be more pragmatic!) and equally easy to see why, in the face of considerable protests against it, he never abandoned its use.

82. See especially Andrew Reck, "Royce's Metaphysics," *Revue Internationale de Philosophie* 21, nos. 79/80 (1967): 11.

83. An interesting account of this course and its relation to Royce's metaphysic is given in Stephen Tyman, "Royce and the Destiny of Idealism," *The Personalist Forum* 15 (Spring 1999): 45–58.

84. Review by H. W. Wright, *American Journal of Theology* 16 (Apr. 1912): 315–16.

85. Reviews and comments by Paul Carus, Radoslav Tsanoff, C. D. Broad, Ellen Bliss Talbott, André Lalande, and Charles B. Vibbert are reprinted in John Elof Boodin, *Truth and Reality: An Introduction to the Theory of Knowledge*, Early Defenders of Pragmatism, vol. 2, "Responses and Reviews" appendix (Bristol, UK: Thoemmes Press, 2001), 1–30. Because of the sustained nature of Tsanoff's comment, they are cited separately with Boodin's reply.

86. See Radislove Tsanoff, "Professor Boodin on the Nature of Truth," *Philosophical Review* 19 (Nov. 1910): 632–38; John E. Boodin, "The Nature of Truth: A Reply," *Philosophical Review* 20 (Jan. 1911): 59–63. Tsanoff's "Rejoinder" that immediately follows adds little new, only to reiterate past points and prove how far past each other they are speaking.

87. See Paul Carus's editorial comment on Boodin's "The Divine Five-Fold Truth," *Monist* 21 (Apr. 1911): 288–94 and 295–96 of that issue.

88. John Clendening, *The Life and Thought of Josiah Royce* (Madison: University of Wisconsin Press, 1985), 93.

89. Quoted in Henry Thomas, *Biographical Encyclopedia of Philosophy* (Garden City, NY: Doubleday, 1965), 32.

90. Boodin, *Time and Reality*, 13.

91. Boodin, 23.

Chapter 2

1. Donald Palmer, *Looking at Philosophy: The Unbearable Heaviness of Philosophy Made Lighter* (Mountain View, CA: Mayfield Publishing, 1988), 57, 78.

2. Palmer, 102.

3. Anthony Kenny, *A New History of Philosophy: In Four Parts* (Oxford: Clarendon Press, 2010), 249.

4. John M. Riddle, *A History of the Middle Ages, 300–1500*, 2nd ed. (Lanham, MD: Rowman & Littlefield, 2016), 47.

5. Frederick Copleston, *Greece and Rome*, vol. 1 of *A History of Philosophy*, Bellarmine Series, 9 (London: Burns, Oates, and Washbourne, 1947), 467.

6. Roger Hancock, s.v. "Metaphysics, History of," in *The Encyclopedia of Philosophy*, vol. 5 (New York: Macmillan, 1967).

7. Kenny, 646–52.

8. D. W. Hamlin, s.v. "Metaphysics, History of," in *The Oxford Companion to Philosophy*, 2nd ed. (Oxford: Oxford University Press, 2005).

9. Glenn Alexander Magee makes a compelling case for Hegel's fascination for and engagement with occult mysticism in *Hegel and the Hermetic Tradition* (Ithaca, NY: Cornell University Press, 2001); and his "Hegel on the Paranormal: Altered States of Consciousness in the Philosophy of Subjective Spirit," *Aries* 8 (2008): 21–36.

10. E. J. Lowe, s.v. "Metaphysics, Opposition to," in *Oxford Companion*.

11. Bernard Phillips, "Logical Positivism and the Function of Reason," *Philosophy* 23 (Oct. 1948): 346.

12. See, for example, the essays from 1951 to 1974 republished in Donald C. Williams, *The Elements and Patterns of Being: Essays in Metaphysics*, ed. A. R. J. Fisher (Oxford: Oxford University Press, 2018); A. R. J. Fisher, "Donald C. Williams's Defense of Real Metaphysics," *British Journal for the History of Philosophy* 25, no. 2 (2017): 332–55; and Brian G. Henning, "Recovering the Adventure of Ideas: In Defense of Metaphysics as Revisable, Systematic, Speculative Philosophy," *Journal of Speculative Philosophy* 29, no. 4 (2015): 437.

13. See the preface to the second edition, *A Realistic Universe*, rev. ed. (New York: Macmillan, 1931), xi. Unless otherwise stated, all references to this work are from the revised edition, the most important of which is the addition of a lengthy introduction. Although Boodin claims he made changes in chapters 2 and 3, a

careful examination of both editions show only modest but substantive changes to chapter 2. All the others (including chapter 3) remain identical in content and pagination to the original.

14. Boodin, *A Realistic Universe*, xxxv. A similar "overlapping genius" in our own generation is the Princeton mathematician/physicist Freeman Dyson. Only recently deceased, Dyson shared his immense knowledge of the physical sciences in several books aimed at an intelligent lay public. His essays remain exemplary examples of popular science done exceptionally well. See, for example, *The Scientist as Rebel* (2006), *A Many-Colored Glass: Reflections on the Place of Life in the Universe* (2007), *Dreams of Earth and Sky* (2015), and *Birds and Frogs: Selected Papers, 1990–2014* (2015).

15. I. Bernard Cohen, *Revolution in Science* (Cambridge, MA: Belknap Press [Harvard University Press], 1985), 405.

16. Boodin, *A Realistic Universe*, xxvii. Similarly, Herman Weyl (whom Boodin was fond of referencing), a key architect of the relativity and quantum revolutions by establishing symmetry-groups in particle physics in 1928, told a young Freeman Dyson, "I always try to combine the true with the beautiful, but when I have to choose one or the other, I usually choose the beautiful." See Freeman Dyson, *Dreams of Earth and Sky* (New York: New York Review of Books, 2015), 192. Weyl later applied this scientific concept to a broader ranging metaphysical work in *Symmetry* (1952), still available from Princeton University Press.

17. Spencer Weart, "Trend-Spotting: Physics in 1931 and Today," *Physics Today* 59, no. 6 (June 2006): 32, https://doi.org/10.1063/1.2218552.

18. Boodin, *A Realistic Universe*, xxxv.

19. Boodin, 10.

20. Charles H. Nelson, *John Elof Boodin: Philosopher-Poet* (New York: Philosophical Library, 1987), 71.

21. Boodin, 27.

22. Boodin, 41.

23. For an excellent review, see Andrew Rex, "Maxwell's Demon—A Historical Review," *Entropy* 19, no. 6 (2017): 240, https://doi.org/10.3390/e19060240.

24. Although likely to be misconstrued in today's parlance, Thomson's designation was well chosen, suggesting an inherently teleological entity. J. Lemprière defines *dæmon* as "a kind of spirit which, as the ancients supposed, presided over the actions of mankind, gave them private counsels, and carefully watched over their most secret intentions." See *Lempriere's Classical Dictionary* (1788; repr., London: Bracken Books, 1984), 214.

25. Fred Hoyle, *The Intelligent Universe* (New York: Holt, Rinehart, and Winston, 1983), 37.

26. See, for example, Boodin, *A Realistic Universe*, xx, 10, 56, 186, 327, 338, 341, 343, 345, 367–69, 381.

27. Boodin, 61.

28. Boodin, 73.
29. Boodin, 111.
30. Boodin, 127.
31. Boodin, 128. Boodin could easily have been aware of this rising theory of psychology. Watson first presented his ideas in an address to Columbia University in 1913 (published that same year as "Psychology as the Behaviorist Views It" in the *Psychological Review*, the same journal where Boodin published his monograph *Time and Reality* nine years earlier). Watson would elaborate upon his ideas in *Behavior: An Introduction to Comparative Psychology* (1914). It is fair to say that at Boodin's writing behaviorism was a "hot topic" and even more so in 1931.
32. Bertalanffy despised behaviorism and set his sights on everything and anything B. F. Skinner wrote. According to Bertalanffy, Watson and Skinner were responsible for making "a large part of modern psychology . . . a sterile and pompous scholasticism which, with the blinders of preconceived notions or superstitions on its nose, doesn't see the obvious; which covers the triviality of its results and ideas with preposterous language bearing no resemblance either to normal English or normal scientific theory; and which provides modern society with the techniques for the progressive stultification of mankind." He characterized behaviorism as "a positivistic-mechanistic-reductionist approach which can be epitomized as *the robot model of man*." See his *Robots, Man and Minds: Psychology in the Modern World* (New York: George Braziller, 1967), 6–7.
33. Boodin, *A Realistic Universe*, 143.
34. Quoted in Rupert Sheldrake, *Morphic Resonance: The Nature of Formative Causation*, 4th ed. (Rochester, VT: Park Street Press, 2009), xiii.
35. Deborah Blum, *Ghost Hunters: William James and the Search for Scientific Proof of Life After Death* (New York: Penguin Books, 2006), 223.
36. Quoted in Robert D. Richardson, *William James: In the Maelstrom of American Modernism: A Biography* (Boston: Houghton Mifflin, 2006), 261.
37. Kenneth Ray Sutton, "John Elof Boodin's Creationism" (PhD diss., University of New Mexico, 1969), 23. From Mollenhauer's personal communication to the author.
38. Boodin, autobiographical statement in *Contemporary American Philosophy*, 142.
39. Boodin, *A Realistic Universe*, 178.
40. Boodin, 212–13.
41. Boodin, 241.
42. Dyson, *Dreams of Earth and Sky*, 242.
43. Boodin, *A Realistic Universe*, 313.
44. Boodin, 320.
45. Boodin, 323.
46. For a detailed discussion, see Alan Sokal and Jean Bricmont, *Fashionable Nonsense: Postmodern Intellectuals' Abuse of Science* (New York: Picador, 1998). They

offer some additional comments and correctives in their "Defense of a Modest Scientific Realism," in *Knowledge and the World: Challenges Beyond the Science Wars*, eds. Martin Carrier, Johannes Roggenhofer, Günter Küppers, and Philippe Blanchard (Berlin: Springer, 2004).

    47. Boodin, *A Realistic Universe*, 350.

    48. Raymond Tallis, *Aping Mankind: Neuronmania, Darwinitis and the Misrepresentation of Humanity* (London: Routledge, 2016), 225–26.

    49. Max Horkheimer, *Eclipse of Reason* (1947; repr., Mansfield Center, CT: Martino Publishing, 2013), 124–25.

    50. Stephen Reynolds, trans., *Gregory of Nazianzus: Five Theological Orations* (Toronto: Stephen Reynolds, 2011), 18. This translation was part of a course given at Trinity College in the University of Toronto, 2001–2009.

    51. Frank Wilczek, *A Beautiful Question: Finding Nature's Deep Design* (New York: Penguin Books, 2015), 8.

    52. Boodin, *A Realistic Universe*, 365.

    53. Boodin, 367.

    54. Boodin, 333.

    55. R. F. A. Hoernlé, review of *A Realistic Universe*, by John Elof Boodin, *Harvard Theological Review* 12 (Jan. 1919): 129.

    56. Radislav A. Tsaniff, reviews, *Philosophical Review* 26 (Nov. 1917): 660–65; and *Philosophical Review* 43 (Jan. 1934): 90–92.

    57. E. C. Wilm, review, *International Journal of Ethics* 30 (July 1920): 464–67.

    58. Carl Haessler, review, *Journal of English and Germanic Philology* 16 (Oct. 1917): 617–20.

    59. Joseph Brent, *Charles Sanders Peirce: A Life*, rev. ed. (Bloomington: University of Indiana Press, 1998), 207–8.

    60. M. T. McClure, review, *Journal of Philosophy, Psychology and Scientific Methods* 14 (Dec. 6, 1917): 693–95.

    61. Boodin, "Cosmic Attributes," *Philosophy of Science* 10 (Jan. 1943): 1–12.

Chapter 3

    1. Janet Browne, *Charles Darwin: The Power of Place* (Princeton, NJ: Princeton University Press, 2002), 177.

    2. Adrian Desmond and James Moore, *Darwin* (New York: W. W. Norton, 1991), 12.

    3. Desmond and Moore, 34.

    4. Janet Browne, *Charles Darwin: Voyaging* (Princeton, NJ: Princeton University Press, 1995), 83.

    5. For more on the formative influences bearing on Darwin's youth, see Michael A. Flannery, *Nature's Prophet: Alfred Russel Wallace and His Evolution from*

Natural Selection to Natural Theology (Tuscaloosa: University of Alabama Press, 2018), 43; and Michael A. Flannery, *Intelligent Evolution: How Alfred Russel Wallace's "World of Life" Challenged Darwinism* (Nashville, TN: Erasmus Press, 2020), 6–10.

6. Karl W. Giberson, *Saving Darwin: How to Be a Christian and Believe in Evolution* New York: HarperOne, 2008), 19–20.

7. Alister E. McGrath, *Darwinism and the Divine: Evolutionary Thought and Natural Theology* (Oxford: Wiley-Blackwell, 2011), 159–60.

8. Michael Ruse, *On Purpose* (Princeton, NJ: Princeton University Press, 2018), 88.

9. Browne, *The Power of Place*, 427.

10. Browne, *Voyaging*, x–xi.

11. Phillip R. Sloan, "'The Sense of Sublimity': Darwin on Nature and Divinity," *Osiris* 16 (2001): 257.

12. Historians underestimate Darwin's tendency to self-promotion at their own peril. Those so inclined should read John Angus Campbell, "The Invisible Rhetorician: Charles Darwin's 'Third Party' Strategy," *Rhetorica: A Journal of the History of Rhetoric* 7 (Winter 1989): 55–85; and Randy Moore, "The Persuasive Mr. Darwin," *BioScience* 47 (Feb. 1997): 107–14.

13. For references to Auguste Comte, see *Charles Darwin's Notebooks, 1836–1844*, eds. Paul H. Barrett et al. (London: Natural History Museum, 1987), 535, 539, 553, 566, 608.

14. Maurice Mandelbaum, "Darwin's Religious Views," *Journal of the History of Ideas* 19 (June 1958): 376.

15. See Edward Aveling, *The Religious Views of Charles Darwin* (1883), reproduced in full at Darwin Online: http://darwin-online.org.uk/content/frameset?pageseq=1&itemID=A234&viewtype=text.

16. Richard Dawkins, *River Out of Eden: A Darwinian View of Life* (New York: Basic Books, 1995), 133.

17. Thomas Henry Huxley, *Collected Essays*, vol. 2 of *Darwiniana* (London: Macmillan, 1899), 82.

18. Jacques Barzun, *Darwin, Marx, Wagner: Critique of a Heritage*, 2nd ed. (1958; repr., Chicago: University of Chicago Press, 1981), 63–64.

19. Examples abound. See John Elof Boodin, *Cosmic Evolution* (New York: Macmillan, 1925), 18–459 passim.

20. Boodin, 58, 67.

21. Any doubt that Wallace viewed natural selection as an eliminative force has been thoroughly dispelled in Charles H. Smith's "Alfred Russell Wallace and the Elimination of the Unfit," *Journal of Biosciences* 37 (June 2012): 203–5. Significantly, Ernst Mayr, one of the leading architects of the neo-Darwinian synthesis of the 1930s and 1940s, agrees. He defines natural selection as, "The process by which in every generation individuals of lower fitness are *removed* [emphasis added] from the population." See his *What Evolution Is* (New York: Basic Books, 2001), 288.

22. Boodin, *Cosmic Evolution*, 83.

23. On the recantation, see Karl Popper, "Natural Selection and the Emergence of Mind," *Dialectica* 32, no. 3/4 (1978): 339–55. Mehmet Elgin and Elliott Sober insist that Popper merely changed the subject but not his mind. See their "Popper's Shifting Appraisal of Evolutionary Theory," *HOPOS* 7 (Spring 2017): 31–55.

24. John Elof Boodin, *Time and Reality*, *Psychological Review*, monographic suppl., vol. 6, no. 3 (New York: Macmillan, 1904), 294.

25. James K. Feibleman, "The Metaphysics of Logical Positivism," *Review of Metaphysics* 5 (Sept. 1951): 59.

26. Robert Henry Peters, "Tautology in Evolution and Ecology," *American Naturalist* 110 (Jan./Feb. 1978): 1–12.

27. Robert Henry Peters, "Predictable Problems with Tautology in Evolution and Ecology," *American Naturalist* 112 (July/Aug. 1978): 761.

28. Edward S. Reed, "The Lawfulness of Natural Selection," *American Naturalist* 118 (July 1981): 61–71; and Henry C. Byerly, "Natural Selection as a Law: Principles and Processes," *American Naturalist* 121 (May 1983): 739–45.

29. Ronald H. Brady, "Natural Selection and the Criteria by which a Theory is Judged," *Systematic Zoology* 28 (Dec. 1979): 600–21.

30. Tam Hunt, "Reconsidering the Logical Structure of the Theory of Natural Selection," *Communicative & Integrative Biology* 7, no. 6 (2014): e972848, https://www.ncbi.nlm.nih.gov/pmc/articles/PMC4594354/.

31. Guilhem Doulcier, Peter Takacs, and Pierrick Bourrat, "Taming Fitness: Organism-Environment Interdependencies Preclude Long-Term Fitness Forecasting," *BioEssays* (2020), https://onlinelibrary.wiley.com/doi/abs/10.1002/bies.202000157.

32. John C. Greene, *Darwin and the Modern World* (Baton Rouge: Louisiana State University Press, 1961), 44.

33. Quoted in John C. Greene, "The Interaction of Science and World View in Sir Julian Huxley's Evolutionary Biology," *Journal of the History of Biology* 23 (Spring 1990): 43.

34. The letter is reproduced in its 1871 version in Flannery, *Nature's Prophet*, 178–87.

35. John Elof Boodin, *A Realistic Universe: An Introduction to Metaphysics*, rev. ed. (New York: Macmillan, 1931), 354.

36. It wasn't just Darwin's analogy that was dismissed, but Darwinian theory in general, especially in France where it was largely ignored. The Positivists considered the theory metaphysical and too speculative to receive serious notice while others, like Paul Janet (an idealist who took his moral philosophy seriously), rejected any evolutionary theory that eliminated final causes. One French correspondent is said to have exclaimed, "Will there not be found in British science a man of eminence to fight the battle of good sense and of the facts, against the monstrous imaginations of Darwin? If such a man comes out, he will find a powerful assistant in our *Quatrefages*, our *Blanchard*, and our *Janet*," a clear reference to Janet's *Les Causes*

Finales (1876). Even when evolution was accepted, it was of a very un-Darwinian kind, as in the case of Bergson's *L'Évolution créatrice* (1907). In 1959 it was not Darwin's centennial of *Origin* that was celebrated in France, but the sesquicentennial of Lamarck's *Philosophie zoologique*. For details, see Robert E. Stebbins essay on France in *The Comparative Reception of Darwinism*, ed. Thomas F. Glick (Austin: University of Texas Press, 1972), 117–63.

37. Rémy Collin, *Evolution*, trans. J. Tester (New York: Hawthorne Books, 1959), 71.

38. Pierre-P. Grassé, *Evolution of Living Organisms: Evidence for a New Theory of Transformation* (New York: Academic Press, 1977), 122.

39. Bert Theunissen, "Darwin and His Pigeons: The Analogy Between Artificial and Natural Selection Revisited," *Journal of the History of Biology* 45 (Summer 2012): 179–212.

40. Darwin, *Variation of Animals and Plants Under Domestication*, 2nd ed., 2 vol. (New York: D. Appleton, 1883), 2:496.

41. Jean Gayon, *Darwinism's Struggle for Survival: Heredity and the Hypothesis of Natural Selection*, trans. Matthew Cobb (Cambridge: Cambridge University Press, 1998), 59.

42. David N. Reznick, *The "Origin" Then and Now: An Interpretive Guide to the "Origin of Species"* (Princeton, NJ: Princeton University Press, 2010), 79.

43. Arthur Koestler, *Janus: A Summing Up* (London: Pan Books, 1979), 192.

44. It has been alleged that Darwin never argued that *Homo sapiens* descended from the ape. Actually, this is untrue. He wrote, "The Simiadæ then branched off into two great stems, the New World and the Old World monkeys; and from the latter, at a remote period, Man, the wonder and glory of the Universe, proceeded. Thus we have given to man a pedigree of prodigious length, but not, it may be said, of noble quality." See Charles Darwin, *Descent of Man* (1871; repr. New York: Barnes & Noble, 2004), 141.

45. For an excellent overview of the facts of the case and insightful analysis, see Shala Barczewska, "*Inherit the Wind*: The Movie and the Myth in American Cultural Memory," in *Confluence of Literature, History and Cinema*, eds. Paweł Kaptur and Agnieszka Szwach (Kielce, Poland: Jan Kochanowski University, 2020), 109–30. See also Peter J. Bowler, *Monkey Trials and Gorilla Sermons: Evolution and Christianity from Darwin to Intelligent Design* (Cambridge, MA: Harvard University Press, 2007), 182–86.

46. Frederick Lewis Allen, *Only Yesterday: An Informal History of the 1920s* (New York: Perennial Library, 1964), 168.

47. Allen, 171.

48. See Barczewska, "*Inherit the Wind*," 118; and Tom Arnold-Foster, "Rethinking the Scopes Trial: Cultural Conflict, Media Spectacle, and Circus Politics," *Journal of American Studies* (May 12, 2021): 6, https://doi.org/10.1017/S0021875821000529.

49. Quoted in Barczewska, "*Inherit the Wind*," 113.

50. Barczewska, 127.

51. Hugh Ross, *A Matter of Days: Resolving a Creation Controversy* (Colorado Springs, CO: NavPress, 2004), 31–32.

52. For a complete listing and descriptive summaries, see Tom McIver, *Anti-Evolution: An Annotated Bibliography* (Jefferson, NC: McFarland, 1988), 232–34.

53. Raymond Dart, "*Australopithecus africanus*: The Man-Ape of South Africa," *Nature* 115, no. 2884 (Feb. 7, 1925): 195–99.

54. Robert Jurmain et al., *Introduction to Physical Anthropology* (Belmont, CA: Wadsworth, Cengage Learning, 2014), 296.

55. Frank Sulloway, "Darwin and His Finches: The Evolution of a Legend," *Journal of the History of Biology* 15 (Spring 1982): 1–53.

56. V. B. Smocovitis, "The Evolutionary Synthesis and Evolutionary Biology," *Journal of the History of Biology* 25 (Spring 1992): 1–65.

57. Vassiliki Betty Smocovitis, "The 1959 Darwin Centennial Celebration in America," *Osiris* 14, Commemorative Practices in Science: Historical Perspectives on the Politics of Collective Memory (1999): 321.

58. Smocovitis, 323.

59. Ron Amundson, *The Changing Role of the Embryo in Evolutionary Thought: Roots of Evo-Devo* (Cambridge: Cambridge University Press, 2007), 11.

60. Amundson, 13, 25.

61. Amundson, 81.

62. See Günter Wagner, "How Wide and How Deep is the Divide Between Population Genetics and Developmental Evolution?," *Biology and Philosophy* 22 (2007): 145–53; Mary P. Winsor, "The Creation of the Essentialism Story: An Exercise in Metahistory," *History and Philosophy of the Life Sciences* 28, no. 2 (2006): 149–74.

63. Wagner, 151.

64. Mark B. Adams, "Little Evolution, BIG Evolution: Rethinking the History of Darwinism, Population Genetics, and the 'Synthesis,'" in *Natural Selection: Revisiting Its Explanatory Role in Evolutionary Biology*, ed. Richard G. Delisle (Cham, Switzerland: Springer, 2021), 215.

65. Adams, 224–25.

66. Baily D. McKay and Robert M. Zink, "Sisyphean Evolution in Darwin's Finches," *Biological Reviews* 90 (2015): 689–98.

67. William H. Werkmeister, *A History of Philosophical Ideas in America* (New York: Ronald Press, 1949), 512–18.

68. Boodin, *Cosmic Evolution*, 33.

69. Boodin, 35.

70. Boodin, 38.

71. R. Broom, *The Coming of Man: Was It Accident or Design?* (London: H. F. & G. Witherby, 1933), 220–21.

72. Broom, 416.

73. Broom, 49.

74. Quoted in Rupert Sheldrake, *Science Set Free: 10 Paths to New Discovery* (New York: Deepak Chopra Books, 2012), 65.

75. Boodin, *Cosmic Evolution*, 57.

76. Michael J. Behe, "Experimental, Evolution, Loss-of-Function Mutations, and 'The First Rule of Adaptive Evolution,'" *Quarterly Review of Biology* 85 (Dec. 2010): 419–45.

77. Michael J. Behe, *Darwin Devolves: The New Science About DNA that Challenges Evolution* (New York: HarperOne, 2019).

78. Boodin, *Cosmic Evolution*, 61.

79. Boodin, 115.

80. Boodin, 125.

81. Boodin, 163.

82. Boodin, 178.

83. Boodin, 207.

84. Boodin, 268.

85. John C. Eccles, *Evolution of the Brain: Creation of the Self* (London: Routledge, 1989), 241.

86. Eccles, 237.

87. John C. Eccles, "Do Mental Events Cause Neural Events Analogously to the Probability Fields of Quantum Mechanics?," *Proceedings of the Royal Society of London: Series B, Biological Sciences* 227, no. 1249 (1986): 411–28.

88. Boodin, *Cosmic Evolution*, 240. Like Eccles, Boodin also gives importance to the "birth of a soul," 240–53.

89. Charles H. Nelson, *John Elof Boodin: Philosopher-Poet* (New York: Philosophical Library, 1987), 133.

90. Boodin, *Cosmic Evolution*, 316–17.

91. Alfred North Whitehead, *Science and the Modern World* (1925; repr. New York: Free Press, 1967), 54.

92. Boodin, *Cosmic Evolution*, 332.

93. Boodin, 396.

94. Boodin, 109.

95. Paul J. Steinhardt and Neil Turok, *Endless Universe: Beyond the Big Bang* (New York: Doubleday, 2007). See also Charles Seife, "Eternal-Universe Idea Comes Full Circle," *Science*, New Series 296, no. 5568 (Apr. 26, 2002): 639.

96. Plamena Marcheva and Stoil Ivanov, "On the Geodesics in Bondi-Gold-Hoyle Universe Model," *Journal of Physics and Technology* 1, no. 1 (2017): 3–5.

97. João Barbosa, "*Creation*: a Multifaceted and *Thematic* Concept in the Construction of Modern Cosmology—from Friedmann's *Creation of the Universe* to the Steady-State's *Continuous Creation*," *Philosophy and Cosmology* 27 (2021): 31.

98. Boodin, *Cosmic Evolution*, 467.

99. Boodin, 454–55.

100. Boodin, 263.

101. Roy Wood Sellars, "What is the Correct Interpretation of Critical Realism?," *Journal of Philosophy* 24 (Apr. 28, 1927): 238.

102. Naturalism is a philosophically two-fold term. Specifically, it can be applied to ethics, while more generally it applies to the study of phenomena and the methods of studying. It can be reductive or non-reductive. Essentially it is the idea that everything belongs to the observable, testable world we sense and perceive. If there are exceptions, these can for the naturalist be typically explained away. Metaphysically, it need not be strictly physicalist but it often has affinities with materialism. Naturalism, however, does insist that nature is a single realm without incursions of outside forces (spirits, souls, special entities, entelechies, or abstract universals). Naturalism can be quite nuanced as in Hilary Putnam's "liberal naturalism," in which he rules out appeals to the sorts of outside forces described above, but at the same time he is a non-reductionist, whereby scientific explanations do not require physicalist interpretations or conclusions. For Putnam, reason lies beyond natural explanation such that science always relies on that which is not entirely within the bounds of science itself. By comparison, Sellars's materialism could hardly comport to what could be considered "liberal" in Putnam's sense, nor in the *process* sense in which all nature is coextensive (see glossary). See the essays in *Naturalism, Realism, and Normativity: Hilary Putnam*, ed. Mario De Caro (Cambridge, MA: Harvard University Press, 2016); Alan Lacey, s.v. "Naturalism," in *Oxford Companion to Philosophy*, ed. Ted Honderich, 2nd ed. (Oxford: Oxford University Press, 2005); and David Ray Griffin, *Reenchantment without Supernaturalism: A Process Philosophy of Religion* (Ithaca, NY: Cornell University Press, 2001).

103. Roy Wood Sellars, *Evolutionary Naturalism* (Chicago: Open Court, 1922), 18.

104. See George H. Sabine, review of *Evolutionary Naturalism*, by Roy Wood Sellars, *Philosophical Review* 32 (Jan. 1923): 93–95; and Maurice Picard, review of *Evolutionary Naturalism, Journal of Philosophy* 19 (Oct. 12, 1922): 582–87.

105. See, for example, the numerous scientists subscribing to "The Third Way: Evolution in the Era of Genomics and Epigenetics," *The Third Way*, accessed Jan. 21, 2022, https://www.thethirdwayofevolution.com/.

106. For details, see Sydney Eisen, "Huxley and the Positivists," *Victorian Studies* 7 (1964): 337–58.

107. Will Durant, *The Mansions of Philosophy: A Survey of Human Life and Destiny* (Garden City, NY: Garden City Publishing, 1929), 601.

108. Robert J. Kreyche, "The Naturalism of Roy Wood Sellars" (PhD diss., University of Ottawa, 1950), 170.

109. Kreyche, 188.

110. John C. Greene, *Debating Darwin: Adventures of a Scholar* (Claremont, CA: Regina Books, 1999), 217.

111. Whitehead, *Science and the Modern World*, 107.

112. Boodin, *Cosmic Evolution*, 72.

113. Denis Noble, *Dance to the Tune of Life: Biological Relativity* (Cambridge: Cambridge University Press, 2017), 181.
114. Noble, 175–76.
115. Noble, 249.
116. Ruse, *On Purpose*, 235.
117. Boodin, *Realistic Universe*, 381–82.
118. Rupert Sheldrake, *Science and Spiritual Practices* (Berkeley, CA: Counterpoint, 2017), 63–65.
119. Joanna Leidenhag, "How to Be a Theological Panpsychist, but Not a Process Theologian," *Philosophy, Theology and the Sciences* 7, no. 1 (2020): 10–29, https://research-repository.st-andrews.ac.uk/handle/10023/23541.
120. Michael B. Foster, review, *Journal of Philosophical Studies* 4 (Apr. 1929): 255–257.
121. A. K. Stout, review, *Mind*, New Series 36 (Oct. 1927): 496–99.
122. R. F. A. Hoernlé, review, *Journal of Philosophy* 24 (Mar. 17, 1927): 160–63.
123. G. W. T. Whitney, review, *Social Science* 2 (Aug.–Sept. 1927): 442–45.

Chapter 4

1. John Elof Boodin, *Three Interpretations of the Universe* (New York: Macmillan, 1934), 27–28.
2. Boodin, 28.
3. David C. Lindberg, *The Beginnings of Western Science: The European Scientific Tradition in Philosophical, Religious, and Institutional Context, Prehistory to A.D. 1450*, 2nd ed. (Chicago: University of Chicago Press, 2007), 198–99.
4. Boodin, *Three Interpretations*, 31.
5. Jonathan Loose, s.v. "Emergence," in *Dictionary of Christianity and Science*, ed. Paul Copan et al. (Grand Rapids, MI: Zondervan, 2017).
6. Categorizing emergent evolution (the idea that entirely new properties such as mind and consciousness can spring or develop from the unpredictable rearrangement of existing phenomena) can be quite difficult. For example, C. Lloyd Morgan, most often associated with the term, is sometimes regarded as a teleologist who proposed an expressly theistic version of evolution. But it has been correctly pointed out that Morgan's evolutionary process was neither mechanistic nor finalistic; it rejected teleology. He believed that a transcendent "God" or "spirit" must be assumed, but only for purely heuristic purposes. See T. A. Goudge, s.v. "Morgan, C. Lloyd," in *The Encyclopedia of Philosophy*, vol. 5 (New York: Macmillan, 1965). Boodin himself admitted difficulty in placing Morgan's evolutionary theory, but in the end he determined that Lloyd gave a complete statement of naturalistic emergence that reduced God to a completely monistic interpretation (*Three Interpretations*, 134). But Boodin recognized that Samuel Alexander's brand of emergence—the emergence

of successive stages—as stated in his *Space, Time, and Deity* (1920) could have a non-naturalistic aspect since in Alexander's view "God is the whole drive of the process." But in this case Boodin suggests that Alexander's theory according to his schema is more preformationist (*Three Interpretations*, 131). In any case, Boodin thought little of Alexander's metaphysical balancing act (see chapter 3).

7. Rupert Sheldrake, *Science Set Free: 10 Paths of New Discovery* (New York: Deepak Chopra Books, 2012), 106.

8. Boodin, *Three Interpretations*, 13.

9. Boodin, 123.

10. Anthony Kenny, *A New History of Western Philosophy: In Four Parts* (Oxford: Clarendon Press, 2010), 83.

11. Boodin, *Three Interpretations*, 99.

12. Boodin, 111.

13. Quoted in Boodin, 116.

14. Boodin, 143.

15. Boodin, 153.

16. Boodin, 186.

17. Boodin, 189.

18. Whether or not origin of life hypotheses have reached a crisis is a matter of interpretation. See Iris Fry's chapter "Crisis—Real or Fictitious" in *The Emergence of Life on Earth: A Historical and Scientific Overview* (New Brunswick, NJ: Rutgers University Press, 2000), 112–34.

19. See, for example, the following peer-reviewed literature: J. T. Trevers and D. L. Abel, "Chance and Necessity Do Not Explain the Origin of Life," *Cell Biology International* 28, no. 11 (2004): 729–739; David L. Abel, "The Capabilities of Chaos and Complexity," *International Journal of Molecular Science* 2009, no. 10 (2009): 247–91; and Vera Vasas, Eörs Szathmáry, and Mariop Santos, "Lack of Evolvability in Self-Sustaining Auotcatalytic Networks Constraints Metabolism-First Scenarios for the Origin of Life," *PNAS* 107, no. 4 (2010): 1470–1475.

20. Johnjoe McFadden and Jim Al-Kalili, *Life on the Edge: The Coming Age of Quantum Biology* (New York: Crown Publishers, 2014), 99.

21. McFadden and Al-Kalili, 286.

22. See, for example, Adriana Marais, et al., "The Future of Quantum Biology," *Journal of the Royal Society: Interface* 15 (Nov. 14, 2018), https://doi.org/10.1098/rsif.2018.0640.

23. Boodin, *Three Interpretations*, 231.

24. Although Spinoza is usually considered a monist, Boodin argues that his parallelism of the physical and the psychical amounts to dualism. Wolfgang Köhler, who helped found Gestalt psychology, falls prey to this dualism as Boodin explains. Boodin, 252–53.

25. Boodin, 256.

26. Boodin, 217.

27. Boodin, 275.

28. Boodin, 299–98.

29. Boodin, 307.

30. Boodin, 323.

31. Harvey, Boyle, and Maupertuis are thoroughly discussed in Jeffrey K. McDonough, "Not Dead Yet: Teleology and the 'Scientific Revolution,'" in *Teleology: A History*, ed. Jeffrey K. McDonough (New York: Oxford University Press, 2020), 150–79.

32. In this regard, see also Emerson Thomas McMullen, "A Barren Virgin? Teleology in the Scientific Revolution" (PhD diss., University of Indiana, 1989).

33. Boodin, *Three Interpretations*, 377.

34. John M. Riddle, *A History of the Middle Ages, 300–1500*, 2nd ed. (Lanham, MD: Rowman & Littlefield, 2016), 43.

35. Frederick Copleston, *Greece and Rome: A History of Philosophy*, vol. 1. Bellarmine Series 9 (London: Burns, Oates, and Washbourne, 1946), 464.

36. Boodin, *Three Interpretations*, 422.

37. Boodin, 422–23.

38. Boodin, 426–27.

39. Alfred North Whitehead, *Process and Reality*, eds. David Ray Griffin and Donald W. Sherburne (New York: Free Press, 1978), 342. Corrected ed.

40. Boodin, *Three Interpretations*, 487.

41. Francis M. Sanchez, "A Coherent Resonant Cosmology Approach and its Implications in Microphysics and Biophysics," in *Quantum Systems in Physics, Chemistry, and Biology*, eds. Alia Tadjer et al. (Cham, Switzerland: Springer, 2017), 377.

42. See, for example, Paul J. Steinhardt and Neil Turok's *Endless Universe: Beyond the Big Bang* (New York: Doubleday, 2007). Of some relevance here is the fact that Steinhardt and Turok note the ambiguity of the Bible on this issue and admit that "a common [Rabbinic] interpretation is that this creation was from preexisting material, meaning that space, time, matter, and energy existed before the moment of creation" (172), a view clearly echoed by Boodin. As Turok said in an interview, "It's exactly what the steady-state universe people wanted. Our model really realizes their goal." Quoted in Charles Seif, "Eternal-Universe Idea Comes Full Circle," *Science*, New Series 296 (Apr. 26, 2002): 639. Work along these lines continues: Plamena Marcheva and Stoil Ivanov, "On the Geodesics in Bondi-Gold-Hoyle Universe Model," *Journal of Physics and Technology* 1, no. 1 (2017): 3–5; Anna Ijjas and Paul J. Steinhardt, "Bouncing Cosmology Made Simple," *Classical and Quantum Gravity* 35 (June 1, 2018), https://iopscience.iop.org/article/10.1088/1361-6382/aac482.

43. Boodin, *Three Interpretations*, 499.

44. Boodin, 500.

45. See Charles F. Sawhill Virtue, review of *Three Interpretations of the Universe and God*, by John Elof Boodin, *Philosophical Review* 45 (Jan. 1936): 88.

46. R. F. Hoernlé, review of *Three Interpretations and God*, *Mind* 45 (Apr. 1936): 219.

47. See Boodin, *Three Interpretations*, 43–45.

48. Alfred North Whitehead, *Religion in the Making* (1926; repr., New York: Fordham University Press, 1996), 111–12.

49. See Randall E. Auxier, "A Note on Whitehead's View of God in *Religion in the Making*," in Whitehead, *Religion in the Making*, 244–51.

50. Boodin, *Three Interpretations*, 204–5.

51. Boodin, *Realistic Universe: An Introduction to Metaphysics*, rev. ed. (New York: Macmillan,1931), 366. It is worth noting here that Sheldrake agrees with Boodin. For him, Bergson's ideas are very similar to other emergent evolutionists like Darwin and Marx. See his *The Presence of the Past: Morphic Resonance and the Memory of Nature*, rev. ed. (Rochester, VT: Park Street Press, 2012), 373–74.

52. Adrian Desmond, *Archetypes and Ancestors: Paleontology in Victorian London, 1850–1875* (Chicago: University of Chicago Press, 1982), 32–33.

53. The calculated misrepresentations of Owen's positions and views have now been well chronicled by historians. See, for example, Ron Amundson, *The Changing Role of the Embryo in Evolutionary Thought: Roots of Evo-Devo* (Cambridge: Cambridge University Press, 2007), 77; Nicolaas Rupke, *Richard Owen: Biology without Darwin*, rev. ed. (Chicago: University of Chicago Press, 2009), 169; and Ruth Barton, *The X Club: Power and Authority in Victorian Science* (Chicago: University of Chicago Press, 2018), 173–75. Examples of Huxley's rank deception in his dispute with Owen over the hippocampus is thoroughly examined in Christopher E. Cosans, *Owen's Ape & Darwin's Bulldog: Beyond Darwinism and Creationism* (Bloomington: Indiana University Press, 2009).

54. The published version of Owen's 1849 lecture is still available, now enhanced with commentaries by leading modern experts in the field. See Richard Owen, *On the Nature of Limbs: A Discourse*, ed. Ron Amundson (Chicago: University of Chicago Press, 2007). Preface by Brian K. Hall and introductory essays by Ron Amundson, Kevin Padian, Mary P. Winsor, and Jennifer Coggon.

55. Quoted in Rupke, *Richard Owen*, 132.

56. Owen, *On the Nature of Limbs*, 2–3.

57. Michael Ruse, *The Darwinian Revolution: Science Red in Tooth and Claw*, 2nd ed. (Chicago: University of Chicago Press, 1999), 122.

58. Rupke, *Richard Owen*, 133.

59. Kevin F. Doherty, "Location of Platonic Ideas," *Review of Metaphysics* 14 (Sept. 1960): 57–72, esp. at 63–64.

60. Desmond, *Archetypes and Ancestors*, 43.

61. Desmond, 62.

62. For an interesting discussion of these Germanic antecedents, see Andrea Gambarotto, *Vital Forces, Teleology and Organization: Philosophy of Nature and the Rise of Biology in Germany* (Cham, Switzerland: Springer, 2018).

63. Rupke, *Richard Owen*, 179.

64. D'Arcy Wentworth Thompson, *On Growth and Form* (1917; repr., New York: Dover Publications, 1992), 7.

65. Michael Denton first announced his Darwinian doubts with his controversial *Evolution: A Theory in Crisis* (Bethesda, MD: Adler & Adler, 1986). He reiterated that skepticism and added his own commitment to structuralist evolution with his *Evolution: Still a Theory in Crisis* (Seattle: Discovery Institute Press, 2016). The quoted passage on page 281 is really discussed throughout the book at 39–41, 65, 67–69, 279–81, 291.

66. Will Durant, *The Reformation: A History of European Civilization from Wyclif to Calvin, 1300–1564* (New York: Simon & Schuster, 1957), 371.

67. Boodin, *God and Creation: A Cosmic Philosophy of Religion* (New York: Macmillan, 1934), 6.

68. Boodin, 15.

69. Boodin, 34

70. Boodin, 32.

71. The Gaia hypothesis is an idea of cooperative and symbiotic rather than competitive evolution proposed by James Lovelock and Lynn Margulis. Margulis, for example, is prone to talk of "Mother Earth" that expresses itself as a "physiology" of "environmental regulation," conceived as "an emergent property of interaction among organisms." See Margulis, *Symbiotic Planet: A New Look* (New York: Basic Books, 1998), 118–19. Although it has captured the attention of many New Age mystics, Lovelock has been very clear: "Neither Lynn Margulis nor I ever proposed a teleological hypothesis. Nowhere in our writings do we express the idea that planetary self-regulation is purposeful, or involves foresight or planning by the biota." See James E. Lovelock, "Hands Up for the Gaia Hypothesis," *Nature* 344 (Mar. 8, 1990): 100–02. So here we have just another example of Boodin's emergent, mechanistic cosmology—random, blind, and thoroughly reductionist.

72. Boodin, *God and Creation*, 101.

73. Boodin, 113.

74. Boodin, note on 61.

75. Boodin, 49.

76. Boodin, 183.

77. Boodin, 75.

78. Boodin, 129.

79. An interesting discussion of this, including the distinction between healthy and diseased communities, is available in Randall E. Auxier, *Time, Will, and Purpose: Living Ideas from the Philosophy of Josiah Royce* (Chicago: Open Court, 2013), 265–73.

80. Boodin, *Religion of Tomorrow* (New York: Philosophical Library, 1943), 27.

81. Boodin, 46.

82. Boodin, 49.

83. Boodin, 62.

84. Boodin, 74, 77, 85.

85. Boodin, 101.

86. Boodin, 137.

87. Boodin, 177.

88. Clifford Barrett, review of *Three Interpretations* and *God*, *Journal of Philosophy* 32 (Mar. 1935): 157–60.

89. Boodin, "Fictions in Science and Philosophy. II," *Journal of Philosophy* 40 (Dec. 23, 1943): 702.

90. Boodin, 704.

91. Boodin, 712.

92. Roland Faber, "Three Hundred Years of Whitehead: Halfway," *Process Studies* 41, no. 1 (2012): 19. The entire seminar is available at "Letter of Alfred North Whitehead," Library of Congress, February 17, 2011, https://www.loc.gov/item/webcast-5200.

93. Faber, 20.

94. R. F. Alfred Hoernlé, review of *Three Interpretations* and *God*, *Mind* 45 (Apr. 1936): 218.

95. Hoernlé, 227.

96. L. J. Russell, review of *Three Interpretations*, *Philosophy* 43 (July 1936): 359–60.

97. J. H. Tufts, review of *Three Interpretations* and *God*, *International Journal of Ethics* 45 (July 1935): 466–68.

98. Harold Bosley, review of *Religion of Tomorrow*, *Journal of Religion* 24 (Apr. 1944): 143–44.

99. Oliver L. Reiser, review of *Religion of Tomorrow*, *Philosophy and Phenomenological Research* 4 (June 1944): 571–72.

100. Boodin, *Religion*, 96.

101. Boodin, *The Social Mind* (New York: Macmillan, 1939), 583.

102. Charles Hartshorne, review of *Religion of Tomorrow*, *Ethics* 54 (Apr. 1944): 233–34.

Chapter 5

1. William James, *The Varieties of Religious Experience* (1902; repr., New York: Penguin Books, 1982), 31.

2. See Jonathan Bricklin, *The Illusion of Will, Self, and Time: William James's Reluctant Guide to Enlightenment* (Albany: State University Press of New York, 2015), 64.

3. Robert D. Richardson, *William James: In the Maelstrom of American Modernism: A Biography* (Boston: Houghton-Mifflin, 2006), 19.

4. John Clendening, *The Life and Thought of Josiah Royce* (Madison: University of Wisconsin Press, 1985), 23.

5. Randall E. Auxier, *Time, Will, and Purpose: Living Ideas from the Philosophy of Josiah Royce* (Chicago: Open Court, 2013), 265.

6. The former position is indicated in M. L. Brody, "Community in Royce: An Interpretation," *Transactions of the Charles S. Peirce Society* 5 (Fall 1969): 224–42; the latter Joseph L. Blau, "Royce's Theory of Community," *Journal of Philosophy* 53 (Feb. 2, 1956): 92–98. Both essays offer sound introductions to Royce's concept of community.

7. Boodin, *The Social Mind* (New York: Macmillan, 1939), 167.

8. Royce, *The Philosophy of Loyalty* in *The Basic Writings of Josiah Royce*, ed. John J. McDermott (Chicago: University of Chicago Press, 1969), 2:1012.

9. Royce, "Provincialism," in *The Basic Writings*, 2:1069.

10. Boodin, *The Social Mind*, 300.

11. Royce, *The Philosophy of Loyalty* in *The Basic Writings*, 2:953.

12. The revolt of Northern industrialism and defense of segregation in an idealized South is embodied in *I'll Take My Stand: The South and the Agrarian Tradition* (1930; repr., Baton Rouge: Louisiana State University Press, 1977). The compilation of essays by twelve Southerners is best considered a manifesto of post–Civil War grievance, protesting the capitalist domination of the region by Northern corporate interests against the Lost Cause mythology. Notable authors include the poets Allen Tate and John Crowe Ransom, novelist Robert Penn Warren, and historian Frank Owsley.

13. Gary Herstein, "The Roycean Roots of the Beloved Community," *Pluralist* 4 (Summer 2009): 91–107. As Herstein convincingly argues, it was not loyalty that drove King's Beloved Community but a more overtly Christian *agape* love. Nonetheless, it was catalyzed by a Roycean spirit. Not only did King read Royce directly, but he was introduced to Royce through his dissertation director L. Harold DeWolf and especially through E. S. Brightman's *Religious Values*.

14. Boodin, *Studies in Philosophy: The Posthumous Papers of John Elof Boodin* (Los Angeles: University of California Press, 1957), 101.

15. Boodin, *The Social Mind*, 229. For more on Royce's triadic thinking, see M. L. Briody, "Community in Royce: An Interpretation," *Transactions of the Charles S. Peirce Society* 5, no. 4 (1969): 228.

16. Boodin, *The Social Mind*, 315–16.

17. Boodin, 134.

18. Boodin, 358.

19. Boodin, 460–61.

20. Boodin, 53–56.

21. Boodin, viii.

22. Boodin, 120.

23. Boodin, 139–40.

24. Boodin, 189.

25. *Travaux du IXe Congrès International de Philosophie* (1937). For particulars on Bloch and Boodin at that conference, see D. S. Robinson, "The Value Studies of the Ninth International Congress of Philosophy," *Ethics* 48 (Apr. 1938): 423–32.

26. For more on Ariès and *mentalité*, see Peter Burke, *The French Historical Revolution: The Annales School, 1929–1989* (Stanford, CA: Stanford University Press, 1990), 67–69, 115.

27. Burke, 71–72.

28. Marc Bloch, *The Historian's Craft*, trans. Peter Putnam (New York: Vintage Books, 1953), 189.

29. Lucien Fabre and Marc Bloch both utilized *mentalité* as an operative historical force, but each were drawn from different perspectives. Fabre's was more psychological; Bloch's was more sociological and influenced by Émile Durkheim. Insofar as there is *mentalité* in Boodin's concept of "the social mind," it is more akin to Bloch than Fabre. For a more complete discussion, see André Burguière, *The Annales School: An Intellectual History*, trans. Jane Marie Todd (Ithaca, NY: Cornell University Press, 2009).

30. Boodin, *The Social Mind*, 235.

31. Boodin, 269.

32. Boodin, 311.

33. Boodin, 316, 351.

34. Boodin, 356.

35. Boodin, 369.

36. For an interesting examination of the influences of Schopenhauer on Hitler, see especially Richard Weikart, *Hitler's Religion: The Twisted Beliefs That Drove the Third Reich* (Washington, DC: Regnery History, 2016).

37. Boodin, *The Social Mind*, 370.

38. For a complete discussion, see Oren Harman, *The Price of Altruism: George Price and the Search for the Origins of Kindness* (New York: W. W. Norton, 2010).

39. See William F. Basener and John C. Sanford, "The Fundamental Theorem of Natural Selection with Mutations," *Journal of Mathematical Biology* 76 (2018): 1589–1622. Basener and Sanford argue that Fisher's theorem "disregards mutations, and because it is invalid beyond one instant in time, it has limited biological relevance." As they say, "What is often overlooked is that without a constant supply of new mutations, selection can only increase fitness by reducing genetic variance (i.e., selecting away undesirable alleles, eventually reducing their frequencies to zero). This means that given enough time, selection must reduce genetic variance all the way to zero, apart from new mutations" (1592). Besides this, kin selection is by no means a settled theory. For a summary of kin selection skepticism, see B. J. Williams, "Kin Selection in Human Populations: Theory Reconsidered," *Human Biology* 77 (Aug. 2005): 421–31; and William T. Lynch, "Between Kin Selection and Cultural Relativism: Cultural Evolution and the Origin of Inequality," *Perspectives on Science* 27 (Mar.–Apr. 2019): 278–315.

40. Oren Harmon, *The Price of Altruism: George Price and the Search for the Origins of Kindness* (New York: W. W. Norton, 2010), 358–59.

41. Boodin, *The Social Mind*, 389.
42. Boodin, 458.
43. Boodin, 561–62.
44. Auxier, *Time, Will, and Purpose*, 176, 341.
45. Quoted in Charles H. Nelson, *John Elof Boodin: Philosopher-Poet* (New York: Philosophical Library, 1987), 115–16.
46. Quoted in Jonathan Rogers, *The Terrible Speed of Mercy: A Spiritual Biography of Flannery O'Connor* (Nashville, TN: Thomas Nelson, 2012), 16–17.
47. These reviews are summarized in Nelson, 119–20.
48. A. D. Ritchie, review of *The Social Mind*, by John Elof Boodin, *Philosophy* 16 (Apr. 1941): 214–15.
49. See reviews of *The Social Mind* in H. M. Kallen, *Social Research* 8 (May 1941): 253–54; and Newel L. Sims, *American Sociological Review* 5 (June 1940): 441–42.
50. J. O. Hertzler, review of *The Social Mind*, *American Journal of Sociology* 46 (Sept. 1940): 257–58.
51. Joseph L. Blau, *Men and Movements in American Philosophy* (Englewood Cliffs, NJ: Prentice-Hall, 1952), 262.
52. See, for example, the following important book-length studies: Maurice Natanson's *Social Dynamics of George H. Mead* (1956), Paul E. Pfeutz's *Self, Society, Existence: Human Nature and Dialogue in the Thought of George Herbert Mead and Martin Buber* (1961), David L. Miller's *George Herbert Mead: Self, Language and the World* (1973), Hans Joas's *G. H. Mead: A Contemporary Re-Examination of His Thought* (1985), Sandra B. Rosenthal and Patrick L. Bourgeois's *Mead and Merleau-Ponty: Toward a Common Vision* (1991), and John D. Baldwin's *George Herbert Mead: A Unifying Theory for Sociology* (2002), to name a few. Eugene C. Holmes, to be discussed later, did provide an analysis of Mead and Boodin's social theory in 1942, but as a self-published dissertation it was not widely available.
53. Herbert Blumer, "Sociological Implications of the Thought of George Herbert Mead," *American Journal of Sociology* 71 (Mar. 1966): 537–39.
54. See Michael J Carter and Celene Fuller, "Symbolic Interactionism," *Sociopedia.isa* (2015), https://www.researchgate.net/publication/303056565_Symbolic_Interactionism.
55. Quoted in Boodin, *The Social Mind*, 9.
56. Boodin, 25.
57. Boodin, 45–46, 329, 549.
58. R. F. Alfred Hoernlé, review of *The Social Mind*, *Mind* 50 (Oct. 1941): 395.
59. Boodin, *The Social Mind*, 580.
60. Admittedly, Boodin moves rather lightly over canalization in *The Social Mind* (see 264–69). He develops canalization with more clarity in his "Value and Social Interpretation," *American Journal of Sociology* 21 (July 1915): 65–103. Although Boodin cites this paper in *The Social Mind* as forming his chapter 7 on

the "Organization of Values," he apparently reorganized and rewrote portions to specifically fit the book.

61. According to Philip G. Fathergill, "Straight-line evolution, or the idea of definite trends or directions in evolution, has long been recognized, and is now an accepted theory." See his *Historical Aspects of Organic Evolution* (New York: Philosophical Library, 1953), 274. For a detailed discussion, see also pages 166–70. Others (especially French biologists) have agreed. See Emile Guyénot, *The Origin of Species*, trans. C. J. Cameron (New York: Walker and Co., 1964), 28–30, 120–23. First published 1944 by *Presses Universitaires de France*. Andrée Tétry also favored orthogenesis. See her section on biology in *Science in the Twentieth Century*, ed. René Taton, trans. A. J. Pomerans (London: Thames & Hudson, 1966), esp. 444–45. Even later, Pierre-Paul Grassé was sympathetic to orthogenesis. See his *Evolution of Living Organisms* (New York: Academic Press, 1977).

62. See, for example, Julian Huxley, "Darwinism To-Day," *Virginia Quarterly Review* 19 (Winter 1943): 107–20; George Gaylord Simpson, "Evolutionary Determinism and the Fossil Record," *Scientific Monthly* 71 (Oct. 1950): 262–67; and Ernst Mayr, "The Idea of Teleology," *Journal of the History of Ideas* 53 (Jan.–Mar. 1992): 117–35.

63. Most recently, see Igor Popov, *Orthogenesis versus Darwinism*, trans. Natalia Lentsman (Cham, Switzerland: Springer, 2018).

64. Wendell Berry, *The Unsettling of America: Culture & Agriculture* (San Francisco: Sierra Club Books, 1977), 19.

65. Berry, 176.

66. Gracy Olmstead, "Wendell Berry's Right Kind of Farming," *New York Times*, Oct. 1, 2018, https://www.nytimes.com/2018/10/01/opinion/Wendell-berry-agriculture-farm-bill.html.

67. Berry is a well-known novelist, short story writer, and poet who takes nature and God seriously. See *Wendell Berry and Religion: Heaven's Earthly Life*, eds. Joel James Shuman and L. Roger Owens (Lexington: University of Kentucky Press, 2009).

68. Herbert W. Schneider, review of *The Social Mind*, *Journal of Philosophy* 37 (Mar. 28, 1940): 187–90.

69. John Elof Boodin, *Man in His World* (Berkeley: University of California Press, 1939), 10–11.

70. Eugene Clay Holmes, *Social Philosophy and the Social Mind: A Study of the Genetic Methods of J. M. Baldwin, G. H. Mead, and J. E. Boodin* (New York: The Author, 1942), 1. This is a self-published version of the author's dissertation completed at Columbia University under Herbert Wallace Schneider.

71. Holmes, 54–55.

72. John H. McClendon, "Dr. Holmes, the Philosopher Rebel," *Freedomways* 2, no 1 (1982): 32, 37.

73. John H. McClendon III and Stephen C. Ferguson II, *African American Philosophers and Philosophy: An Introduction to the History, Concepts, and Contemporary Issues* (London: Bloomsbury Academic. 2019), 201. It is interesting to note that the self-published version of his dissertation carries a 1942 copyright stamp on the verso of the title page reading: "Eugene Clay Holmes, Howard University, Wash., D. C."

74. Holmes, *Social Philosophy*, 50.

75. Holmes, 61.

76. Holmes, 50.

77. Holmes, 3, 59, 63.

78. For details, see D. G. Charlton, *Positivist Thought in France During the Second Empire, 1852–1870* (Oxford: Clarendon Press, 1959), 104–23.

79. W. M. Simon, *European Positivism in the Nineteenth Century: An Essay in Intellectual History* (Ithaca, NY: Cornell University Press, 1963), 267.

80. Hilary Putnam, "Putnam on Pragmatism and Positivism," interview Oct. 30, 2013, YouTube video, https://www.youtube.com/watch?v=jNsMP-Jj5Pw.

81. Diana L. Hayes, "A Great Cloud of Witnesses: Martin Luther King Jr.'s Roots in the African American Religious and Spiritual Tradition," in *Revives My Soul Again: The Spirituality of Martin Luther King Jr.*, eds. Lewis V. Baldwin and Victor Anderson (Minneapolis, MN: Fortress Press, 2018), 39–59. See also in that volume Beverly J. Lanzetta, "The Heart of a World Citizen: Martin Luther King Jr. as Social Mystic," 239–69. More radical activists like Holmes called for liberation rather than integration, but in the context of the times from the postwar years to the tragic end of King's ministry in 1968 there was little interest in Marxist solutions to African-Americans' problems. A careful study of black journals and newspapers of the period as well as statements by virtually all black community leaders reveal a distrust, resistance, and hostility to this more extreme activism. This was not, as many historians have supposed, due to Cold War fears and political self-interest but was, in fact, principled and practical. See Eric Arnesen, "The Traditions of African-American Anti-Communism," *Twentieth Century Communism* 6 (2014): 124–48.

82. John Elof Boodin, *Religion of Tomorrow* (New York: Philosophical Library, 1943), 27.

83. Martin Luther King Jr., "Three Essays on Religion," Martin Luther King Jr. Research and Education Institute, accessed Aug. 14, 2022, https://kinginstitute.stanford.edu/king-papers/documents/three-essays-religion.

84. Additional examples are provided in Michael A. Flannery, "Strong and Weak Teleology in the Life Sciences Post-Darwin," *Religions* 11, no. 6 (2020): 298, https://doi.org/10.3390/rel11060298. On teleology, orthogenesis, and their social/scientific applications, see also Inna Dyachenko, "Directionality of the Living: Evolution, Forms and Consequences of Manifestation (Filomenological Aspects)," *Scientific Journal of Polonia University* 38, no. 1 (2020): 254–65.

85. Holmes, *Social Philosophy*, 50.

86. Holmes, 52–53.

87. A more recent version of the anthropic principle may be found in John Gribben, *Alone in the Universe: Why Our Planet is Unique* (Hoboken, NJ: John Wiley & Sons, 2011).

88. Friedel Weinert, *Copernicus, Darwin, & Freud: Revolutions in the History and Philosophy of Science* (Malden, MA: Wiley-Blackwell, 2009), 86–87.

Chapter 6

1. John S. Haller Jr., *Fictions of Certitude: Science, Faith, and the Search for Meaning, 1840–1920* (Tuscaloosa: University of Alabama Press, 2020), 6.

2. Haller, *Fictions of Certitude*, 164–65.

3. Thaddeus J. Kozinski, *Modernity as Apocalypse: Sacred Nihilism and the Counterfeits of Logos* (Brooklyn, NY: Angelico Press, 2019), 2.

4. Charles H. Nelson, *John Elof Boodin: Philosopher-Poet* (New York: Philosophical Library, 1987), 148.

5. G. K. Chesterton, *Orthodoxy* (1908; repr., New York: Barnes & Noble, 2007), 28.

6. John Elof Boodin, *Religion of Tomorrow* (New York: Philosophical Library, 1943), 188–89.

7. Richard McDonough, s.v. "Roy Wood Sellars," in *Internet Encyclopedia of Philosophy*, https://iep.utm.edu/sella-rw/#H2.

8. Roy Wood Sellars, *Principles of Emergent Realism*, comp. and ed. W. Preston Warren (St. Louis, MO: Warren H. Green, 1970), 148.

9. Sellars, 193.

10. Sellars, 149.

11. Pouwel Slurink, "Back to Roy Wood Sellars: Why His Evolutionary Naturalism Is Still Worthwhile," *Journal of the History of Philosophy* 34 (July 1996): 425–49.

12. R. Broom, *The Coming of Man: Was It Accident or Design?* (London: H. F. & G. Witherby, 1933), 21.

13. George A. Coe, review of *The Next Step in Religion*, by Roy Wood Sellars, *Journal of Philosophy* 16 (Apr. 1919): 248–50.

14. Roy Wood Sellars, *Religion Coming of Age* (New York: Macmillan, 1928), 275.

15. W. de Burgh, review of *Religion Coming of Age*, by Roy Wood Sellars, *Journal of Philosophical Studies* 4 (Apr. 1929): 271–73.

16. Balázs Gimes, "The Materialism of Roy Wood Sellars," *HOPOS: Journal of the International Society for the History of Philosophy of Science* 11 (Spring 2021): 166–82.

17. Austin L. Hughes, "The Folly of Scientism," *New Atlantis* (2012): 32–50, https://www.thenewatlantis.com/publications/the-folly-of-scientism.

18. For more on Shailer Mathews, see John S. Haller Jr., "Shailer Mathews: A Modernist's Search for the Christian God," OpenSIU, Southern Illinois University, 2021, https://opensiuc.lib.siu.edu/histcw_sm.

19. J. E. Boodin, "The Meaning of History," *American Journal of Theology* 21 (Oct. 1917): 628.

20. Kenneth Ray Sutton, "John Elof Boodin's Creationism" (PhD diss., University of New Mexico, 1969), 44.

21. Sutton, 233.

22. James Wayne Dye, "John Elof Boodin's Theory of Consciousness," *Southern Journal of Philosophy* 12, no. 3 (1974): 313–32.

23. Dye, 326.

24. Matthias Neuber, "Between Pragmatism and Realism: The Philosophy of John Elof Boodin," *European Journal of Pragmatism and American Philosophy*, XI-1. Posted July 19, 2019, http://journals.openedition.org/ejpap/1547.

25. Sellars, *Evolutionary Naturalism*, 43.

26. John Elof Boodin, *Cosmic Evolution: Outlines of Cosmic Idealism* (New York: Macmillan, 1925), 261.

27. Sellars, *Evolutionary Naturalism*, 44.

28. Boodin, "Functional Realism," *Philosophical Review* 43 (Mar. 1934): 147.

29. John Kuiper, "Roy Wood Sellars on the Mind-Body Problem," *Philosophy and Phenomenological Research* 15 (Sept. 1954): 59.

30. Boodin, *Cosmic Evolution*, 222.

31. Boodin, 229–230.

32. Boodin, *The Social Mind*, 138.

33. Boodin, 140.

34. Boodin, 269, 289.

35. Quoted in Nelson, *John Elof Boodin*, 95.

36. Nelson, 116.

37. Boodin, *Cosmic Evolution*, 221.

38. See H. O. Mounce, *The Two Pragmatisms: From Peirce to Rorty* (London: Routledge, 1997), 126–27. For more on this, Mounce's discussion of Dewey in chapters 8–10 is highly recommended.

39. John Dewey, *The Influence of Darwin on Philosophy: And Other Essays in Contemporary Thought* (New York: Henry Holt, 1910).

40. Martin Fichman, "Alfred Russel Wallace's North American Tour: Transatlantic Evolutionism," *Endeavor* 25, no. 2 (2001): 74.

41. Flannery, *Nature's Prophet: Alfred Russel Wallace and His Evolution from Natural Selection to Natural Theology* (Tuscaloosa: University of Alabama Press, 2018), 74–75.

42. Fichman, "Alfred Russel Wallace's," 76. As Barton points out, it was more than just these three X Clubbers who opposed Wallace's teleological evolution. Even the most silent members tacitly endorsed the thoroughly naturalistic program. See her *X Club: Power and Authority in Victorian Science* (Chicago: University of Chicago Press, 2018), 422–23.

43. Joseph Brent, *Charles Sanders Peirce: A Life*, rev. ed. (Bloomington: Indiana University Press, 1998), 254.

44. Ralph Barton Perry, *The Thought and Character of William James*, 2 vols. (Boston: Little, Brown, 1935), 2:515.

45. Robert D. Richardson, *William James: In the Maelstrom of American Modernism: A Biography* (Boston: Houghton-Mifflin, 2006), 486.

46. Richardson, 136; Brent, *Charles Sanders Peirce*, 60.

47. See the interview with Hilary Putnam, "Putnam on Pragmatism and Positivism," interview Oct. 30, 2013, YouTube video, 2:00–2:55, https://www.youtube.com/watch?v=jNsMP-Jj5Pw.

48. See Timothy Mccune, "Dewey's Dilemma: Eugenics, Education, and the Art of Living," *Pluralist* 7 (Fall 2012): 96–106.

49. Gertrude Himmelfarb, *Darwin and the Darwinian Revolution* (1962; repr., Chicago: Elephant Books, 1996), 425.

50. Neuber, "Between Pragmatism and Realism," 11, para. 48.

51. Nelson, *John Elof Boodin*, 133.

52. P. Hicks, s.v. "Teilhard de Chardin, Pierre," in *New Dictionary of Christian Apologetics*, eds. W. C. Campbell-Jack and C. Stephen Evans (Downers Grove, IL: Intervarsity Press, 2006).

53. Wolfgang Smith, *Theistic Evolution: The Teilhardian Heresy* (Tacoma, WA: Angelicus Press, 2012), 243.

54. Ronny Desmet and Andrew Irvine, s.v. "Alfred North Whitehead," in *The Stanford Encyclopedia of Philosophy* (Fall 2018), ed. Edward N. Zalta, https://plato.stanford.edu/archives/fall2018/entries/whitehead/.

55. Quoted in Erik L. Peterson, *The Life Organic: The Theoretical Biology Club and the Roots of Epigenetics* (Pittsburgh, PA: University of Pittsburgh Press, 2016), 6–7.

56. Peterson, 60.

57. Peterson, 143.

58. Peterson, 137.

59. Quoted in Peterson, 192.

60. Ernst Mayr, "Darwin's Impact on Modern Thought," *Proceedings of the American Philosophical Society* 139 (Dec. 1995): 317–25. Mayr would surely balk at my suggestion that this essay represented a full blown excursion into metaphysical realms, but like all positivists Mayr could easily ignore the plank in his own eye. Of course, this is not to suggest that Mayr's metaphysics was in any sense *good* metaphysics.

61. The correspondence and discussions between Mayr and Greene make for fascinating and illuminating reading. See John C. Greene, *Debating Darwin: Adventures of a Scholar* (Claremont, CA: Regina Books, 1999), 150.

62. Quoted in Matthew Allen, "Compelled by the Diagram: Thinking through C. H. Waddington's Epigenetic Landscape," *Contemporaneity* 4, no. 1 (2015), https://pdfs.semanticscholar.org/f4d6/e588d304b7a0c7c2c21ec5de8f69b307a5e9.pdf.

63. C. H. Waddington, *Tools for Thought: How to Understand and Apply the Latest Scientific Techniques of Problem Solving* (New York: Basic Books, 1977), 16.

64. Rupert Sheldrake, *Morphic Resonance: The Nature of Formative Causation*, 4th ed. (Rochester, VT: Park Street Press, 2009), xi.

65. Rupert Sheldrake, *Presence of the Past: Morphic Resonance and the Memory of Nature*, rev. ed. (Rochester, VT: Park Street Press, 2012), xvi–xvii.

66. Sheldrake, 441.

67. Sheldrake, 380.

68. Sheldrake, *Morphic Resonance*, 73.

69. Sheldrake, 199.

70. Alex Gomez-Marin, "Facing Biology's Open Questions: Rupert Sheldrake's 'Heretical' Hypothesis Turns 40," *BioEssays* (Feb. 18, 2021), https://doi.org/10.1002/bies.202100055.

71. Boodin, *Three Interpretations of the Universe* (New York: Macmillan, 1934), 217.

72. Boodin, 256.

73. Sheldrake, *The Presence of the Past*, 273.

74. Boodin, *Cosmic Evolution*, 218.

75. Boodin, *Studies in Philosophy: The Posthumous Papers of John Elof Boodin*, comp. Donald Ayres Piatt (Los Angeles: University of California Press, 1957), 65, 85, 100, 125.

76. Michael J. Behe, *Darwin's Black Box: The Biochemical Challenge to Evolution* (New York: Free Press, 1996), 70–73.

77. Sheldrake, *Science Set Free: 10 Paths to New Discovery* (New York: Deepak Chopra Books, 2012), 37–38.

78. Erkki Vesa Rope Kojonen, *The Intelligent Design Debate and the Temptation of Scientism* (London: Routledge, 2016), 6.

79. Casey Luskin, "Why Intelligent Design Is Science: A Reading List," *Evolution News & Science Today*, Nov. 12, 2012, https://evolutionnews.org/2012/11/why_intelligent1/.

80. Stephen C. Meyer, *The Signature in the Cell: DNA and the Evidence for Design* (New York: HarperOne, 2009), 4.

81. Behe, *Darwin's Black Box*, 212–16.

82. Greene, *Debating Darwin*, 249.

83. Boodin, "Philosophy of History," in *Twentieth Century Philosophy: Living Schools of Thought*, ed. Dagobert D. Runes (New York: Philosophical Library, 1947), 104–5.

Epilogue

1. David Bohm, *Wholeness and the Implicit Order* (1980; repr., London: Routledge Classics, 2002).

2. Bernard d'Espagnot, *On Physics and Philosophy* (Princeton, NJ: Princeton University Press, 2006), 463–64.

3. George Levine, "Paradox: The Art of Scientific Naturalism," in *Victorian Science Naturalism: Community, Identity, Continuity*, eds. Gowan Dawson and Bernard Lightman (Chicago: University of Chicago Press, 2014), 84.

4. Although seldom cast in this way, the idea is not without precedence. While aspects of Darwinian theory are clearly testable, the theory in its broadest and most popular representation as "survival of the fittest" is little more than an unverifiable explanatory framework—the epitome of a tautology—of metaphysical proportions. See Momme von Sydow, "Darwinian Metaphysics," in *Encyclopedia of Sciences and Religion*, eds. A. Runehov and L. Oviedo (Dordrecht: Springer, 2013): 1306–1314.

5. See the following by Hilde Hein: "Mechanism and Vitalism as Meta-Theoretical Commitments," *Philosophical Forum* 1, no. 2 (1968): 185–205; "Molecular Biology vs. Organicism: The Enduring Dispute Between Mechanism and Vitalism," *Synthese* 20 (Aug. 1969): 238–53; and "The Endurance of the Mechanism-Vitalism Controversy," *Journal of the History of Biology* 5 (Spring 1972): 159–88.

6. For Erik L. Peterson's discussion of Hilde Hein, see his *The Life Organic: The Theoretical Biology Club and the Roots of Epigenetics* (Pittsburgh, PA: University of Pittsburgh Press, 2016), 247, 248, 262 n. 15.

7. Hein, "Mechanism and Vitalism," 186.

8. Hein, "Molecular Biology vs. Organicism," 238.

9. Michael A. Flannery, "Toward a New Evolutionary Synthesis," *Theoretical Biology Forum* 110, no. 1/2 (2017): 47–61.

10. Peterson, *The Life Organic*, 32.

11. Irwin Edman, "Eighteenth Annual Meeting of the American Philosophical Association," *Journal of Philosophy* 16 (Feb. 27, 1919): 131.

12. Edman, 130.

13. Robert J. Kreyche, *The Betrayal of Wisdom: The Challenge to Philosophy Today* (Staten Island, NY: Alba House, 1972), 6, 13, 100, 58.

14. Kreyche, 30.

15. Randall E. Auxier, *Time, Will, and Purpose: Living Ideas from the Philosophy of Josiah Royce* (Chicago: Open Court, 2013) chap. 10.

16. I will just name two of many. First is R. Paul Thompson, *A Remarkable Journey: The Story of Evolution* (London: Reaktion Books, 2015). Thompson's Darwinian triumphalism knows no bounds, declaring that "science" has put biblical literalism in its death throes. This seems doubtful for two reasons: (1) the embarrassing popularity of creationists like Ken Ham as witnessed in the attendance of his Creationist Museum and Ark theme parks and (2) the slippery phrase "biblical literalism." Ham's literalism is certainly not Hugh Ross's literalism, and besides it is not necessary to sign on to any kind of biblical literalism in order to question Darwinian evolution (i.e., Thompson's "science"). Second is Chris Moody and Sheryl Kirschenbaum's *Unscientific America: How Scientific Illiteracy Threatens Our Future* (New York: Basic Books, 2009). This polemical ode to scientism suggests the imminent collapse of America for being so "unscientific." The book is really a front for the National Center for Science Education, a lobbying group for Darwinian evolution in school education and practically everywhere else.

17. Peterson, *The Life Organic*, 242.

18. Thomas Nagel, *Mind & Cosmos: Why the Materialist Neo-Darwinian Conception of Nature is Almost Certainly False* (Oxford: Oxford University Press, 2012), 10.

19. Randall Auxier, "The Death of Darwinism and the Limits of Evolution," *Philo* 9 (Fall–Winter 2006): 216.

20. Alvin Plantinga, *Where the Conflict Really Lies: Science, Religion, and Naturalism* (Oxford: Oxford University Press, 2011), 113–25.

21. James A. Shapiro, "What We Have Learned About Evolutionary Genome Change in the Past 7 Decades," *Biosystems* 215–16 (June 2022), https://www.sciencedirect.com/science/article/abs/pii/S0303264722000594#.

22. J. Scott Turner, *Purpose & Desire: What Makes Something "Alive" and Why Darwinism His Failed to Explain It* (New York: HarperOne, 2017), 207.

23. Wendell Berry, *Life is a Miracle: An Essay Against Modern Superstition* (Washington, DC: Counterpoint, 2000), 93.

Appendix

1. Both steel plows and wooden plows were used, the former for breaking the turf, the latter for stirring the soil.

Bibliography

Books

Adams, Mark B. "Little Evolution, BIG Evolution: Rethinking the History of Darwinism, Population Genetics, and the 'Synthesis.'" In *Natural Selection: Revisiting Its Explanatory Role in Evolutionary Biology*, edited by Richard G. Delisle. Cham, Switzerland: Springer, 2021.

Al-Khalili, Jim. *The World According to Physics*. Princeton, NJ: Princeton University Press, 2020.

Allen, Frederick Lewis. *Only Yesterday: An Informal History of the 1920s*. New York: Perennial Library, 1964.

Amundson, Ron. *The Changing Role of the Embryo in Evolutionary Thought: Roots of Evo-Devo*. Cambridge: Cambridge University Press, 2007.

Aronowitz, Stanley. *Science as Power: Discourse and Ideology in Modern Society*. Minneapolis: University of Minnesota Press, 1988.

Auxier, Randall E., s.v. "Boodin, John Elof." In *The Bloomsbury Encyclopedia of Philosophers from 1600 to Present*, edited by John R. Shook. London: Bloomsbury Academic, 2016.

———. *The Dictionary of Modern American Philosophers*. Vol. 1. Bristol, UK: Thoemmes Continuum, 2006.

———. Introduction to *Truth and Reality: An Introduction to the Theory of Knowledge*, by John Elof Boodin. Early Defenders of Pragmatism, edited by John R. Shook, vol. 2. 1911. Reprint, Bristol, UK: Thoemmes Press, 2001.

———. *Time, Will, and Purpose: Living Ideas from the Philosophy of Josiah Royce*. Chicago: Open Court Press, 2013.

Baldwin, Lewis V., and Victor Anderson, eds. *Revives My Soul Again: The Spirituality of Martin Luther King Jr.* Minneapolis, MN: Fortress Press, 2018.

Barczewska, Shala. "*Inherit the Wind*: The Movie and the Myth in American Cultural Memory." In *Confluence of Literature, History and Cinema*, edited by Paweł Kaptur and Agnieszka Szwach. Kielce, Poland: Jan Kochanowski University, 2020.

Barrett, Clifford. Review of *Three Interpretations of the Universe* and *God and Creation*, by John Elof Boodin. *Journal of Philosophy* 32 (Mar. 1935): 157–60.
Barton, Ruth. *The X Club: Power and Authority in Victorian Science*. Chicago: University of Chicago Press, 2018.
Barzun, Jacques. *Darwin, Marx, Wagner: Critique of a Heritage*. 2nd ed. 1958. Reprint, Chicago: University of Chicago Press, 1981.
Behe, Michael J. *Darwin Devolves: The New Science About DNA that Challenges Evolution*. New York: HarperOne, 2019.
———. *Darwin's Black Box: The Biochemical Challenge to Evolution*. New York: Free Press, 1996.
Bertalanffy, Ludwig von. *Robots, Man and Minds: Psychology in the Modern World*. New York: George Braziller, 1967.
Berry, Wendell. *Life is a Miracle: An Essay Against Modern Superstition*. Washington, DC: Counterpoint, 2000.
———. *The Unsettling of America: Culture & Agriculture*. San Francisco: Sierra Club Books, 1977.
Blau, Joseph L. *Men and Movements in American Philosophy*. Englewood Cliffs, NJ: Prentice-Hall, 1952.
Bloch, Marc. *The Historian's Craft*. Translated by Peter Putnam. New York: Vintage Books, 1953.
Blum, Deborah. *Ghost Hunters: William James and the Search for Scientific Proof of Life After Death*. New York: Penguin Books, 2006.
Bohm, David. *Wholeness and the Implicit Order*. 1980. Reprint, London: Routledge Classics, 2002.
Boodin, John Elof. Autobiographical Statement and "Nature and Reason." In *Contemporary American Philosophy*, edited by George P. Adams and William Pepperell Montague. New York: Russell & Russell, 1962.
———. *Cosmic Evolution: Outlines of Cosmic Idealism*. New York: Macmillan, 1925.
———. *God and Creation: A Cosmic Philosophy of Religion*. New York: Macmillan, 1934.
———. *Man in His World*. Berkeley: University of California Press, 1939.
———. "Philosophy of History." In *Twentieth Century Philosophy: Living Schools of Thought*, edited by Dagobert D. Runes. New York: Philosophical Library, 1947.
———. *A Realistic Universe: An Introduction of Metaphysics*. New York: Macmillan, 1916. Rev. ed., 1931.
———. *Religion of Tomorrow*. New York: Philosophical Library, 1943.
———. *The Social Mind*. New York: Macmillan, 1939.
———. *Studies in Philosophy: The Posthumous Papers of John Elof Boodin*. Compiled by Donald Ayres Piatt. Los Angeles: University of California Press, 1957.
———. *Time and Reality*. Psychological Review, Monographic Supplements, vol. 6, no. 3. New York: Macmillan, 1904.

———. *Three Interpretations of the Universe*. New York: Macmillan, 1934.

———. *Truth and Reality: An Introduction to the Theory of Knowledge*. Early Defenders of Pragmatism, ed. John R. Shook, vol. 2. 1911. Reprint, Bristol, UK: Thoemmes Press, 2001.

———. "William James as I Knew Him." In *William James Remembered*, by Linda Simon. Lincoln: University of Nebraska Press, 1996.

Bowler, Peter J. *Darwin Deleted: Imagining a World without Darwin*. Chicago: University of Chicago Press, 2013.

———. *Monkey Trials and Gorilla Sermons: Evolution and Christianity from Darwin to Intelligent Design*. Cambridge, MA: Harvard University Press, 2007.

Brent, Joseph. *Charles Sanders Peirce: A Life*. Rev. ed. Bloomington: University of Indiana Press, 1998.

Bricklin, Jonathan. *The Illusion of Will, Self, and Time: William James's Reluctant Guide to Enlightenment*. Albany: State University of New York Press, 2015.

Briggs, John P., and F. David Peat. *Looking Glass Universe: The Emerging Science of Wholeness*. New York: Simon & Schuster, 1984.

Browne, Janet. *Charles Darwin: The Power of Place*. Princeton, NJ: Princeton University Press, 2002.

———. *Charles Darwin: Voyaging*. Princeton, NJ: Princeton University Press, 1995.

Burguière, André. *The Annales School: An Intellectual History*. Translated by Jane Marie Todd. Ithaca, NY: Cornell University Press, 2009.

Burke, Peter. *The French Historical Revolution: The Annales School, 1929–1989*. Stanford, CA: Stanford University Press, 1990.

Butterfield, Herbert. *The Origins of Modern Science*. Rev. ed. New York: Macmillan, 1962.

———. *The Whig Interpretation of History*. 1931. Reprint, New York: W. W. Norton, 1965.

Carroll, Peter N., and David W. Noble. *The Free and the Unfree: A New History of the United States*. 2nd ed. New York: Penguin Books, 1988.

Charlton, D. G. *Positivist Thought in France During the Second Empire, 1852–1870*. Oxford: Clarendon Press, 1959.

Chesterton, G. K. *Orthodoxy*. 1908. Reprint, New York: Barnes & Noble, 2007.

Clendening, John. *The Life and Thought of Josiah Royce*. Madison: University of Wisconsin Press, 1985.

Cohen, I. Bernard. *Revolution in Science*. Cambridge, MA: Belknap Press of Harvard University, 1985.

Collin, Rémy. *Evolution*. Translated by J. Tester. New York: Hawthorne Books, 1959.

Copleston, Frederick. *Greece and Rome: A History of Philosophy*. Vol. 1. Bellarmine Series 9. London: Burns, Oates, and Washbourne, 1946.

Cosans, Christopher E. *Owen's Ape & Darwin's Bulldog: Beyond Darwinism and Creationism* Bloomington: Indiana University Press, 2009.

Darwin, Charles. *Charles Darwin's Notebooks, 1836–1844.* Edited by Paul H. Barrett, Peter J. Gautrey, Sandra Herbert, David Kohn, and Sydney Smith. London: Natural History Museum, 1987.

———. *Descent of Man and Selection in Relation to Sex.* 1871. Reprint, New York: Barnes & Noble, 2004.

———. *Variation of Animals and Plants Under Domestication.* 2nd ed. 2 vols. New York: D. Appleton, 1883.

Dawkins, Richard. *River Out of Eden: A Darwinian View of Life.* New York: Basic Books, 1995.

de Burgh, W. Review of *Religion Coming of Age*, by Roy Wood Sellars. *Journal of Philosophical Studies* 4 (Apr. 1929): 272–73.

De Caro, Mario, ed. *Naturalism, Realism, and Normativity: Hilary Putnam.* Cambridge, MA: Harvard University Press, 2016.

Denton, Michael. *Evolution: A Theory in Crisis.* Bethesda, MD: Adler & Adler, 1986.

———. *Evolution: Still a Theory in Crisis.* Seattle: Discovery Institute, 2016.

Desmond, Adrian. *Archetypes and Ancestors: Paleontology in Victorian London, 1850–1875.* Chicago: University of Chicago Press, 1982.

Desmond, Adrian, and James Moore. *Darwin.* New York: W. W. Norton, 1991.

Durant, Will. *The Lessons of History.* New York: Simon & Schuster, 1968.

———. *The Mansions of Philosophy: A Survey of Human Life and Destiny.* Garden City, NY: Garden City Publishing, 1929.

———. *The Reformation: A History of European Civilization form Wycliffe to Calvin, 1300–1564.* New York: Simon & Schuster, 1957.

Dewey, John. *The Influence of Darwin on Philosophy: And Other Essays in Contemporary Thought.* New York: Henry Holt, 1910.

Dyson, Freeman. *Dreams of Earth and Sky.* New York: New York Review of Books, 2015.

———. *A Many-Colored Glass: Reflections on the Place of Life in the Universe.* Charlottesville: University of Virginia Press, 2007.

Eccles, John C. *Evolution of the Brain: Creation of the Self.* London: Routledge, 1989.

Fathergill, Philip G. *Historical Aspects of Organic Evolution.* New York: Philosophical Library, 1953.

Fischer, David Hackett. *Historians' Fallacies: Toward a Logic of Historical Thought.* New York: Harper Torchbooks, 1970.

Flannery, Michael A. *Intelligent Evolution: How Alfred Russel Wallace's "World of Life" Challenged Darwinism.* Nashville, TN: Erasmus Press, 2020.

———. *Nature's Prophet: Alfred Russel Wallace and His Evolution from Natural Selection to Natural Theology.* Tuscaloosa: University of Alabama Press, 2018.

Fry, Iris. *The Emergence of Life on Earth: A Historical and Scientific Overview.* New Brunswick, NJ: Rutgers University Press, 2000.

Gambarotto, Andrea. *Vital Forces, Teleology and Organization: Philosophy of Nature and the Rise of Biology in Germany.* Cham, Switzerland: Springer, 2018.

Gayon, Jean. *Darwinism's Struggle for Survival: Heredity and the Hypothesis of Natural Selection*. Translated by Matthew Cobb. Cambridge: Cambridge University Press, 1998.

Giberson, Karl W. *Saving Darwin: How to Be a Christian and Believe in Evolution*. New York: HarperOne, 2008.

Gillespie, Neal C. *Charles Darwin and the Problem of Creation*. Chicago: University of Chicago Press, 1979.

Gimes, Balázs. "The Materialism of Roy Wood Sellars." *HOPOS: The Journal of the International Society for the History of Philosophy of Science* 11 (Spring 2021): 166–82.

Grassé, Pierre-P. *Evolution of Living Organisms: Evidence for a New Theory of Transformation*. New York: Academic Press, 1977.

Greene, John C. *Darwin and the Modern World*. Baton Rouge: Louisiana State University Press, 1961.

———. *Debating Darwin: Adventures of a Scholar*. Claremont, CA: Regina Books, 1999.

———. *Science, Ideology, and World View: Essays in the History of Evolutionary Biology*. Berkeley: University of California Press, 1981.

Gregory of Nazianzus: Five Theological Orations. Translated with an introduction and notes by Stephen Reynolds. [Toronto]: Stephen Reynolds, 2011.

Griffin, David Ray. *Reenchantment without Supernaturalism: A Process Philosophy of Religion*. Ithaca, NY: Cornell University Press, 2001.

Guelac, Suzanne. *Thinking in Time: An Introduction to Henri Bergson*. Ithaca, NY: Cornell University Press, 2006.

Guyénot, Emile. *The Origin of Species*. Translated by C. J. Cameron. New York: Walker and Co., 1964.

Haller, John S. Jr. *Fictions of Certitude: Science, Faith, and the Search for Meaning, 1840–1920* Tuscaloosa: University of Alabama Press, 2020.

Hamlin, D. W., s.v. "Metaphysics, History of." In *The Oxford Companion to Philosophy*, edited by Ted Honderich, 2nd ed. Oxford: Oxford University Press, 2005.

Hancock, Roger, s.v. "Metaphysics, History of." In *The Encyclopedia of Philosophy*, vol. 5. New York: Macmillan, 1967.

Hannam, James. *The Genesis of Science: How the Christian Middle Ages Launched the Scientific Revolution*. Washington, DC: Regnery Publishing, 2011.

Hare, Peter H., s.v. "Sellars, Roy Wood." In *The Oxford Companion to Philosophy*, edited by Ted Honderich, 2nd ed. Oxford: Oxford University Press, 2005.

Harman, Oren. *The Price of Altruism: George Price and the Search for the Origins of Kindness*. New York: W. W. Norton, 2010.

Heisenberg, Werner. *Natural Law and the Structure of Matter*. London: Rebel Press, 1970.

———. *Physics and Philosophy: The Revolution in Modern Science*. 1958. Reprint, New York: Harper Perennial Modern Thought, 2007.

Hicks, P., s.v. "Teilhard de Chardin, Pierre." In *New Dictionary of Christian Apologetics*, edited by W. C. Campbell-Jack and C. Stephen Evans. Downers Grove, IL: Intervarsity Press, 2006.

Himmelfarb, Gertrude. *Darwin and the Darwinian Revolution*. 1962. Reprint, Chicago: Elephant Books, 1996.

Holmes, Eugene Clay. *Social Philosophy and Social Mind: A Study of the Genetic Method of J. M. Baldwin, G. H. Mead and J. E. Boodin*. New York: The Author, 1942.

Hooper, Walter. *C. S. Lewis: A Companion and Guide*. New York: HarperSanFrancisco, 1996.

Horkheimer, Max. *Eclipse of Reason*. 1947. Reprint, Mansfield Center, CT: Martino Publishing, 2013.

Hoyle, Fred. *The Intelligent Universe*. New York: Holt, Rhinehart, and Winston, 1983.

Huxley, Julian. "Darwinism To-Day." *Virginia Quarterly Review* 19 (Winter 1943): 107–20.

Huxley, Thomas Henry. *Collected Essays*. In *Darwiniana*, vol. 2. London: Macmillan, 1899.

James, William. *The Varieties of Religious Experience*. 1902. Reprint, New York: Penguin Books, 1982.

Johnson, Curtis. *Darwin's Dice: The Idea of Chance in the Thought of Charles Darwin*. Oxford: Oxford University Press, 2015.

Jurmain, Robert et al., *Introduction to Physical Anthropology*. Belmont, CA: Wadsworth, Cengage Learning, 2014.

Kenny, Anthony. *A New History of Philosophy: In Four Parts*. Oxford: Clarendon Press, 2010.

Koestler, Arthur. *Janus: A Summing Up*. London: Pan Books, 1979.

———. *The Sleepwalkers: A History of Man's Changing Vision of the Universe*. 1959. Reprint, London: Arkana, 1989.

Kojonen, Erkki Vesa Rope. *The Intelligent Design Debate and the Temptation of Scientism*. London: Routledge, 2016.

Kozinski, Thaddeus J. *Modernity as Apocalypse: Sacred Nihilism and the Counterfeits of Logos*. Brooklyn, NY: Angelico Press, 2019.

Kreyche, Robert J. *The Betrayal of Wisdom: The Challenge to Philosophy Today*. Staten Island, NY: Alba House, 1972.

Kuhn, Thomas S. *The Structure of Scientific Revolutions*. 2nd ed. Chicago: University of Chicago Press, 1970.

Kuklick, Bruce. *A History of Philosophy in America, 1720–2000*. Oxford: Clarendon Press, 2001.

Larson, Edward J. *Evolution: The Remarkable History of a Scientific Theory*. New York: Modern Library, 2004.

Lempriere, J. *Lempriere's Classical Dictionary*. 1788. Reprint, London: Bracken Books, 1984.

Levine, George. "Paradox: The Art of Scientific Naturalism." In *Victorian Science Naturalism: Community, Identity, Continuity*, edited by Gowan Dawson and Bernard Lightman. Chicago: University of Chicago Press, 2014.

Lowe, E. J., s.v. "Metaphysics, Opposition to." In *The Oxford Companion to Philosophy*, 2nd ed. Oxford: Oxford University Press, 2005.

Magee, Glenn Alexander. *Hegel and the Hermetic Tradition*. Ithaca, NY: Cornell University Press, 2001.

Margulis, Lynn. *Symbiotic Planet: A New Look*. New York: Basic Books, 1998.

Mayr, Ernst. *What Evolutions Is*. New York: Basic Books, 2001.

McFadden, Johnjoe, and Jim Al-Kalili. *Life on the Edge: The Coming Age of Quantum Biology*. New York: Crown Publishers, 2014.

McGrath, Alister E. *Darwinism and the Divine: Evolutionary Thought and Natural Theology*. Oxford: Wiley-Blackwell, 2011.

McIver, Tom. *Anti-Evolution: An Annotated Bibliography*. Jefferson, NC: McFarland, 1988.

Meyer, Stephen C. *The Signature in the Cell: DNA and the Evidence for Design*. New York: HarperOne, 2009.

Moody, Chris, and Sheryl Kirschenbaum. *Unscientific America: How Scientific Illiteracy Threatens Our Future*. New York: Basic Books, 2009.

Mumford, Lewis. *Technics & Civilization*. 1934. Reprint, Chicago: University of Chicago Press, 2010.

Nagel, Thomas. *Mind & Cosmos: Why the Materialist Neo-Darwinian Conception of Nature is Almost Certainly False*. Oxford: Oxford University Press, 2012.

Nelson, Charles H. *John Elof Boodin: Philosopher-Poet*. New York: Philosophical Library, 1987.

Noble, Denis. *Dance to the Tune of Life: Biological Relativity*. Cambridge: Cambridge University Press, 2017.

Owen, Richard. *On the Nature of Limbs: A Discourse*. Edited by Ron Amundson. Preface by Brian K. Hall, introductory essays by Ron Amundson, Kevin Padian, Mary P. Winsor, and Jennifer Coggon. Chicago: University of Chicago Press, 2007.

Palmer, Donald. *Looking at Philosophy: The Unbearable Heaviness of Philosophy Made Lighter*. Mountain View, CA: Mayfield Publishing, 1988.

Perry, Ralph Barton. *The Thought and Character of William James*. 2 vols. Boston: Little, Brown, 1935.

Peterson, Erik L. *The Life Organic: The Theoretical Biology Club and the Roots of Epigenetics*. Pittsburgh, PA: University of Pittsburgh Press, 2016.

Plantinga, Alvin. *Where the Conflict Really Lies: Science, Religion, and Naturalism*. Oxford: Oxford University Press, 2011.

Popper, Karl. "Who Killed Logical Positivism?" In *Unended Quest: An Intellectual Autobiography*, 98–101. 1974. New ed., London: Routledge, 1992.

Putnam, Hilary. *The Collapse of the Fact/Value Dichotomy and Other Essays*. Cambridge, MA: Harvard University Press, 2002.

Putnam, Hilary, and Ruth Anna Putnam. *Pragmatism as a Way of Life: The Lasting Legacy of William James and John Dewey*. Cambridge, MA: Belknap Press of Harvard University Press, 2017.
Reck, Andrew J., s.v. "Boodin, John Elof." In *The Encyclopedia of Philosophy*, vol. 1. New York: Macmillan, 1967.
———. "The Cosmic Philosophy of John Elof Boodin." In *Recent American Philosophy: Studies of Ten Representative Thinkers*, by Andrew J. Reck. New York: Pantheon Books, 1964.
Reznick, David N. *The "Origin" Then and Now: An Interpretive Guide to the "Origin of Species."* Princeton, NJ: Princeton University Press, 2010.
Richardson, Robert D. *William James: In the Maelstrom of American Modernism: A Biography*. Boston: Houghton-Mifflin, 2006.
Riddle, John M. *A History of the Middle Ages, 30–1500*. 2nd ed. Lanham, MD: Rowman & Littlefield, 2016.
Rogers, Jonathan. *The Terrible Speed of Mercy: A Spiritual Biography of Flannery O'Connor*. Nashville, TN: Thomas Nelson, 2012.
Ross, Hugh. *A Matter of Days: Resolving a Creation Controversy*. Colorado Springs, CO: NavPress, 2004.
Roth, John K. Introduction to *The Philosophy of Josiah Royce*. New York: Thomas Y. Crowell, 1971.
Royce, Josiah. *The Basic Writings of Josiah Royce*. Edited by John J. McDermott. 2 vols. Chicago: University of Chicago Press, 1969.
———. *The Letters of Josiah Royce*. Edited with an introduction by John Clendening. Chicago: University of Chicago Press, 1970.
———. *The Philosophy of Josiah Royce*, edited with an introduction by John K. Roth. New York: John Y. Crowell, 1971.
Rupke, Nicolaas. *Richard Owen: Biology without Darwin*. Rev. ed. Chicago: University of Chicago Press, 2009.
Ruse, Michael. *The Darwinian Revolution: Nature Red in Tooth and Claw*. 2nd ed. Chicago: University of Chicago Press, 1999.
———. *On Purpose*. Princeton, NJ: Princeton University Press, 2018.
Sanchez, Francis M. "A Coherent Resonant Cosmology Approach and its Implications in Microphysics and Biophysics." In *Quantum Systems in Physics, Chemistry, and Biology*, edited by Alia Tadjer et al. Cham, Switzerland: Springer, 2017.
Schlereth, Thomas J. *Victorian America: Transformations in Everyday Life*. New York: HarperPerennial, 1991.
Schneider, Herbert W. Review of *The Social Mind*, by John Elof Boodin. *Journal of Philosophy* 37 (Mar. 28, 1940): 187–90.
Sellars, Roy Wood. *Evolutionary Naturalism*. Chicago: Open Court, 1922.
———. *Principles of Emergent Realism*. Compiled and edited by W. Preston Warren. St. Louis, MO: Warren H. Green, 1970.
———. *Religion Coming of Age*. New York: Macmillan, 1928.

Shapin, Stephen. *Never Pure: Historical Studies of Science as if It Was Produced by People with Bodies, Situated in Time, Space, Culture, and Society, and Struggling for Credibility and Authority.* Baltimore, MD: Johns Hopkins University Press, 2010.

Sheldrake, Rupert. *Morphic Resonance: The Nature of Formative Causation.* 4th ed. Rochester, VT: Park Street Press, 2009.

———. *The Present of the Past: Morphic Resonance and the Memory of Nature.* Rev. ed. Rochester, VT: Park Street Press, 2012.

———. *Science and Spiritual Practices.* Berkeley, CA: Counterpoint, 2017.

———. *Science Set Free: 10 Paths to New Discovery.* New York: Deepak Chopra Books, 2012.

Shuman, Joal James, and L. Roger Owens, eds. *Wendell Berry and Religion: Heaven's Earthly Life.* Lexington: University of Kentucky Press, 2009.

Simon, W. M. *European Positivism in the Nineteenth Century: An Essay in Intellectual History.* Ithaca, NY: Cornell University Press, 1963.

Slotten, Ross A. *A Heretic in Darwin's Court.* New York: Columbia University Press, 2004.

Smith, Wolfgang. *Theistic Evolution: The Teilhardian "Heresy."* Tacoma, WA: Angelicus Press, 2012.

Snyder, Laura J. *The Philosophical Breakfast Club.* New York: Broadway Books, 2011.

Sokal, Alan, and Jean Bricmont. "Defense of a Modest Scientific Realism." In *Knowledge and the World: Challenges Beyond the Science Wars*, edited by Martin Carrier, Johannes Roggenhofer, Günter Küppers, and Philippe Blanchard. Berlin: Springer, 2004.

———. *Fashionable Nonsense: Postmodern Intellectuals' Abuse of Science.* New York: Picador, 1998.

Stebbins, Robert E. "France." In *The Comparative Reception of Darwinism*, edited by Thomas F. Glick. Austin: University of Texas Press, 1972.

Steinhardt, Paul J., and Neil Turok. *Endless Universe: Beyond the Big Bang.* New York: Doubleday, 2007.

Tallis, Raymond. *Aping Mankind: Neuronmania, Darwinitis and the Misrepresentation of Humanity.* London: Routledge, 2016.

Thomas, Henry, s.v. "Comte, Auguste." In *Biographical Encyclopedia of Philosophy.* Garden City, NY: Doubleday, 1965.

Thompson, D'Arcy Wentworth. *On Growth and Form.* 1917. Reprint, New York: Dover, 1992.

Thompson, R. Paul. *A Remarkable Journey: The Story of Evolution.* London: Reaktion Books, 2015.

Tuchman, Barbara. *Practicing History: Selected Essays.* New York: Alfred A. Knopf, 1981.

Turner, J. Scott. *Purpose & Desire: What Makes Something "Alive" and Why Modern Darwinism Has Failed to Explain It.* New York: HarperOne, 2017.

von Sydow, Momme. "Darwinian Metaphysics." In *Encyclopedia of Sciences and Religion*, edited by A. Runehov and L. Oviedo. Dordrecht: Springer, 2013.

Waddington, C. H. *Tools for Thought: How to Understand and Apply the Latest Scientific Techniques of Problem Solving*. New York: Basic Books, 1977.

Wallace, Alfred Russel. *The World of Life: A Manifestation of Creative Power, Directive Mind and Ultimate Purpose*. London: Chapman and Hall, 1910.

Weikart, Richard. *Hitler's Religion: The Twisted Beliefs That Drove the Third Reich*. Washington, DC: Regnery History, 2016.

Weinert, Friedel. *Copernicus, Darwin, & Freud: Revolutions in the History and Philosophy of Science*. Malden, MA: Wiley-Blackwell, 2009.

Werkmeister, William H. *A History of Philosophical Ideas in America*. New York: Ronald Press, 1949.

Whitehead, Alfred North. *Adventures in Ideas*. 1933. Reprint, New York: Free Press, 1967.

———. *Process and Reality*. Corrected ed. Edited by David Ray Griffin and Donald W. Sherburne. New York: Free Press, 1978.

———. *Religion in the Making*. 1926. Reprint, New York: Fordham University Press, 1996.

———. *Science and the Modern World*. 1925. Reprint, New York: Free Press, 1967.

Wilczek, Frank. *A Beautiful Question: Finding Nature's Deep Design*. New York: Penguin Books, 2015.

Williams, Donald C. *The Elements and Patterns of Being: Essays in Metaphysics*. Edited by A. R. J. Fisher. Oxford: Oxford University Press, 2018.

Dissertations

Kreyche, Robert J. "The Naturalism of Roy Wood Sellars." PhD diss., University of Ottawa, 1950.

McMullen, Emerson Thomas. "A Barren Virgin? Teleology in the Scientific Revolution." PhD diss., University of Indiana, 1989.

Sutton, Kenneth Ray. "John Elof Boodin's Creationism." PhD diss., University of New Mexico, 1969.

Print Journals

Abel, David L. "The Capabilities of Chaos and Complexity." *International Journal of Molecular Science* 2009, no. 10 (2009): 247–91.

Arnesen, Eric. "The Traditions of African-American Anti-Communism." *Twentieth Century Communism* 6 (2014): 124–48.

Auxier, Randall. "The Death of Darwinism and the Limits of Evolution." *Philo* 9 (Fall–Winter 2006): 193–220.

Barbosa, João. "*Creation*: A Multifaceted and *Thematic* Concept in the Construction of Modern Cosmology—from Friedmann's *Creation of the Universe* to the Steady-State's *Continuous Creation.*" *Philosophy and Cosmology* 27 (2021): 22–33.
Basener, William F., and John C. Sanford. "The Fundamental Theorem of Natural Selection with Mutations." *Journal of Mathematical Biology* 76 (2018): 1589–1622.
Behe, Michael J. "Experimental, Evolution, Loss-of-Function Mutations, and 'The First Rule of Adaptive Evolution.'" *Quarterly Review of Biology* 85 (Dec. 2010): 419–45.
Biesta, Gert J. J. "How the Use Pragmatism Pragmatically? Suggestions for the Twenty-First Century." *Education and Culture* 25, no. 2 (2009): 34–45.
Blau, Joseph L. "Royce's Theory of Community." *Journal of Philosophy* 53 (Feb. 2, 1956): 92–98.
Blumer, Herbert. "Sociological Implications of the Thought of George Herbert Mead." *American Journal of Sociology* 71 (Mar. 1966): 535–44.
Bode, B. H. Review of *Time and Reality*, by John Elof Boodin. *Philosophical Review* 14 (Nov. 1905): 730–31.
Boodin, John Elof. "The Concept of Time." *Journal of Philosophy, Psychology and Scientific Methods* 2 (July 6, 1905): 365–72.
———. "Cosmic Attributes." *Philosophy of Science* 10 (Jan. 1943): 1–12.
———. "The Existence of Social Minds." *American Journal of Sociology* 19 (July 1913): 1–47.
———. "Fictions in Science and Philosophy. I." *Journal of Philosophy* 40 (Dec. 9, 1943): 673–82.
———. "Fictions in Science and Philosophy. II." *Journal of Philosophy* 40 (Dec. 23, 1943): 701–16.
———. "Functional Realism." *Philosophical Review* 43 (Mar. 1934): 147–78.
———. "The Meaning of History." *American Journal of Theology* 21 (Oct. 1917): 624–28.
———. "The Nature of Truth: A Reply." *Philosophical Review* 20 (Jan. 1911): 59–63.
———. "The New Realism." *Journal of Philosophy, Psychology and Scientific Methods* 4 (Sept. 26, 1907): 533–42.
———. "The Ought and Reality." *International Journal of Ethics* 17 (July 1907): 454–74.
———. "The Reality of the Ideal with Special Reference to the Religious Ideal." *The Unit* of Iowa College [Grinnell College] 5 (1900): 97–109.
———. "The Reinstatement of Teleology." *Harvard Theological Review* 6 (Jan. 1913): 76–99.
———. "Value and Social Interpretation." *American Journal of Sociology* 21 (July 1915): 65–103.
Bosley, Harold. Review of *Religion of Tomorrow*, by John Elof Boodin. *Journal of Religion* 24 (Apr. 1944): 143–44.

Brady, Ronald H. "Natural Selection and the Criteria by which a Theory is Judged." *Systematic Zoology* 28 (Dec. 1979): 600–21.
Briody, M. L. "Community in Royce: An Interpretation." *Transactions of the Charles S. Peirce Society* 5 (Fall 1969): 224–42.
Brown, Frank Burch. "The Evolution of Darwin's Theism." *Journal of the History of Biology* 19 (Spring 1986): 1–45.
Buxton, Michael. "The Influence of William James on John Dewey's Early Work." *Journal of the History of Ideas* 45 (July–Sept. 1984): 451–63.
Byerly, Henry C. "Natural Selection as a Law: Principles and Processes." *American Naturalist* 121 (May 1983): 739–45.
Campbell, John Angus. "The Invisible Rhetorician: Charles Darwin's 'Third Party' Strategy." *Rhetorica: A Journal of the History of Rhetoric* 7 (Winter 1989): 55–85.
Čapek, Millič. "Time and Eternity in Royce and Bergson." *Revue Internationale de Philosophie* 79, nos. 79/80 (1967): 22–45.
Carus, Paul. "Editorial Comment." *Monist* 21 (Apr. 1911): 295–96.
Coe, George A. Review of *The Next Step in Religion*. *Journal of Philosophy* 16 (Apr. 1919): 248–50.
Cotkin, George. "William James and the Cash-Value Metaphor." *ETC: A Review of General Semantics* 42 (Spring 1985): 37–46.
Dart, Raymond. "*Australopithecus africanus*: The Man-Ape of South Africa." *Nature* 115, no. 2884 (Feb. 7, 1925): 195–99.
Doherty, Kevin F. "Location of Platonic Ideas." *Review of Metaphysics* 14 (Sept. 1960): 57–72.
Dyachenko, Inna. "Directionality of the Living: Evolution, Forms and Consequences of Manifestation (Filomenological Aspects)." *Scientific Journal of Polonia University* 38, no. 1 (2020): 254–65.
Dye, James Wayne. "John Elof Boodin's Theory of Consciousness." *Southern Journal of Philosophy* 12, no. 3 (1974): 313–32.
Eccles, John C. "Do Mental Events Cause Neural Events Analogously to the Probability Fields of Quantum Mechanics?" *Proceedings of the Royal Society of London. Series B, Biological Sciences* 227, no. 1249 (1986): 411–28.
Edman, Irwin. "Eighteenth Annual Meeting of the American Philosophical Association." *Journal of Philosophy* 16 (Feb. 27, 1919): 127–32.
Eisen, Sydney. "Huxley and the Positivists." *Victorian Studies* 7 (1964): 337–58.
Elgin, Mehmet, and Elliott Sober. "Popper's Shifting Appraisal of Evolutionary Theory." *HOPOS: The Journal of the International Society for the History of Philosophy of Science* 7 (Spring 2017): 31–55.
Faber, Roland. "Three Hundred Years of Whitehead: Halfway." *Process Studies* 41, no. 1 (2012): 5–20.
Feibleman, James K. "The Metaphysics of Logical Positivism." *Review of Metaphysics* 5 (Sept. 1951): 55–82.
Fichman, Martin. "Alfred Russel Wallace's North American Tour: Transatlantic Evolutionism." *Endeavor* 25, no. 2 (2001): 74–78.

Fisher, A. R. J. "Donald C. Williams's Defense of Real Metaphysics." *British Journal for the History of Philosophy* 25, no. 2 (2017): 332–55.

Flannery, Michael A. "Toward a New Evolutionary Synthesis." *Theoretical Biology Forum* 110, no. 1/2 (2017): 47–61.

Foster, Michael B. Review of *Cosmic Evolution*, by John Elof Boodin. *Journal of Philosophical Studies* 4 (Apr. 1929): 255–57.

Freire, Olival, and Christoph Lehner. " 'Dialectical Materialism and Modern Physics,' an Unpublished Text by Max Born." *Notes and Records of the Royal Society of London* 64 (June 20, 2010): 155–62.

Greene, John C. "The Interaction of Science and World View in Sir Julian Huxley's Evolutionary Biology." *Journal of the History of Biology* 23 (Spring 1990): 39–55.

Haessler, Carl. Review of *A Realistic Universe*, by John Elof Boodin. *Journal of English and Germanic Philology* 16 (Oct. 1917): 617–20.

Hartshorne, Charles. Review of *Religion of Tomorrow*, by John Elof Boodin. *Ethics* 54 (Apr. 1944): 233–34.

Hein, Hilde. "The Endurance of the Mechanism: Vitalism Controversy." *Journal of the History of Biology* 5 (Spring 1972): 159–88.

———. "Mechanism and Vitalism as Meta-Theoretical Commitments." *Philosophical Forum* 1, no. 2 (1968): 185–205.

———. "Molecular Biology vs. Organicism: The Enduring Dispute Between Mechanism and Vitalism." *Synthese* 20 (Aug. 1969): 238–53.

Henning, Brian G. "Recovering the Adventure of Ideas: In Defense of Metaphysics as Revisable, Systematic, Speculative Philosophy." *Journal of Speculative Philosophy* 29, no. 4 (2015): 437–56.

Herstein, Gary. "The Roycean Roots of the Beloved Community." *Pluralist* 4 (Summer 2009): 91–107.

Hertzler, J. O. Review of *The Social Mind*, by John Elof Boodin. *American Journal of Sociology* 46 (Sept. 1940): 257–58.

Hoernlé, R. F. A. Review of *Cosmic Evolution*, by John Elof Boodin. *Journal of Philosophy* 24 (Mar. 17, 1927): 160–63.

———. Review of *A Realistic Universe*, by John Elof Boodin. *Harvard Theological Review* 12 (Jan. 1919): 128–32.

———. Review of *The Social Mind*, by John Elof Boodin. *Mind* 50 (Oct. 1941): 393–401.

———. Review of *Three Interpretations of the Universe* and *God and Creation*, by John Elof Boodin. *Mind* 45 (Apr. 1936): 217–29.

Hughes, Percy. Review of *The Social Mind*, by John Elof Boodin. *Mind* 50 (Oct. 1941): 393–401.

———. Review of *Time and Reality*, by John Elof Boodin. *Journal of Philosophy, Psychology and Scientific Methods* 2 (Apr. 13, 1905): 218–20.

James, William. "The Perception of Time." *Journal of Speculative Philosophy* 20 (Oct. 1886): 374–407.

Kallen, H. M. Review of *The Social Mind*, by John Elof Boodin. *Social Research* 8 (May 1941): 253–54.

Kuiper, John. "Roy Wood Sellars on the Mind-Body Problem." *Philosophy and Phenomenological Research* 15 (Sept. 1954): 48–64.

Levine, George. "By Knowledge Possessed: Darwin, Nature, and Victorian Narrative." *New Literary History* 24 (Spring 1993): 363–91.

Lindberg, David C. *The Beginnings of Western Science: The European Scientific Tradition in Philosophical, Religious, and Institutional Context, Prehistory to A.D. 1450*. 2nd ed. Chicago: University of Chicago Press, 2007.

Lovelock, James E. "Hands Up for the Gaia Hypothesis." *Nature* 344 (Mar. 8, 1990): 100–02.

Lynch, William T. "Between Kind Selection and Cultural Relativism: Cultural Evolution and the Origin of Inequality." *Perspectives on Science* 27 (Mar.–Apr. 2019): 278–315.

Mandelbaum, Maurice. "Darwin's Religious Views." *Journal of the History of Ideas* 19 (June 1958): 363–78.

Marcheva, Plamena, and Stoil Ivanov. "On the Geodesics in Bondi-Gold-Hoyle Universe Model." *Journal of Physics and Technology* 1, no. 1 (2017): 3–5.

Mayr, Ernst. "Darwin's Impact on Modern Thought." *Proceedings of the American Philosophical Society* 139 (Dec. 1995): 317–25.

———. "The Idea of Teleology." *Journal of the History of Ideas* 53 (Jan.–Mar. 1992): 117–35.

McClendon, John H. "Dr. Holmes, the Philosopher Rebel." *Freedomways* 22, no. 1 (1982): 32–40.

McClure, M. T. Review of *A Realistic Universe*, by John Elof Boodin. *Journal of Philosophy, Psychology and Scientific Methods* 14 (Dec. 6, 1917): 693–95.

Mccune, Timothy. "Dewey's Dilemma: Eugenics, Education, and the Art of Living." *Pluralist* 7 (Fall 2012): 96–106.

McDonough, Jeffrey K. "Not Dead Yet: Teleology and the 'Scientific Revolution.' " In *Teleology: A History*, edited by Jeffrey K. McDonough. New York: Oxford University Press, 2020.

McKay, Baily D., and Robert M. Zink. "Sisyphean Evolution in Darwin's Finches." *Biological Reviews* 90 (2015): 689–98.

Moore, Randy. "The Persuasive Mr. Darwin." *BioScience* 47 (Feb. 1997): 107–14.

Mounce, H. O. *The Two Pragmatisms: From Peirce to Rorty*, 126–27. London: Routledge, 1997.

Peters, Robert Henry. "Predictable Problems with Tautology in Evolution and Ecology." *American Naturalist* 112 (July/Aug. 1978): 759–62.

———. "Tautology in Evolution and Ecology." *American Naturalist* 110 (Jan./Feb. 1978): 1–12.

Phillips, Bernard. "Logical Positivism and the Function of Reason." *Philosophy* 23 (Oct. 1948): 346–60.

Picard, Maurice. Review of *Evolutionary Naturalism*, by Roy Wood Sellars. *Journal of Philosophy* 19 (Oct. 12, 1922): 582–87.

Popov, Igor. *Orthogenesis versus Darwinism*. Translated by Natalia Lentsman. Cham, Switzerland: Springer, 2018.

Popper, Karl. "Natural Selection and the Emergence of Mind." *Dialectica* 32, no. 3/4 (1978): 339–55.

Putnam, Hilary. "Pragmatism." *Proceedings of the Aristotelian Society*. New Series 95 (1995): 291–306.

Radislove, Tsanoff. "Professor Boodin on the Nature of Truth. *Philosophical Review* 19 (Nov. 1910): 632–38.

———. Review of *A Realistic Universe*, by John Elof Boodin. *Philosophical Review* 26 (Nov. 1917): 660–65; and *Philosophical Review* 43 (Jan. 1934): 90–92.

Reck, Andrew. "Royce's Metaphysics." *Revue Internationale de Philosophie* 21, nos. 79/80 (1967): 8–21.

Reed, Edward S. "The Lawfulness of Natural Selection." *American Naturalist* 118 (July 1981): 61–71.

Reiser, Oliver L. Review of *Religion of Tomorrow*, by John Elof Boodin. *Philosophy and Phenomenological Research* 4 (June 1944): 571–72.

Ritchie, A. D. Review of *The Social Mind*, by John Elof Boodin. *Philosophy* 16 (Apr. 1941): 214–15.

Robinson, D. S. "The Value Studies of the Ninth International Congress of Philosophy." *Ethics* 48 (Apr. 1938): 423–32.

Rosenblith, Walter A. "On Cybernetics and the Human Brain." *American Scholar* 35 (Spring 1966): 243–48.

Russell, L. J. Review of *Three Interpretations of the Universe*, by John Elof Boodin. *Philosophy* 43 (July 1936): 359–60.

Sabine, George H. Review of *Evolutionary Naturalism*, by Roy Wood Sellars. *Philosophical Review* 32 (Jan. 1923): 93–95.

Schweber, Silvan S. "The Young Darwin." *Journal of the History of Biology* 12 (Spring 1979): 175–92.

Seife, Charles. "Eternal-Universe Idea Comes Full Circle." *Science*. New Series 296, no. 5568 (Apr. 26, 2002): 639.

Sellars, Roy Wood. "What is the Correct Interpretation of Critical Realism?" *Journal of Philosophy* 24 (Apr. 28, 1927): 238–41.

Simpson, George Gaylord. "Evolutionary Determinism and the Fossil Record." *Scientific Monthly* 71 (Oct. 1950): 262–67.

Sims, Newel L. Review of *The Social Mind*, by John Elof Boodin. *American Sociological Review* 5 (June 1940): 441–42.

Sloan, Philip R. "'The Sense of Sublimity': Darwin on Nature and Divinity." *Osiris* 16 (2001): 251–69.

Slurink, Pouwel. "Back to Roy Wood Sellars: Why His Evolutionary Naturalism Is Still Worthwhile." *Journal of the History of Philosophy* 34 (July 1996): 425–49.

Smith, Charles H. "Alfred Russell Wallace and the Elimination of the Unfit." *Journal of Biosciences* 37 (June 2012): 203–5.

Smocovitis, V. B. "The 1959 Darwin Centennial Celebration in America." *Osiris* 14. Commemorative Practices in Science: Historical Perspectives on the Politics of Collective Memory (1999): 274–23.

———. "The Evolutionary Synthesis and Evolutionary Biology." *Journal of the History of Biology* 25 (Spring 1992): 1–65.

Stout, A. K. Review of *Cosmic Evolution*, by John Elof Boodin. *Mind.* New Series 36 (Oct. 1927): 496–99.

Sulloway, Frank J. "Darwin and His Finches: The Evolution of a Legend." *Journal of the History of Biology* 15 (Spring 1982): 1–53.

Theunissen, Bert. "Darwin and His Pigeons: The Analogy Between Artificial and Natural Selection Revisited." *Journal of the History of Biology* 45 (Summer 2012): 179–212.

Trevers, J. T., and D. L. Abel. "Chance and Necessity Do Not Explain the Origin of Life." *Cell Biology International* 28, no. 11 (2004): 729–39.

Tufts, J. H. Review of *Three Interpretations of the Universe* and *God and Creation*, by John Elof Boodin. *International Journal of Ethics* 45 (July 1935): 466–68.

Tyman, Stephen. "Royce and the Destiny of Idealism." *Personalist Forum* 15 (Spring 1999): 45–58.

Vasas, Vera, Eörs Szathmáry, and Mariop Santos. "Lack of Evolvability in Self-Sustaining Autocatalytic Networks Constraints Metabolism-First Scenarios for the Origin of Life." *Proceedings of the National Academy of Sciences* (*PNAS*) 107, no. 4 (2010): 1470–1475.

Virtue, Charles F. Sawhill. Review of *Three Interpretations of the Universe* and *God*, by John Elof Boodin. *Philosophical Review* 45 (Jan. 1936): 88–90.

Wagner, Günter. "How Wide and How Deep is the Divide Between Population Genetics and Developmental Evolution?" *Biology and Philosophy* 22 (2007): 145–53.

Weiner, Philip Paul. "Some Metaphysical Assumptions and Problems of Neo-Positivism." *Journal of Philosophy* 32 (Mar. 28, 1935): 174–81.

Whitney, G. W. T. Review of *Cosmic Evolution*, by John Elof Boodin. *Social Science* 2 (Aug.–Sept. 1927): 442–45.

Williams, B. J. "Kin Selection in Human Populations: Theory Reconsidered." *Human Biology* 77 (Aug. 2005): 421–31.

Williams, Rosalind. "Lewis Mumford's *Technics and Civilization*." *Technology and Culture* 43 (Jan. 2002): 139–49.

Wilm, E. C. Review of *A Realistic Universe*, by John Elof Boodin. *International Journal of Ethics* 30 (July 1020): 464–67.

Winsor, Mary P. "The Creation of the Essentialism Story: An Exercise in Metahistory." *History and Philosophy of the Life Sciences* 28, no. 2 (2006): 149–74.

Woelfel, James. "Challengers of Scientism Past and Present: William James and Marilynne Robinson." *American Journal of Theology & Philosophy* 34 (May 2013): 175–87.

Wright, H. W. Review of *Time and Reality*, by John Elof Boodin. *American Journal of Theology* 16 (Apr. 1912): 315–16.
Yeager, Jeffrey Wayne. "The Social Mind: John Elof Boodin's Influence on John Steinbeck's Phalanx Writings, 1935–1942." *Steinbeck Review* 10, no. 1 (2013): 31–46.

Online Sources

Allen, Matthew. "Compelled by the Diagram: Thinking through C. H. Waddington's Epigenetic Landscape." *Contemporaneity* 4, no. 1 (2015). https://pdfs.semanticscholar.org/f4d6/e588d304b7a0c7c2c21ec5de8f69b307a5e9.pdf.
Arnold-Foster, Tom. "Rethinking the Scopes Trial: Cultural Conflict, Media Spectacle, and Circus Politics." *Journal of American Studies* (May 12, 2021): 1–25. https://doi.org/10.1017/S0021875821000529.
Aveling, Edward. *The Religious Views of Charles Darwin*. 1883. Reproduced at *Darwin Online*. Accessed Jan. 12, 2022. http://darwin-online.org.uk/content/frameset?pageseq=1&itemID=A234&viewtype=text.
Carter, Michael J., and Celene Fuller. "Symbolic Interactionism." *Sociopedia.isa*. 2015. https://www.researchgate.net/publication/303056565_Symbolic_Interactionism.
Darwin, Charles. Letter to William Graham, July 3, 1881. "Letter no. 13230." Darwin Project. University of Cambridge. https://www.darwinproject.ac.uk/letter/?docId=letters/DCP-LETT-13230.xml&query=william%20Graham%203%20July%201881.
Desmet, Ronny, and Andrew Irvine, s.v. "Alfred North Whitehead." In *Stanford Encyclopedia of Philosophy: Fall 2018*, edited by Edward N. Zalta. Revised Sept. 4, 2018. https://plato.stanford.edu/archives/fall2018/entries/whitehead/.
Disaster Center. "United States Crime Rates 1960–2019." Accessed Dec. 11, 2021. https://www.disastercenter.com/crime/uscrime.htm.
Doulcier, Guilhem, Peter Takacs, and Pierrick Bourrat. "Taming Fitness: Organism-Environment Interdependencies Preclude Long-Term Fitness Forecasting." *BioEssays* 2020. https://onlinelibrary.wiley.com/doi/abs/10.1002/bies.202000157.
Flannery, Michael A. "The Process Theology of John Elof Boodin." *Religions* 14, no. 2 (2023). https://www.mdpi.com/2077-1444/14/2/238.
———. "Strong and Weak Teleology in the Life Sciences Post-Darwin." *Religions* 11, no. 6 (2020): 298. https://doi.org/10.3390/rel11060298.
Gomez-Marin, Alex. "Facing Biology's Open Questions: Rupert Sheldrake's 'Heretical' Hypothesis Turns 40." *BioEssays* (Feb. 18, 2021). https://doi.org/10.1002/bies.202100055.
Haller, John S. Jr. "Shailer Mathews: A Modernist's Search for the Christian God." OpenSIU-Southern Illinois University, 2021. https://opensiuc.lib.siu.edu/histcw_sm.

Hilary Putnam. "Putnam on Pragmatism and Positivism." Interview Oct. 30, 2013. YouTube video. https://www.youtube.com/watch?v=jNsMP-Jj5Pw.

Hughes, Austin L. "The Folly of Scientism." *New Atlantis* (2012): 32–50. https://www.thenewatlantis.com/publications/the-folly-of-scientism.

Hunt, Tam. "Reconsidering the Logical Structure of the Theory of Natural Selection." *Communicative & Integrative Biology* 7, no. 6 (2014). https://www.ncbi.nlm.nih.gov/pmc/articles/PMC4594354/.

Iijas, Anna, and Paul J. Steinhardt. "Bouncing Cosmology Made Simple." *Classical and Quantum Gravity* 35 (June 1, 2018). https://iopscience.iop.org/article/10.1088/1361-6382/aac482.

King, Martin Luther Jr. "Three Essays on Religion." Martin Luther King Jr. Research and Education Institute. Stanford University. Accessed Aug. 14, 2022. https://kinginstitute.stanford.edu/king-papers/documents/three-essays-religion.

Leidenhag, Joanna. "How to Be a Theological Panpsychist, but Not a Process Theologian." *Philosophy, Theology and the Sciences* 7, no. 1 (2020). https://research-repository.st-andrews.ac.uk/handle/10023/23541.

Library of Congress. "Letter of Alfred North Whitehead." Symposium on Feb. 17, 2011. Video. https://www.loc.gov/item/webcast-5200.

Luskin, Casey. "Why Intelligent Design Is Science: A Reading List." *Evolution News & Science Today*. Nov. 12, 2012. https://evolutionnews.org/2012/11/why_intelligent1/.

Marais, Adriana, Betony Adams, Andrew K. Ringsmuth, Marco Ferretti, J. Michael Gruber, Ruud Hendrikx, Maria Schuld et al. "The Future of Quantum Biology." *Journal of the Royal Society: Interface* 15 (Nov. 14, 2018). https://doi.org/10.1098/rsif.2018.0640.

McDonough, Richard, s.v. "Roy Wood Sellars." In *Internet Encyclopedia of Philosophy*. https://iep.utm.edu/sella-rw#H2.

Neuber, Matthias. "Between Pragmatism and Realism: The Philosophy of John Elof Boodin." *European Journal of Pragmatism and American Philosophy*, XI-1. Posted online July 19, 2019. http://journals.openedition.org/ejpap/1547.

Olmstead, Gracy. "Wendell Berry's Right Kind of Farming." *New York Times*, Oct. 1, 2018. https://www.nytimes.com/2018/10/01/opinion/Wendell-berry-agriculture-farm-bill.html.

Rex, Andrew. "Maxwell's Demon—A Historical Review." *Entropy* 19, no. 6 (2017). https://doi.org/10.3390/e19060240.

Shapiro, James A. "What We Have Learned About Evolutionary Genome Change in the Past 7 Decades." *Biosystems* (Mar. 28, 2022). https://www.sciencedirect.com/science/article/abs/pii/S0303264722000594#.

Tanaka, Stefan. *History Without Chronology*. Ann Arbor, MI: Lever Press (Amherst College Press and Michigan Publishing), 2019. https://doi.org/10.3998/mpub.11418981.

"The Third Way: Evolution in the Era of Genomics and Epigenetics." *The Third Way*. Accessed Jan. 21, 2022. https://www.thethirdwayofevolution.com/.

Weart, Spencer. "Trend-Spotting: Physics in 1931 and Today." *Physics Today* 59 (June 2006). https://doi.org/10.1063/1.2218552.

Index

Absolute, the, xvi, xvii, xxiii, 18, 20, 24, 26, 28, 29, 35, 46, 65, 140, 148–49, 173, 174, 231n27
Adams, Mark B., 85–86
Agassiz, Louis, 85, 122
Alexander, Samuel, 91, 119, 245n6
Amundson, Ron, 85
anagenesis, 75, 84
Anaxagoras, 38, 114, 223
Andronicus, 46
Annales school, 19, 145–46, 171
anthropic principle, 93, 168, 256n87
Aquinas, Thomas, 117, 119, 125
Aristotle, 3, 19, 30, 38, 46, 51, 59, 92, 108, 114, 115–16, 117, 119, 124–25, 163, 186, 223
Aronowitz, Stanley, 10
Auxier, Randall E., xxii–xxiii, 24, 30, 183, 209, 222, 226n15, 226n19, 227n37, 232n39

Barbosa, João, 96–97
Barrett, Clifford, 133, 154
Barzun, Jacques, 71
Beak of the Finch (Weiner), 86
Behe, Michael, 90–91, 200, 201, 212
Beloved Community, 107, 128, 130, 132, 138–42, 161, 162, 167, 169, 173, 210, 211, 213, 251n13

Bergson, Henri, xxvii, 17, 22, 23, 26–27, 30, 31, 53, 63, 64, 87, 88, 91, 92, 119, 121, 128, 153, 193, 194, 195, 198, 220, 232n38, 248n51
Berkeley, George, 47, 148
Berry, Wendell, 161–62, 210–11
Bertalanffy, Ludwig von, 55, 194, 207, 217, 237n32
big bang cosmology, 96–97, 118
Bloch, Marc, 19, 145, 230n5, 252n29
Blumer, Herbert, 156–57, 158
Bode, B. H., 32, 42
Bohm, David, 203
Böhme, Jakob, 119, 120
Bohr, Neils, 2, 25, 50, 51
Boodin, John Elof: boyhood home and early life, xi–xii, 142, 158, 213–215; as Christian, xii, 130, 135, 166; as poet, xii, xx, 14–16, 51, 98, 133, 225n5; at Carleton College, xviii; at Grinnell College, xvi, xxi; at Harvard, xiii, xv–xvi, xvii, xxi, 40, 172; at the University of Kansas, xvii, xviii; at UCLA, xvii, xviii–xix, xxi, xxvi, 13, 135, 180, 194; death, xii; influence of James, xiii–xvi, xvii, 40, 174, 225n6; influence of Royce, xvi–xvii, 40, 42, 138–41;

283

Boodin, John Elof *(continued)*
  professional isolation, xvi, xix–xxi;
  relations with colleagues, xviii, 108,
  153–54, 177, 180, 187–88
Boole, George, 42
Bosanquet, Bernard, xx, 48
Bosley, Harold, 135
Bowler, Peter J., 84
Boyle, Robert, 47, 94, 115
Brady, Ronald H., 74
Brightman, E. S., 140
Broom, Robert, 83, 88–89, 90, 176
Bruno, Giordano, 119, 120
Bryan, William Jennings, 79–82
Butler, Samuel, 72–73
Byerly, Henry C., 74

Čapek, Milič, 26
Cartesian, 92, 94, 181, 222. *See also* Descartes
Carus, Paul, 41–42
cash-value, 40, 45, 53, 59, 98, 173, 190, 234n81
Chesterton, G. K., 74
Christianity: Catholicism, 6, 125–26, 129, 131, 166, 167; early period, 11, 108, 117, 125; modern/contemporary, 174; perennialist, 136, 198; Protestantism, 125–26, 131, 135, 166–67; trinity, 109
chronological snobbery, xxvi, 227n41
Cobb, John B. Jr., 107, 194, 222
Coe, George A., 177
Comte, Auguste, 3–4, 8, 48, 69, 101, 166, 178, 181, 220–21
Conybeare, William Daniel, 123
cosmic evolution, 62, 65, 87, 92, 93, 98, 143, 165, 197, 207, 217
Cosmic Evolution (Boodin), xviii, xix, xx, xxvi, 53, 62, 67, 87–104, 114, 150, 181, 184, 186, 188, 212; reviews of 104–106

creation(ism), 46, 80, 82, 86–87, 96–98, 180–81, 195
creativity, xvii, 51, 92–93, 98, 118, 127, 130, 133, 136, 143–44, 152, 156–57, 159, 160, 169, 197–98, 200
critical realism, 99, 100, 175–76, 183–84, 217–18, 220, 222
Cusanus, Nicolaus, 119, 120

Darrow, Clarence, 79–81
Dart, Raymond, 83, 88
Darwin, Charles, xxvi, 4–6, 7, 9, 60–61, 67–79, 109–10, 112, 115, 119, 122, 123, 129, 165, 171, 188–89, 190, 217, 218, 228n3, 232n46, 240n36, 241n44, 248n51
Darwin, Erasmus (grandfather), 4, 68, 111
Darwinian theism(ists), 71, 205, 207
Darwinism, 1, 7, 61, 67, 77, 84, 86, 98, 176–77, 179, 181, 188, 191–92, 205–206, 218. *See also* neo-Darwinism
Darwinism (Wallace), 72, 189
Dawkins, Richard, xxiv, 2, 48, 71, 171, 219
Dayton, Tennessee, 79, 81
de Burgh, W., 178
de Chardin, Teilhard, 48, 87, 109, 110, 168, 192–93, 220
de Vries, Hugo, 73, 83
Democritus, xxiv, 29, 57, 110, 128, 219
Denton, Michael, 125, 249n65
Descartes, René, 46, 94, 114. *See also* Cartesian
Descent of Man (Darwin), 4, 67
Desmond, Adrian, 68, 124
Dewey, John, xiii, xvi, xxii, 12–13, 45, 100, 147, 153, 154, 155, 164, 165, 180, 186–88, 190–92, 208, 221, 225n7

dialectical materialism, 111, 164–65
Dobzhansky, Theodosius, 75, 84, 85–86, 177
Driesch, Hans, 63, 110, 119, 121, 168, 194–95
Duns, John, 77, 123
Dye, James Wayne, 181–83
Dyson, Freeman, 10, 39, 57–58, 229n25, 236n14, 236n16

Eccles, John C., 92–93, 168, 243n88
Einstein, Alfred, 2, 25, 49–50, 51, 172, 204
élan vital, 26, 53, 88, 121, 193, 195
embryology, 124, 162–63, 195, 196
emergence, 108–15, 117, 119, 121, 126, 195, 205, 245n6
Engels, Friedrich, 111, 165
epigenetics, 102, 124, 195–96, 210
epistemology, xxvi, xxvii, 12, 14, 17, 27–43 passim, 46, 47, 58, 99, 154, 185, 191, 205, 218–19
Erigena, 108–109, 117, 119, 120–21
evolution (structuralist), 122–25, 249n65
Evolutionary Naturalism (Sellars), 99, 100, 177, 184

Faber, Roland, 134–35
fallacy of misplaced concreteness, 20–21, 24–26, 31, 42, 152, 195
Feuerbach, Ludwig, 111, 119
Fichte, Johann Gottlieb, 47, 57, 115
Filipchenko, Yuri, 85–86
Fisher, Ronald, 84, 150, 177, 252n39
formative causation, 197–99, 207. *See also* Sheldrake, Rupert
Foster, Michael B., 104
Foucault, Michel, 25, 48
Frost, Robert, xix
fundamentalism (religious), xxiv, 2, 67, 81–82, 96, 98

Gaia hypothesis, 103, 127, 217, 249n71
genetics, 76, 83, 84–86, 162, 195, 205
God and Creation (Boodin), xvii, 105, 106, 107, 125–29, 164, 166, 177; reviews of, 133–36
Greene, John C., 1, 5, 7, 9, 70, 75–76, 101, 195, 201
Griffin, David Ray, 107, 222

Haessler, Carl, 65
Haller, John S. Jr., 172–73
Ham, Ken, xxiv, 2, 82, 261n16
Hartshorne, Charles, xvii, 26, 27, 51, 92, 97, 103–104, 107, 136, 194, 220, 222, 226n15
Harvey, William, 115
Hawking, Stephen, xxiv, 2, 10, 48
Hegel, Wilhelm Friedrich, 27, 29–30, 32, 47–48, 109, 140, 149, 181, 186, 235n9
Hein, Hilde, 206–208
Heisenberg, Werner, 2, 10, 25, 50, 51, 112, 172, 176, 229n24
Heraclitus, 222
Herschel, John, 75, 124, 223
Hertzler, J. O., 154–55
Hocking, William Ernest, xxiii, 226n15, 226n19
Hoernlé, R. F. Alfred, xix–xx, xxi, 63–64, 105, 119, 135, 158–63, 173, 187, 208
Holmes, Eugene C., 164–69, 253n52, 255n81
Horkheimer, Max, 61
Howison, George Holmes, xxiii, 27
Hoyle, Fred, 52, 118
Hughes, Percy, 32, 42
Hume, David, 39, 47, 48, 59, 176, 221
Hunt, Tam, 74
Huxley, Julian, 9, 75–76, 84, 177

Huxley, Thomas Henry, 5–6, 71, 101, 113, 122, 123, 166, 190, 205, 248n53

idealism(ists), xiii, xvi–xvii, xxiii, 26, 27, 35, 37, 46–47, 53–54, 58, 62, 64–65, 87, 92, 104, 105, 111, 112, 116, 126, 136, 142, 143, 148–49, 151, 154, 165, 169, 173, 179, 180–81, 190, 209, 217, 219, 220, 221
Inherit the Wind, 81, 82, 241n45
intelligent design, 82, 90–91, 200–201, 206, 209, 212

James, William, xiii–xxvi passim, 12, 13, 17–18, 20, 22–24, 26–28, 35, 37, 40, 53, 55–56, 65, 82, 138, 152, 172, 173, 190–91, 221, 225n6, 229n25, 234n81

Kallen, H. M., 154
Kant, Immanuel, xiii, 28–29, 47–48, 57, 174, 181, 232n38
King, Martin Luther, 140–41, 143, 151, 167, 211, 251n13, 255n81
Koestler, Arthur, 1, 7, 78, 109, 197
Kojonen, Erkki Vesa Rope, 200
Kreyche, Robert J., 101, 208, 218
Kuhn, Thomas, 1, 6–7
Kuklick, Bruce, xxii

Lack, David, 84, 86
Lamarck, Jean Baptise, 4, 7, 102. *See also* neo-Lamarckian
Levine, George, 205
Lewis, C. S., xxvi–xxvii, 227n41
Lindberg, David C., 108
Locke, John, 47, 94
Lovejoy, Arthur O., xix, 153, 177, 187, 220
Lovelock, James, 217, 249n71

Lucretius, 110–11, 116, 119, 128

Man in His World (Boodin), xxvi
Mandelbaum, Maurice, 70, 205
Man's Place in the Universe (Wallace), 72, 168, 189
Margulis, Lynn, 217, 249n71
Marx, Karl, 9, 21, 48, 109, 111, 119, 164–65, 180, 248n51, 255n81
materialism(ists), xxvi, 27, 37–39, 54–55, 69, 90, 93, 99, 101, 103–105, 110, 111–12, 126, 130–31, 164–65, 167, 169, 175–77, 184–85, 191, 196, 198, 212, 219, 220, 221, 244n102. *See also* physicalism
Maxwell, James Clerk, 52, 63
Mayr, Ernst, 84, 85, 110, 195–96, 205, 239n21, 258n60
McClure, M. T., 65
McKenna, Terrance, 55, 90
Mead, George Herbert, xxii, 155–58, 164, 165
mentalité, 85, 145–46, 171, 174, 230n5, 252n29
Mill, John Stuart, 3, 109
Miller, Ken, 71
Miller-Urey experiment, 113
Mollenhauer, Bernard, 56, 180
Montague, William Pepperell, 167, 185, 220
Morgan, C. Lloyd, 109, 119, 165, 245n6
morphic resonance, 197, 207. *See also* Sheldrake, Rupert
Mounce, Howard O., 188, 191
Mumford, Lewis, 18–19

Nagel, Thomas, 48, 103, 209
natural selection, 4, 60–63, 72–79, 83–86, 90, 103, 150, 188–91, 196, 200, 210, 218, 228n3, 239n21

naturalism, 99–101, 126, 154, 176–77, 188, 190, 192, 205, 208, 218, 219–20, 244n102
Needham, Joseph, 194–95
neo-Darwinism(ists), xxiv, 6, 61, 75, 76, 79, 84–87, 91, 102, 150, 195, 205–206, 209–10, 239n21. *See also* Darwinism
neo-Lamarckian, 83, 189, 195. *See also* Lamarck
neo-Platonism(ists), 46, 108. *See also* Platonism
Neuber, Matthias, 183–85, 191, 192
Newton, Isaac, 19, 23, 25, 57, 89. *See also* Newtonian
Newtonian, 19–20, 25, 49, 94, 115, 231n15. *See also* Newton
Next Step in Religion (Sellars), 177–78

O'Connor, Flannery, 153–54
ontology, 3, 12, 32, 36, 40, 41, 53, 55, 60, 67, 99, 101, 108, 109, 129, 150, 174, 178, 182, 220
organicism, 194–95, 197, 207
Origen, 117, 119
Origin of Species (Darwin), 1, 67, 71, 84, 177, 190
orthogenesis, 88, 160, 168, 207, 254n61, 255n84
Osborn, Henry Fairfield, 88
Owen, Richard, 119, 122–25, 223, 248n53

Paley, William, 62, 69, 200–201, 212, 228n3
Palmer, Donald, 46
Palmer, George Herbert, xv, xxi
panentheism, 51, 91, 92, 96, 97, 107, 115, 131, 171, 173, 179, 207, 220
pangenesis, 83, 189
panpsychism(ists), 103–104, 108, 128, 148, 182

pantheism(ists), 39, 46, 91, 108, 111, 115, 136, 193, 220
Parmenides, 57
Peabody Model School (PMS), 153–54
Peirce, Charles Sanders, xiii, 37, 42, 45, 65, 181, 186, 188, 190–91, 221, 231n19
Peters, Robert Henry, 73
Peterson, Erik, 207, 209
Phalanx theory, xix
Phillips, Bernard, 48, 221
physicalism, 109, 176, 178, 198, 219. *See also* materialism
Piatt, Donald Ayers, xxvi, 14, 180, 187
Picard, Maurice, 100
Piper, Leonora, 55–56
Plantinga, Alvin, 39, 48, 210
Plato, xxv, 3, 10, 19, 29, 38, 46, 51, 53, 108, 112, 114–17, 119, 124, 127, 128, 129, 132, 149, 159, 173, 176, 223
Platonism(ists), 10, 29, 46, 51, 85, 103, 109, 117, 118–19, 122–24, 220. *See also* neo-Platonism
Plotinus, 46, 93, 108, 111, 116, 119
Poincaré, Henri, 49, 172, 181
Popper, Karl, 12, 73, 221
positivism, xxv–vi, 1, 3–9, 11, 12, 13, 30, 68, 69, 84, 98, 101, 128, 134, 166, 174, 178, 188, 191, 205–206, 220–21, 222
pragmatic energism, 45, 51, 65
pragmatism(ists), xiii–xxiii passim, 13, 24–25, 35–37, 41–42, 45, 51, 57, 59, 128, 164–65, 174, 185–88, 190–92, 208, 221
preformation, 108–11, 117, 119, 120–21, 124, 126, 246n6
Price, George, 149–50
process philosophy, xvii, 26, 27, 32, 67, 87, 93, 103, 104, 124, 128,

process philosophy *(continued)* 143, 169, 173, 183, 193–94, 205, 221–22, 223, 231n19, 232n39
progress (historical), 151–53
psychic phenomena, 56
Putnam, Hilary, 8, 11, 12, 48, 191, 221, 223n77, 244n102

quantum biology, 112–13
quantum physics, 2, 20, 50, 113, 172

Realistic Universe, A (Boodin), xvii, xxi, xxvi, 14, 33, 38, 45, 49–62, 137, 163, 168, 181, 218, 235n13; reviews of, 63–65
Reck, Andrew, xx, xxiii, xxv, 203
Reed, Edward S., 74
Reichenbach, Hans, xxii, 12, 191
Reiser, Oliver L., 135
Religion Coming of Age (Sellars), 177–78
Religion of Tomorrow (Boodin), xvii, xxi, xxvi, 106, 107, 129–32, 141, 166, 167, 173, 175, 177, 194, 211; reviews of, 135–36
Rimmer, Harry, 82, 180
Ritchie, A. D., 154
Ritter, William E., 194
Rosenblith, Walter, 32–33
Royce, Josiah, xvi–viii, xxi, xxiii, 13, 17–18, 22, 24–27, 28, 30, 31, 40, 42–43, 58, 65–66, 107, 128, 130, 132, 137, 138–41, 142, 152, 172–73, 210, 220, 226n15, 231n19, 231n27, 232n39, 234n83, 251n13
Rupke, Nicolaas, 124
Ruse, Michael, 69, 70–71, 103, 123–24, 219
Russell, Bertrand, 48, 63, 193
Russell, L. J., 135

Sabine, George H., 100

Santayana, George, xvi, xxi, 153, 167, 183, 185
Sartre, John-Paul, 48
Schelling, William Joseph, 109, 119
Schlick, Moritz, 8, 48, 221
Schopenhauer, Arthur, 56, 109, 119, 149, 181, 252n36
Schrödinger, Irwin, 2, 25, 50, 51, 172
Scientism, xxiv, xxv, xxvi, 4, 9–13, 27, 42, 48, 85, 87, 101–102, 116, 171, 178, 190, 191, 200–201, 206, 212, 218, 222, 261n16
Sellars, Roy Wood, 36, 99–102, 112, 167, 175–79, 183–85, 187, 188, 191–92, 208, 217–18, 219, 220, 222, 244n102
Shailer, Mathews, 179
Sheldrake, Rupert, 39, 104, 109–10, 114, 164, 168, 182, 197–200, 206, 207, 210, 212, 248n51. *See also* formative causation. *See also* morphic resonance
Sims, Newel L., 154–55
Sisyphean evolution, 86
Skinner, B. F., 9, 54, 100, 114, 155, 157, 217, 237n32
Smith, Wolfgang, 193
Smocovitis, Betty, 84–85
Social Mind, The (Boodin), xvii, xix, xxi, xxvi, 130, 132, 135, 137, 141–51, 166, 169, 173, 187, 210–11, 253n60; reviews of, 153–55, 158–60, 163–64
Socrates, 46, 136, 160
Spencer, Herbert, 5, 59, 60, 72, 119, 181, 222
Spengler, Oswald, 152, 190
Spinoza, Baruch, 46, 114, 246n24
steady-state cosmology, 96–97, 117–18, 247n42
Steinbeck, John, xix, 180, 226n23
Steinhardt, Paul J., 96

Stoics, Greek, 108, 111, 119
Stout, A. K., 104–105
Studies in Philosophy (Boodin), xxvi, 14
Sutton, Kenneth Ray, 180–81
symbolic interactionism, 156–57

Tanaka, Stefan, 19–20, 25
Taung child, xxvi, 83, 88
tautology, 60, 72–76, 79, 90, 176, 222, 260n4
teleology, xxiv, 60–63, 71, 101, 102, 114–15, 124, 125, 128, 150, 165–68, 172, 173, 189, 195–97, 198, 200–201, 212, 222–23, 228n3, 235n24, 245n6, 249n71, 258n42
Theoretical Biology Club, 194, 196
Thompson, D'Arcy Wentworth, 125
Three Interpretations of the Universe (Boodin), xvii, xxiv, xxvi, 18, 62, 105, 107–20, 130, 163, 168, 208, 212; reviews of, 132–35
Timaeus (Plato), 10, 114–15, 117, 119, 124
Time and Reality (Boodin), xiv, xxi, xxiv, xxvi, 17, 30; reviews of, 32–33, 40–41
Truth and Reality (Boodin), xvii, xxvi, 17, 31, 33–39, 45, 218, 221; reviews of, 40–43
Tsanoff, Radoslave, 41, 64
Tufts, James Hayden, 135, 147, 155, 180, 187–88
Turok, Neil, 96, 247n42

Tyson, Neil deGrasse, 10, 48

Virtue, Charles F., 118, 133
vitalism(ist), 62–63, 75, 91, 121, 168, 194–95, 197, 206, 207–208, 210
Von Hartmann, Eduard, 109, 119

Waddington, C. H. ("Hal"), 194–97
Wallace, Alfred Russel, xxiii, 4, 60, 72, 76, 78, 88, 168, 188–91, 228n3, 239n21, 258n42
Watson, John B., 9, 54–55, 100, 114, 155, 157, 217, 237n31, 237n32
Werkmeister, William ("Werkie"), 87, 155
Weyl, Hermann, 172, 236n16
Whitehead, Alfred North, xvii, xxiv, xxvi, 8, 20, 24–27, 30, 32, 36, 48, 51, 87, 92–95, 97, 101, 103, 104, 107, 109, 115, 117, 118–21, 126, 127, 133–35, 152, 178, 182, 192–97, 220, 222, 223
Whitney, G. W. T., 105
Wiener, Norbert, 32
Wilm, E. C., 64
Wilson, E. O., xxiv, 2, 48
Wittgenstein, Ludwig, 48, 193
Woodger, Joseph, 194
World of Life, The (Wallace), 72, 88, 189
Wright, H. W., 40

X Club, 123, 171, 190, 258n42

www.ingramcontent.com/pod-product-compliance
Lightning Source LLC
Chambersburg PA
CBHW031706230426
43668CB00006B/131